Topology

THE WILEY BICENTENNIAL–KNOWLEDGE FOR GENERATIONS

Each generation has its unique needs and aspirations. When Charles Wiley first opened his small printing shop in lower Manhattan in 1807, it was a generation of boundless potential searching for an identity. And we were there, helping to define a new American literary tradition. Over half a century later, in the midst of the Second Industrial Revolution, it was a generation focused on building the future. Once again, we were there, supplying the critical scientific, technical, and engineering knowledge that helped frame the world. Throughout the 20th Century, and into the new millennium, nations began to reach out beyond their own borders and a new international community was born. Wiley was there, expanding its operations around the world to enable a global exchange of ideas, opinions, and know-how.

For 200 years, Wiley has been an integral part of each generation's journey, enabling the flow of information and understanding necessary to meet their needs and fulfill their aspirations. Today, bold new technologies are changing the way we live and learn. Wiley will be there, providing you the must-have knowledge you need to imagine new worlds, new possibilities, and new opportunities.

Generations come and go, but you can always count on Wiley to provide you the knowledge you need, when and where you need it!

WILLIAM J. PESCE
PRESIDENT AND CHIEF EXECUTIVE OFFICER

PETER BOOTH WILEY
CHAIRMAN OF THE BOARD

Topology

Point-Set and Geometric

Paul L. Shick

John Carroll University
Department of Mathematics and Computer Science
Cleveland, Ohio

WILEY-INTERSCIENCE
A John Wiley & Sons, Inc., Publication

Published by John Wiley & Sons, Inc., Hoboken, New Jersey.
Published simultaneously in Canada.

For general information on our other products and services or for technical support, please contact our Customer Care Department within the United States at (800) 762-2974, outside the United States at (317) 572-3993 or fax (317) 572-4002.

Wiley also publishes its books in a variety of electronic formats. Some content that appears in print may not be available in electronic format. For information about Wiley products, visit our web site at www.wiley.com.

Library of Congress Cataloging-in-Publication Data is available.

ISBN: 978-0-470-09605-5

10 9 8 7 6 5 4 3 2 1

To Sue, with all my love

CONTENTS

FOREWORD

This book is intended as a text for a first course in topology, modeled on a junior/senior level course offered at John Carroll University. This particular course is required of all our mathematics majors, and is generally taken after the students have had a sophomore-level abstract algebra course. After teaching this course countless times over the last 20 years, and after endless discussions with my colleagues about it, I've become convinced that

1. An introductory topology course has to cover enough point-set topology to prepare the students adequately for ideas they're likely to encounter in analysis, geometry and other areas.

2. An introductory topology course can't do *just point-set topology, for two reasons: (a) this leaves out the more intuitive geometric aspects of the field in favor of the more classical point-set areas, ignoring the portions of topology most applicable to other fields of mathematics; and (b) often students completing a point-set topology course feel that they've learned a number of topics, but that the course (and field) lacks a "big theorem" that forms a capstone for their study.*

3. *An introductory topology course should start with the axiomatic definition of a topology on a set, rather than using metric spaces or the topology of subsets of* $\mathbb{R}^n$ *to "ease" the students into the subject, for three reasons: (a) the metric approach leaves out too many important examples (such as function spaces); (b) students who see only the metric approach, or who see this first, tend to develop more simplistic intuition than students who learn the more general definition first; and (c) the more general approach allows the student to learn how to write precise proofs in a brand new context, an invaluable experience for math majors. The important examples or* $\mathbb{R}$ *and* $\mathbb{R}^n$ *(in their usual topologies) should be presented only after the general definitions. Metric spaces should be covered (much) later.*

This text is designed with these beliefs in mind. It's intended to be covered in one semester by typical math majors (as opposed to some "one-semester" texts that are so dense that typical classes can cover only a small portion in one term). However, the text is intended to be quite rigorous, modeling for the students how one writes precise proofs. It covers the essentials of point-set topology in a relatively terse presentation, with lots of examples and motivation along the way. The introductory chapter attempts to explain what topology is in the context of what math majors have usually seen in their previous coursework. Along with the standard point-set topology topics (connected spaces, compact spaces, separation axioms and metric spaces), we include path-connectedness and a chapter on constructing spaces from other spaces (including products, quotients, etc.). Each chapter has a short introduction designed to motivate the ideas and place them into an appropriate context. Each section has a set of exercises, ranging in difficulty from easy to fairly challenging. The text culminates in two "capstone" chapters, each independent of the other, enabling instructors to choose which subject best suits their views and students. These capstone chapters are:

1. The Classification Theorem for Compact, Connected Surfaces.
2. Fundamental groups and classifying spaces, with applications.

As you can see, geometric and algebraic parts of topology are introduced in the two capstone chapters. This is not intended to diminish the importance of these topics, but rather to make the text more flexible.

ACKNOWLEDGMENTS

What appears in this text is strongly influenced by the topologists with whom I've worked or studied over the years. Among these, Bob Kolesar and Mark Mahowald deserve special thanks. Much of my understanding of the topics covered here is due to the classic books of James Munkres and William Massey. My thanks especially go to Sue Simonson Shick for her patient help with the graphics. Bob Kolesar, among others, deserves thanks for patient and careful proofreading. Finally, my heartfelt thanks go to my students, who have helped shape me as well as this book.

P. S.

CHAPTER 1

INTRODUCTION: INTUITIVE TOPOLOGY

1.1 INTRODUCTION: INTUITIVE TOPOLOGY

What is topology? One would hope that a book entitled *Topology* would provide a simple answer to this question as a starting point. Unfortunately, it's not terribly easy to give a brief answer without first building up some background. Here's a first attempt: Topology is the study of the qualitative properties of certain objects (called *topological spaces*) that are invariant under certain kinds of transformation (called *continuous maps*), especially those properties which are invariant under a certain kind of equivalence (called a *homeomorphism*). This answer probably raises more questions than it answers, but it makes a good deal of sense when put into an appropriate mathematical context.

Euclidean geometry, for example, is a branch of mathematics that nearly everyone has spent some time studying. If asked for a short definition of Euclidean geometry, most students would have a difficult time distilling a year-long high school course into a single sentence. However, there is a very simple theme that unites nearly all of the varied topics studied in high school geometry – one considers the properties of certain objects (triangles, squares and other planar figures) that are preserved under congruence or similarity. Precisely, we consider certain transformations of the plane, called *Euclidean transformations*: rotations (of the coordinate axes by any angle), reflections (across any line in the plane) and translations (moving the "origin" of the coordinate system to any point in the plane), and transformations obtained as some combination of these. Two triangles are said to be congruent if one can be transformed into the other under some Euclidean transformation. If we add "rescalings" (scalar multiplication of the plane by any positive number) to our list of possible transformations, we can define the concept of similar triangles in much the same way. So our short definition of Euclidean geometry is as follows: geometry is the study of the properties of planar figures that are invariant under Euclidean transformations. We'll look at a similar explanation of topology shortly.

This idea ties in very nicely to another area of mathematics that most students have studied before encountering topology – abstract algebra. The set of Euclidean transformations of the plane forms a *group* under the operation of composition. In fact, Felix Klein defined a "geometry on a space X" to be the study of the properties of subsets of X that are invariant under the action of some transformation group G. [1]

The various areas of study lumped together as abstract algebra provide another important analogy for understanding what topology is all about. In group theory, for example, one studies groups and group homomorphisms. (See Section 2.6 for a primer on groups.) Precisely, a group is a set G with an operation, $\bullet$, satisfying three properties: $\bullet$ is associative ($(g_1 \bullet g_2) \bullet g_3 = g_1 \bullet (g_2 \bullet g_3)$ for all $g_1, g_2, g_3 \in G$); G contains an identity for the operation $\bullet$ (there exists $e \in G$ such that $g \bullet e = g = e \bullet g$ for every $g \in G$) and each element of G has an inverse in G (for every $g \in G$, there exists $g^{-1} \in G$ such that $g \bullet g^{-1} = e = g^{-1} \bullet g$). For example, the set of integers $\mathbb{Z} = \{\ldots, -2, -1, 0, 1, 2, 3, \ldots\}$ forms a group under $+$, but not under multiplication. A group homomorphism is a function $f : G \to H$ from a group G to a group H that respects the group structures: $f(g_1 \bullet_G g_2) = f(g_1) \bullet_H f(g_2)$ for every $g_1, g_2 \in G$, where $\bullet_G$ indicates the operation in G and $\bullet_H$ the operation in H. Group theory can be summarized as the study of the properties of groups that are preserved under group homomorphisms, especially those preserved by isomorphisms (equivalences between groups). For example, the property of being *abelian*[2] (or commutative: $g_1 \bullet g_2 = g_2 \bullet g_1$ for every $g_1, g_2 \in G$) is preserved by group homomorphisms (both trivial and nontrivial). So if G is abelian and H is not, then G cannot be isomorphic to H. Thus, the cyclic group of order 6, $\mathbb{Z}_{/6}$, cannot be isomorphic to the dihedral group on three letters, D_3, despite the fact that they both have six elements. This idea shows up in other areas of algebra; including semigroup theory, ring theory and field

[1] Felix Klein's (1849–1925) Erlangen program (named after the Universitat Erlange, where he held his chair) defined the concept of geometry entirely in terms of abstract algebra, a notable departure from previous mainstream mathematical thought.

theory. In each case, we have a set plus some other structure, and we study properties preserved by functions that respect this extra structure.

The field of topology works similarly. The objects we study are topological spaces, where a *topological spaceis* a set X with an extra structure, called a *topology* on X, which has to do with how "close" points in X are to each other. A function between spaces that respects these structures is called a *continuous function* or *continuous map*, with an equivalence referred to as a *homeomorphism.* Examples of topological spaces that are commonly studied are curves in $\mathbb{R}^2$ or $\mathbb{R}^3$, surfaces in $\mathbb{R}^3$ or $\mathbb{R}^4$, and spaces of functions between surfaces.

It's rather complicated to give a description of what a topology on a set X entails, but it's much easier to give an intuitive idea of what is meant by a continuous function between spaces. For subsets of $\mathbb{R}^n$, a Euclidean transformation is any composition of rotations (of the coordinate axes by any angle), reflections [across any (codimension 1) hyperplane (or subvector space) in $\mathbb{R}^n$] and translations (moving the "origin" of the coordinate system to any point in $\mathbb{R}^n$). Subsets of $\mathbb{R}^n$ are congruent if one can be transformed into the other by a Euclidean transformation. We add "rescalings" (scalar multiplication of $\mathbb{R}^n$ by any positive number) to the list of Euclidean transformations to define when two sets are similar. These concepts are generalized further in defining topological transformations. We add to the list of Euclidean transformations some new ones; we allow any transformations that involve bending or stretching, but not tearing. For example, an ellipse with eccentricity $e < 1$ is not geometrically equivalent to a circle, because such an ellipse has nonconstant curvature, while the circle's curvature is constant. The two planar curves are topologically equivalent (homeomorphic), however, because a composition of Euclidean transformations and bending/stretching can change one into the other. So, while curvature is a nice geometric property (preserved by Euclidean transfomations), it is not a topological property.

Topological properties are more "qualitative" than curvature. Examples of topological properties can be explained only intuitively at this point, but some of the central ideas can be made clear. *Connectedness* is perhaps the easiest topological property to look at. A topological space X is said to be connected if X is in a single "continuous" piece. For example, the two planar figures illustrated here (an "8" and an "81")

[2]The use of the term *abelian* for commutative groups is in honor of Nils H. Abel (1802–1828), a Danish mathematician whose work on the unsolvability of the quintic polynomial was part of a movement in mathematics that helped start the field of group theory. Abel's work is closely tied into that of Evariste Galois, who also died quite young, but in more romantic circumstances (reputedly in a duel over a barmaid, rather than by tuberculosis). These two exemplify the maxim that a mathematician's greatest work is most often accomplished before age 30. We'll see some striking counterexamples to this statement later.

8 81

cannot be homeomorphic because one is connected, the other disconnected. A second topological property, closely related to connectivity, is that of *genus,* which is quite easy to explain in the setting of planar curves. For a given connected planar curve X, a "cutpoint" of X is a point $x \in X$ such that X becomes disconnected if the point x is removed. A planar curve X is said to have genus 0 if every point of X (except the endpoints, if X has any) is a cut point. More generally, a connected planar curve X has genus n if some set of n points removed from X leaves it connected, but every set of $n+1$ "cuts" disconnects X (again, avoiding endpoints). Here are some examples of planar curves and their genera (plural of genus), with some cuts shown:

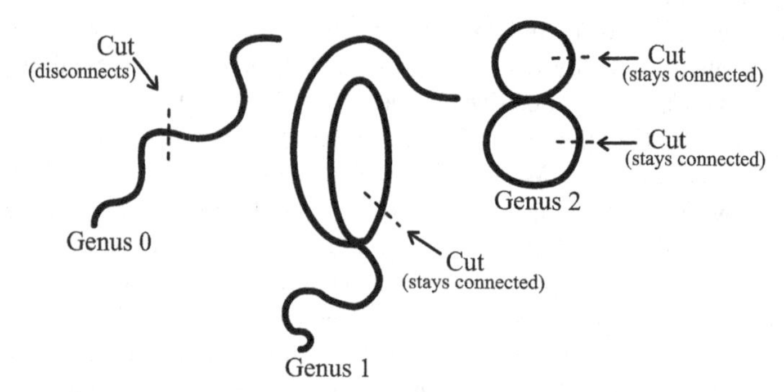

Finally, a very interesting example of a topological property is that of *orientability* in the context of surfaces. A *surface* is a topological space X that is "locally homeomorphic" to the open disk in $\mathbb{R}^2$ (meaning that each point of X has a "neighborhood" that is topologically equivalent to $B^2 = \{(x, y) : x^2 + y^2 < 1\} \subset \mathbb{R}^2$). Some simple examples of surfaces are:

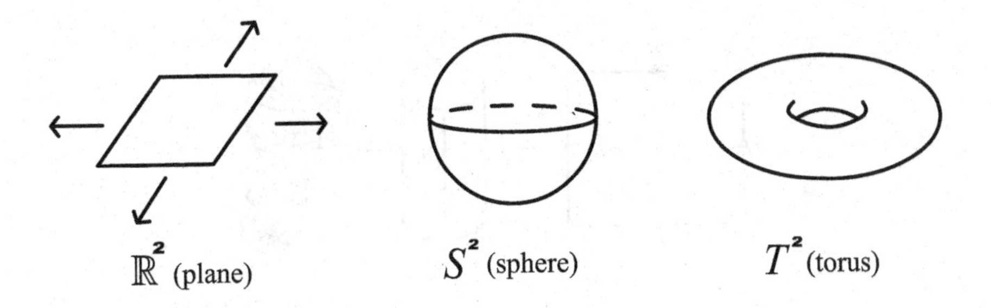

These are all *oriented* surfaces, in that there is a consistently defined "outward" direction on each. Precisely, at each point $x \in X$, we have a neighborhood equivalent to the planar disk, so we have two sides, which we'll call "inward" and "outward." If these directions can be defined *consistently* for each $x \in X$, we say that X is oriented. The key word here is "consistently"; if one takes any simple closed curve on the surface (one forming a simple loop), the "outward" direction must be preserved throughout the curve. It may seem counterintuitive, but it's quite easy to construct nonorientable surfaces. For example, a Moebius band is built by identifying edges of a rectangle as indicated in the following diagrams, which construct a cylinder and a Moebius band:

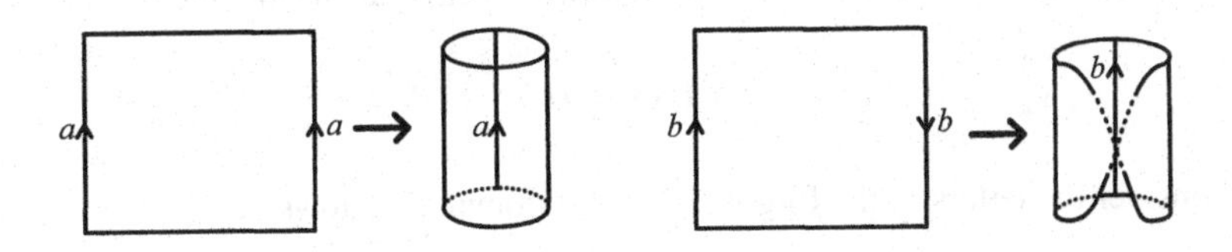

These are not surfaces (since each point on the edge has no neighborhood equivalent to the planar disk) but they are examples of "surfaces with boundary." Note that the cylinder has a consistent orientation, as one can check by looking at the curve that gives the "equator." However, the Moebius band is a "one- sided surface," without a consistently definable outward direction, as one can see by tracing a path along the equator of the band. An example of a true surface that is nonorientable is the *Klein bottle*[3] :

[3]Yes, the same Felix Klein as before. The term "Klein bottle" is used almost universally for this beast, but seems to have been an unintended name. The original German term apparently should have been translated

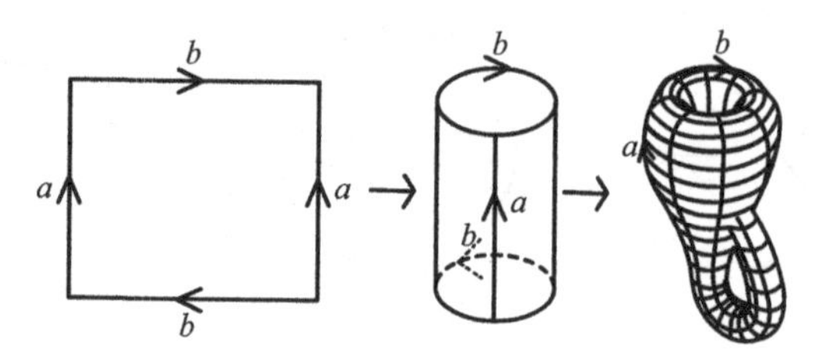

So the Klein bottle cannot be homeomorphic to any oriented surface, despite its similarities to the torus, for example. This is a very good example of how one uses topological properties to tell spaces apart. We'll explore the Klein bottle in more detail in Section 10.2.

In some sense, the goal of topology should be to come up with an exhaustive set of topological properties that would allow one to "classify" all topological spaces. Such a classification program has been completed, for example, in the setting of finite simple groups, where we recall that a simple group is a group with no nontrivial normal subgroups. The Classification Theorem for Finite Simple Groups was completed around 1980, culminating decades of work by scores of mathematicians. Given a finite simple group, one can identify it, up to isomorphism, by knowing its order (the number of elements in the group) and its character table (a complicated invariant, defined by representation theory). Such a classification theorem is nowhere in sight for topological spaces in general, but we do have such a result for connected, compact surfaces (where the term *compact* will be explained in excruciating detail in Chapter 7). The statement and proof of this theorem will occupy us in Chapter 10.

Exercises

Consider the letters of the English alphabet drawn as follows:

A B C D E F G H I J K L M N O P Q R

S T U V W X Y Z

1. Classify the letters by homeomorphism; that is, group the letters together so that all the letters in each set are homeomorphic to each other. You will need a total of nine sets.

as "Klein surface." An inaccurate translation rendered this as "Klein bottle," which has stuck since, for obvious reasons. The surface is often described as a bottle which can hold no water. For more concrete examples, see the website of the Acme Klein Bottle Corporation, `www.kleinbottle.com`.

2. Classify the letters by genus; that is, group the letters together so that all the letters in a set have the same genus. You will need a total of three sets.

CHAPTER 2

BACKGROUND ON SETS AND FUNCTIONS

2.1 SETS

What is a set? In a mathematics course, one is expected to supply precise definitions of the terms one uses, so it seems reasonable to ask for a precise definition of the term "set." However, a class of mathematics majors, when asked to come up with such a definition, typically respond with something like "A set is a collection of elements." When asked to define the term "collection," many reply "a group" or "a gathering." When pressed to be precise, most eventually resort to "you know, a set!"

Given that much of modern mathematics is couched in terms of sets, this seems a rather unfortunate state of affairs. However, the last answer supplied ("you know, a set!") is actually pretty close to being exactly correct. When I'm in a playful mood, and my students ask me for a definition of the term "set," I respond "I can't tell you." Eventually, I explain that we *can't* define the term "set", since it's going to be the basis for nearly all of the definitions to follow. More precisely, the term "set" is part of the starting point for mathematics: a commonly understood notion that we can't define. Similarly, in Euclidean geometry, we cannot define the notions of "point," "line," "plane" or "solid"; we must assume the reader/student has an understanding of this basic idea, and we proceed from there. We can *describe* these ideas, but we can't define them precisely.

We can illustrate this fundamental notion with lots of examples:

- $A = \{a, b, c, \ldots, z\}$, the set of letters in the English alphabet.
- $\emptyset = \{\}$, the empty set.
- $\mathbb{N} = \{1, 2, 3, \ldots\}$, the set of natural numbers.
- $\mathbb{Z} = \{\ldots, -3, -2, -1, 0, 1, 2, 3, \ldots\}$, the set of integers. (The notation comes the German word *zahlen*, or number.)
- $\mathbb{Q} = \{\frac{m}{n} : m \in \mathbb{Z}, n \in \mathbb{N}\}$, the set of rational numbers. This example uses two nice notations: (1) $x \in X$ is pronounced "x is an element of the set X," and (2) we use what's usually called "set-builder notation" to specify a set whose elements would be difficult or impossible to list. Here, we pronounce $\mathbb{Q} = \{\frac{m}{n} : m \in \mathbb{Z}, n \in \mathbb{N}\}$ as "the set of all elements $\frac{m}{n}$, with the property that m is any integer and n is any natural number." Note that we could just as easily have specified that $m, n \in \mathbb{Z}$, with $n \neq 0$.
- $\mathbb{R} =$ the set of all real numbers. We can specify this set, at least intuitively, by saying that $\mathbb{R}$ is the set of all decimals. We'll deal quite extensively with this set later.

We recall that two sets A and B are equal (denoted $A = B$) if they have exactly the same elements. A set C is a subset of A (denoted $C \subset A$) if every element of C is also an element of A. Hence, $A = B$ if and only if $A \subset B$ and $B \subset A$. This observation makes clear why the usual strategy employed in proving two sets are equal is called "double inclusion." Note that our notation for subset ($C \subset A$) does not imply that $C \neq A$. We'll use the notation $C \underset{\neq}{\subset} A$ to indicate that $C \subset A$ and $C \neq A$ (pronounced "C is a proper subset of A"). [*Warning*: Many texts use the convention that $C \subseteq A$ to indicate that C is a (possibly nonproper) subset of A and use $C \subset A$ to indicate inequality (proper inclusion). Our convention is less ambiguous and possibly more standard these days.]

We denote the *complement* (or set difference) of two sets as $X \setminus A = \{x \in X : x \notin A\}$. For example, $\mathbb{Z} \setminus \mathbb{N} = \{\ldots, -3, -2, -1, 0\}$. Note that this is defined even when $A \not\subset X$. For example, let $E = \{1, 2, 3, 4\}$ and $F = \{3, 4, 5, 6\}$. Then $E \setminus F = \{1, 2\}$ and $F \setminus E = \{5, 6\}$. If one is dealing only with subsets of a particular set X, it might be tempting to shorten the notation $X \setminus A$ to A^c, or something like this. However, such notations make sense only in the context of a particular set X, and the notation A^c is getting dangerously close to that of some sort of "set of all sets." We'll be clearer about why this is something to avoid below.

We have two familiar operations on sets, namely, union and intersection. We'll define these, for now, as "binary" operations, but we'll work in more generality shortly. Given sets A and B, we define $A \cup B = \{x : x \in A \text{ or } x \in B\}$, the union of A with B. We define the intersection of A with B as $A \cap B = \{x : x \in A \text{ and } x \in B\}$. For example, for the sets E and F as defined in the previous paragraph, $E \cap F = \{3, 4\}$ and $E \cup F = \{1, 2, 3, 4, 5, 6\}$. We recall that two sets A and B are said to be *disjoint* if $A \cap B = \emptyset$.

We note (a polite way of saying that we won't prove it!) that the operations of $\cup$ and $\cap$ are both commutative and both associative; that is, $A \cup B = B \cup A$ and $A \cap B = B \cap A$, and also $A \cup (B \cup C) = (A \cup B) \cup C$ and $A \cap (B \cap C) = (A \cap B) \cap C$ for any sets A, B and C. Further, the two operations distribute across each other, as we hope: $A \cap (B \cup C) = (A \cap B) \cup (A \cap C)$ and $A \cup (B \cap C) = (A \cup B) \cap (A \cup C)$. We'll prove the first of these identities.

Theorem 2.1.1. *For any sets A, B and C, we have*

$$A \cap (B \cup C) = (A \cap B) \cup (A \cap C).$$

Proof: To show that two sets are equal, the usual approach is to show that they are subsets of each other (a "double inclusion" proof). First, we'll show the left-hand side of the equation is a subset of the right-hand side:
$\subset$) Let $x \in A \cap (B \cup C)$. Then $x \in A$ and $x \in B \cup C$; that is, $x \in A$ and ($x \in B$ or $x \in C$). So we see that ($x \in A$ and $x \in B$) or ($x \in A$ and $x \in C$), another way of saying that $x \in (A \cap B) \cup (A \cap C)$.

The other inclusion:
$\supset$) Let $x \in (A \cap B) \cup (A \cap C)$. Then ($x \in A$ and $x \in B$) or ($x \in A$ and $x \in C$). So it follows that $x \in A$ and ($x \in B$ or $x \in C$), which is another way of saying that $x \in A \cap (B \cup C)$. □

Ə[1] Note that for any set Y, $Y \cup \emptyset = Y$. One might ask, then, whether the "collection" of sets is a group under the operation of union. (See Section 2.6 for a

[1] We're using this symbol Ə as a variant on the roadsign that indicates "dangerous curves ahead," following the tradition of Nicholas Bourbaki. This indicates that the reader should use particular care, because the ideas to follow contain more complexity that one might expect. By the way, Nicholas Bourbaki is not really a person. Bourbaki is the collective pseudonym for a group of (mostly French) mathematicians who began to meet in the 1930s, with the goal of putting much of modern mathematics on a firm logical foundation. The group has published a large number of books, all of which proceed in a very formal definition–theorem–proof format, which give precise proofs of most of the important theorems in algebra, analysis, etc. Ralph Boas, as editor of *Mathematical Reviews*, wrote the annual review of mathematics for *Encyclopedia Brittanica* in the 1950s, and wrote about the Bourbaki program. Ralph later received an

primer on group theory.) Certainly the empty set acts as an identity for this operation. The operation is commutative (since $A \cup B = B \cup A$ for any sets A and B), so the collection, *if* it is a group, is abelian. We observed above that the operation $\cup$ is associative. However, it's easy to see that we have no inverses for nonempty sets. Precisely, if a set A is nonempty, we can never find a set A^{-1} with $A \cup A^{-1} = \emptyset$, since $A \subset A \cup B$ for any set B. So the "collection"[2] of all sets is not a group under $\cup$.

If we're in the situation of dealing with sets that are all subsets of some particular set X, then we have a similar situation: $A \cap X = A$ whenever $A \subset X$. So for subsets of X, the set X acts as an identity under the operation $\cap$. This situation occurs often enough that we adopt the following notation.

Definition 2.1.1. *For a given set X, the power set of X [denoted by $\mathcal{P}(X)$] is the set of all subsets of X.*

irate letter, purportedly from Bourbaki, beginning "You miserable worm, how dare you say that I do not exist?" About this time, the *American Mathematical Society* received an application for membership from Bourbaki, which was rejected on the grounds that the application was for an individual, rather than an institutional membership. The "feud" continued for a while, with members of Bourbaki floating a rumor that Boas did not exist, but was rather a collective pseudonym for the editors of *Mathematical Reviews*. This rumor was perhaps somewhat believable, given how productive Ralph was at that time. The *Seminaire Bourbaki* continues in Paris to this day, bringing rigor to new areas of mathematics (and providing, I'm told, one of the most intimidating seminar audiences around). In any case, we'll use the "dangerous curves ahead" warning from the Bourbaki collective without apology. The details of the Boas/Bourbaki dealings can be found in Ref. [3].

[2]We're avoiding the term "set of all sets" because this is a very problematic term, exposing the difficulty that arises even when dealing with such a simple concept as set. The issue here is what's known as Russell's paradox, which arises when one considers such objects as the $S =$"set of all sets." If such a set S exists, then $S \in S$. (i.e. S is an element of itself.) This property leads to a paradox in the following way: we'll say that a set A is said to be "nice" if $A \notin A$. For example, the set $\mathbb{N}$ is "nice." Let R denote the set of all "nice" sets, which is well defined if we allow the existence of the set S. Now we ask, is R "nice"? If R is "nice," then $R \in R$, since R is the set of all "nice" sets, but this implies R is not "nice." Similarly, if R is not "nice," then $R \notin R$, by the definition of R, so R fulfills the criterion to be in R! So R is "nice" if and only if R is not "nice"!!! We'll stay away from this paradox by scrupulously avoiding any constructions involving a "universal" set.

We can ask the same question: Does the operation $\cap$ give $\mathcal{P}(X)$ the structure of a group? Again, we have an identity and associativity. Do we have inverses? Given $A \subset X$, does there exist a set A^{-1} with $A \cap A^{-1} = X$? The answer is easily seen to be no, since $A \cap B \subset A$ for any set B, so any *proper* subset of X has no inverse.

The familiar set operations of union and intersection relate nicely to complements in what have become known as DeMorgan's[3]laws.

Theorem 2.1.2. *(DeMorgan's laws) For A, B and X any sets, we have*

$$X \setminus (A \cup B) = (X \setminus A) \cap (X \setminus B)$$

and

$$X \setminus (A \cap B) = (X \setminus A) \cup (X \setminus B).$$

We will prove one of these, leaving the other for an exercise.
Proof: $\subset$) Let $x \in X \setminus (A \cup B)$. Then $x \in X$ and $x \notin (A \cup B)$; that is, $x \in X$ and x is in neither A nor B. This means, then, that $x \in (X \setminus A)$ and $x \in (X \setminus B)$, so that $x \in (X \setminus A) \cap (X \setminus B)$, as we wish.
$\supset$) Let $x \in (X \setminus A) \cap (X \setminus B)$. So $x \in (X \setminus A)$ and $x \in (X \setminus B)$. We see, then, that $x \in X$, and that x is in neither A nor B, so that $x \in X \setminus (A \cup B)$. Then $x \in X$ and $x \notin (A \cup B)$, as we had hoped. □

Another useful set construction is the Cartesian product ; for sets A and B, the Cartesian product is defined by

$$A \times B = \{(a, b) : a \in A \text{ and } b \in B\},$$

where (a, b) is our notation for an ordered pair of points. [Unfortunately, this is also the usual notation for "open" intervals in the set $\mathbb{R}$. The meaning of (a, b) should be clear in context.] Cartesian products interact nicely with unions and intersections:

$$A \times (B \cap C) = (A \times B) \cap (A \times C),$$

and so on. The product of k sets is defined analogously:

$$X_1 \times X_2 \times \cdots \times X_k := \{(x_1, x_2, \ldots, x_k) : x_i \in X_i\},$$

[3]Augustus DeMorgan (1806–1871) helped found the British Association for the Advancement of Science. Aside from the laws stated above, he is best known, perhaps, for bringing the famous four-color problem to the attention of the British Royal Society.

the set of ordered k-tuples. We often use the notation $\prod_{i=1}^{k} X_i$ to denote this product.

Finally, we extend the definitions of union and intersection from binary operations to operations on collections of sets. First, an *indexed collection* of sets is a set $\mathcal{C}$ which has as its elements sets, each labeled by an element of an "indexing set" Λ. More precisely, $\mathcal{C} = \{A_\alpha : \alpha \in \Lambda\} = \{A_\alpha\}_{\alpha\in\Lambda}$. This is a nice shorthand notation for dealing with collections of sets that are too complicated to list out. For example, let $B_n = [-n, n]$ for each $n \in \mathbb{N}$, a "nested" sequence of closed intervals. This specifies an indexed collection $\mathcal{B} = \{B_n\}_{n\in\mathbb{N}} = \{[-n, n]\}_{n\in\mathbb{N}}$.

We define the union of an indexed collection $\mathcal{C}$ as

$$\bigcup_{\alpha\in\Lambda} A_\alpha = \{x : x \in A_\beta \text{ for some } \beta \in \Lambda\}.$$

Sometimes we'll denote this set as $\bigcup \mathcal{C}$ as a shorthand notation. For our example $\mathcal{B}$, we see that

$$\bigcup \mathcal{B} = \bigcup_{n\in\mathbb{N}} [-n, n] = \mathbb{R},$$

since every real number is an element of $[-k, k]$ for some $k \in \mathbb{N}$.

Similarly, we define the intersection of an indexed collection $\mathcal{C}$ as

$$\bigcap_{\alpha\in\Lambda} A_\alpha = \{x : x \in A_\beta \text{ for all } \beta \in \Lambda\},$$

with shorthand notation $\bigcap \mathcal{C}$. For our example $\mathcal{B}$, we see that

$$\bigcap \mathcal{B} = \bigcap_{n\in\mathbb{N}} [-n, n] = [-1, 1],$$

since every real number x with $|x| > 1$ fails to be in $[-n, n]$ for $n = 1$.

Note that an indexed collection $\{A_\alpha : \alpha \in \Lambda\}$ can be considerably more complicated than just a sequence of sets, such as $\{B_n : n \in \mathbb{N}\} = \{B_1, B_2, B_3, \ldots\}$. For example, the usual open interval $(0, 1) \subset \mathbb{R}$ can be written as a union of an indexed collection: $(0, 1) = \bigcup_{x\in(0,1)} \{x\}$. This indexed collection is far more complex than a mere sequence of sets (see Theorem 2.5.4).

DeMorgan's laws also hold true for more general unions.

Theorem 2.1.3. *(DeMorgan's laws) For X any set and for any indexed collection $\{A_\alpha : \alpha \in \Lambda\}$ of sets, we have*

$$X \setminus \bigcup_{\alpha\in\Lambda} A_\alpha = \bigcap_{\alpha\in\Lambda} (X \setminus A_\alpha)$$

and

$$X \setminus \bigcap_{\alpha \in \Lambda} A_\alpha = \bigcup_{\alpha \in \Lambda} (X \setminus A_\alpha).$$

Proof: We'll prove the first result, leaving the second as an exercise.
$\subset$) Let $x \in X \setminus \bigcup_{\alpha\in\Lambda} A_\alpha$. Then $x \in X$ and $x \notin \bigcup_{\alpha\in\Lambda} A_\alpha = \{x : x \in A_\beta \text{ for some } \beta \in \Lambda\}$. So we see that $x \in X$ and $x \notin A_\beta$ for every $\beta \in \Lambda$. Hence, for every $\beta \in \Lambda$, x must be an element of $X \setminus A_\beta$. In other words, $x \in \bigcap_{\alpha\in\Lambda}(X \setminus A_\alpha)$, as we wished.
$\supset$) To see the other inclusion, note that each implication in the proof of the inclusion above is actually an equivalence. In other words, one can read the proof from bottom to top to see that $x \in \bigcap_{\alpha\in\Lambda}(X \setminus A_\alpha)$ implies $x \in X \setminus \bigcup_{\alpha\in\Lambda} A_\alpha$. (It is important that you actually check that each of these implications works backward as well.) $\square$

Exercises

1. Prove DeMorgan's law: If X is any set and $\{A_\alpha : \alpha \in \Lambda\}$ is any indexed collection of sets, then $X \setminus \bigcap_{\alpha\in\Lambda} A_\alpha = \bigcup_{\alpha\in\Lambda}(X \setminus A_\alpha)$.

2. Prove that for any sets A, B, C, $A \cup (B \cap C) = (A \cup B) \cap (A \cup C)$.

3. For each of the following subsets of $\mathbb{R}^2 = \mathbb{R} \times \mathbb{R}$, determine whether the set is the Cartesian product of two subsets of $\mathbb{R}$:

   - $A = \{(x, y) : x \text{ is rational.}\}$.
   - $B = \{(x, y) : x \in [0, 2]\}$.
   - $C = \{(x, y) : 1 \leq y < 3\}$.
   - $D = \{(x, y) : x \leq y^2\}$.
   - $E = \{(x, y) : x^2 + y^2 < 3\}$.

4. Sketch the following subsets of $\mathbb{R}^2$, where $A = [1, 2]$, $B = (3, 4)$, $C = [1, 4]$ and $D = \{0, 5, 8\}$:

   - $A \times B$.
   - $B \times D$.
   - $(A \times B) \cup (C \times D)$.
   - $(A \times B) \cap (C \times D)$.

5. For each $n \in \mathbb{N}$, let $A_n = [0, n)$ and $B_n = (-\frac{1}{n}, \frac{1}{n})$. Find

   - $\bigcup_{n\in\mathbb{N}} A_n$.
   - $\bigcap_{n\in\mathbb{N}} A_n$.

- $\bigcup_{n \in \mathbb{N}} B_n$.
- $\bigcap_{n \in \mathbb{N}} B_n$.

6. For each of the following statements, determine whether the statement is true (in full generality) or false. If it's true, prove it. If it's false, provide a precise counterexample.
   - $X \setminus (A \cap B) = (X \setminus A) \cap (X \setminus B)$.
   - If $A \cup C \subset B \cup C$ then $A \subset B$.
   - If $A \cap C \subset B \cap C$ then $A \subset B$.
7. - For every $\gamma \in \Lambda$, prove that $A_\gamma \subset \bigcup_{\alpha \in \Lambda} A_\alpha$.
   - For every $\gamma \in \Lambda$, prove that $\bigcap_{\alpha \in \Lambda} A_\alpha \subset A_\gamma$.
8. - Prove that $B \cup \left(\bigcap_{\alpha \in \Lambda} A_\alpha\right) = \bigcap_{\alpha \in \Lambda} (B \cup A_\alpha)$.
   - Prove that $B \cap \left(\bigcup_{\alpha \in \Lambda} A_\alpha\right) = \bigcup_{\alpha \in \Lambda} (B \cap A_\alpha)$.
9. We have seen that that for any set X, the power set $\mathcal{P}(X)$ does not form a group under intersection or union. Show, instead, that $\mathcal{P}(X)$ is a group with the operation given by the "symmetric difference" of sets: $A \Delta B = (A \setminus B) \cup (B \setminus A)$. (Showing that this operation is associative is rather tedious. You can learn a lot about the personality of your instructor by whether or not he/she requires you to prove that Δ is associative.)

2.2 FUNCTIONS

The concept of a function is one of the most fundamental in all of mathematics. It's also a good example of how the precise definition of the concept is rather different from the way one thinks about it in the "real world."[4]Here's the precise definition:

Definition 2.2.1. *A functionf from a set X to a set Y (denoted $f : X \to Y$) is a subset of $X \times Y$ with the property that if $(x, y_1) \in f$ and $(x, y_2) \in f$ for some $x \in X$, then $y_1 = y_2$.*

[4]By the "real world," I mean the world of mathematics – the world of sets, groups, rings, topological spaces, and so on. Surely no one reading a topology book could have a different meaning of the term "real world" in mind.

To see that this is in fact the usual concept of a function, we recall that the notation $f(x) = y$ is used to indicate that $(x, y) \in f$. However, most mathematicians do not think of a function as a set of ordered pairs. Instead, one usually thinks of a function in the following terms:

Pseudodefinition: A function f from a set X to a set Y (denoted $f : X \to Y$) is a *rule* that assigns to each element $x \in X$ a unique element $f(x) \in Y$.

Note that the property $(x, y_1) \in f$ and $(x, y_2) \in f$ for some $x \in X$ implies $y_1 = y_2$ in the real definition is exactly the same as requiring that $f(x)$ be uniquely defined. We'll refer to the element $f(x) \in Y$ as the *image of x under f*, or, on occasion, as the "value x takes under f." We'll also sometimes refer to a function by the terms *map* or *mapping*.

Of course, for $f : X \to Y$, we refer to the set X as the*domain of f*. We'll use the term *range of f*to refer to the set Y (where some authors prefer the term "codomain.") The subset of Y given by $\{y \in Y : y = f(x) \text{ for some } x \in X\}$ is the *image of f*, denoted by $Im(f) \subset Y$. (Some authors use the term "range" to mean what we're calling the image of the function.)

Most students, when asked for an example of a function, suggest $f(x) = x^2$, meaning, of course, that $f : \mathbb{R} \to \mathbb{R}$. Since most math majors are all too familiar with such functions after three or so semesters of calculus, this sort of example will be a good thing to keep in mind. We can also specify functions by simply exhibiting the image of each element in the domain. For example, consider the function $h : A = \{a, b, c, d\} \to B = \{1, 2, 3\}$ given by

$$\begin{aligned} a &\mapsto 1 \\ b &\mapsto 2 \\ c &\mapsto 3 \\ d &\mapsto 3, \end{aligned}$$

where $x \mapsto y$ denotes $y = h(x)$ (and is read as "x maps to y").

Definition 2.2.2. *A function $f : X \to Y$ is one-to-one if $f(x_1) = f(x_2)$ implies that*

$x_1 = x_2$.

For example, the function $f(x) = x^2$ from $\mathbb{R}$ to $\mathbb{R}$ is not one-to-one, because $f(-1) = 1 = f(1)$. Similarly, the function $h : A \to B$ is not one-to-one, because the elements c and d share the same image. We'll often use the terms *injection* and *monomorphism* as synonyms for one-to-one function.

Definition 2.2.3. *A function $f : X \to Y$ is onto if for every $y \in Y$, there exists*

$x \in X$ such that $y = f(x)$.

Note that this is equivalent to saying that $Im(f) = Y$, where Y is the range of f. In our earlier examples, note that $f : \mathbb{R} \to \mathbb{R}$ given by $f(x) = x^2$ is not onto, since $-1 \notin Im(f)$ (along with many other points in $\mathbb{R}$). However, $h : A \to B$ is onto, since $Im(h) = \{1, 2, 3\} = B$. The terms *surjection* and *epimorphism* are synonymous with onto function. A function that is both one-to-one and onto is called a *bijection*.

We can use certain operations on functions to construct new functions. If, for example, we restrict our attention to functions from $\mathbb{R}$ to $\mathbb{R}$, we can add, subtract, multiply or divide functions (pointwise). For more general sets, we seldom have algebraic structures at our disposal to construct new functions. We can, however, *compose* two functions, whenever the domain of one function equals the range of the other.

Definition 2.2.4. *Given two functions $f : X \to Y$ and $g : Y \to Z$, we define the composition $g \circ f$ as the function $g \circ f : X \to Z$ given by $g \circ f(x) = g(f(x))$.*

Again, note that the operation of composition is defined only when the domain of one function is contained in the range of the other. Even when one considers functions from a set X to *itself*, compositions are not always defined. As a simple example, given the two functions $f, g : \mathbb{R} \to \mathbb{R}$ given by $f(x) = x^2 + 5$ and $g(x) = 3x - 1$, both $g \circ f$ and $f \circ g$ are defined, with

$$(g \circ f)(x) = g(f(x)) = 3(x^2 + 5) - 1 = 3x^2 + 15x - 1$$

and

$$(f \circ g)(x) = f(g(x)) = (3x - 1)^2 + 5 = 9x^2 - 6x + 6.$$

Evidently, then, composition of functions is not commutative, even when we deal only with functions from a set to itself.

One might ask, is the set of functions from a set X to itself [denoted $Fcn(X, X)$] a group under composition? First, it's clear that $f \circ g$ and $g \circ f$ are both well-defined functions in $Fcn(X, X)$, whenever $f, g \in Fcn(X, X)$, so the operation is a good "binary" operation. Next, we observe that $\circ$ is associative on $Fcn(X, X)$. In fact, we'll observe that $\circ$ is associative whenever it makes sense to ask about it, namely, for $f : X \to Y$, $g : Y \to Z$ and $h : Z \to W$, we see that $h \circ (g \circ f) = (h \circ g) \circ f$ as functions from $X \to W$, as we see easily by writing out the definitions. We also have the desired identity for the operation: $i = i_X : X \to X$ given by $i(x) = x$ for all $x \in X$ has the property that $i \circ f = f = f \circ i$ for all $f \in Fcn(X, X)$. So $Fcn(X, X)$ under $\circ$ has all the properties of being a group, except possibly for the existence of inverses. For this property to hold, we would need the following: Given any function $f : X \to X$, there exists a function $f^{-1} : X \to X$ with $f \circ f^{-1}(x) = x = f^{-1} \circ f(x)$, for all $x \in X$; that is, we need $f \circ f^{-1} = i_X = f^{-1} \circ f$. To see that this property does not hold, in general, consider the function $f : \mathbb{R} \to \mathbb{R}$ given by $f(x) = x^2$. This function has no inverse function, when considered as a function from $\mathbb{R}$ to $\mathbb{R}$, because we see that $f(-1) = 1 = f(+1)$, so that the inverse function, if it existed, would have $f^{-1}(1) = 1$ and $f^{-1}(1) = -1$, which a well-defined (unique-valued) function cannot do. The problem here is that our function $f : \mathbb{R} \to \mathbb{R}$ is neither one-to-one nor onto.

More generally, a function $f : X \to Y$ is *invertible* if there exists a function $g : Y \to X$ with $g \circ f = i_X$ and $f \circ g = i_Y$. When this happens, we write $g = f^{-1}$ and $f = g^{-1}$. A nice way to think of this is as

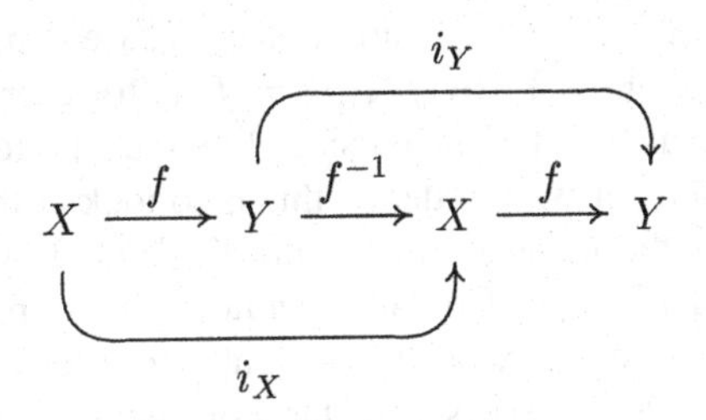

We have the following characterization of invertible functions:

Theorem 2.2.1. *A function $f : X \to Y$ is invertible if and only if f is one-to-one and onto.*

Proof: $\Longrightarrow$) Let $f : X \to Y$ have an inverse function $f^{-1} : Y \to X$. We need to show f is one-to-one and onto. To see that f is onto, let $y \in Y$. Since we have

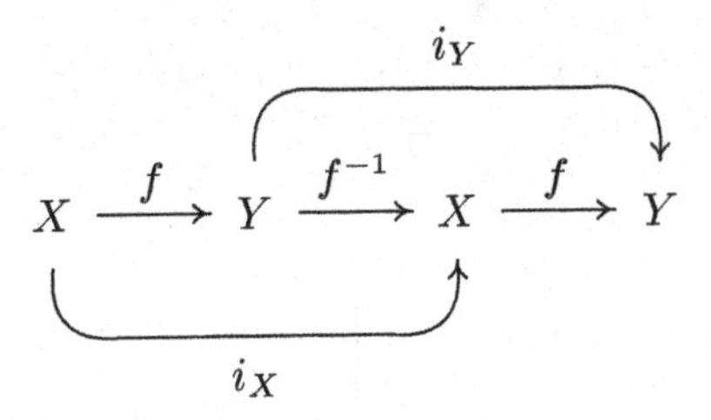

consider $x = f^{-1}(y)$. Then $f(x) = f(f^{-1}(y)) = f \circ f^{-1}(y) = i_Y(y) = y$. Thus $y \in Im(f)$, and since y is an arbitrary point in Y, we see that $Im(f) = Y$. Hence f is onto. To see that f must be an injection, let $f(x_1) = f(x_2)$. Since the function $f^{-1} : Y \to X$ is well defined, we see that $f^{-1}(f(x_1)) = f^{-1}(f(x_2))$, and hence that $x_1 = x_2$, as we wished.

$\Longleftarrow$) Let $f : X \to Y$ be both one-to-one and onto. Then for each $y \in Y$, we define $f^{-1}(y) = x$ if $y = f(x)$. We must show that f^{-1} is well–defined, and that $f \circ f^{-1} = i_Y$ and $f^{-1} \circ f = i_X$. Since f is onto, we know that each $y \in Y$ is the image of some $x \in X$, so we see that $f^{-1}(y)$ makes sense for each $y \in Y$. To see that f^{-1} is well–defined, suppose that $y = f(x_1) = f(x_2)$, so that f^{-1} seems to have two values in X (namely, x_1 and x_2). But the function f is one-to-one, so that $f(x_1) = f(x_2)$, which implies that $x_1 = x_2$, and hence $f^{-1}(y)$ is uniquely defined. Now we need to show that f^{-1} is indeed an inverse for the function f. Let $y \in Y$. Recall that $f^{-1}(y) = x$ exactly when $y = f(x)$. So $f \circ f^{-1}(y) = f(x) = y$, showing that $f \circ f^{-1} = i_Y$. Similarly, for any $x \in X$, we see that $f^{-1} \circ f(x) = f^{-1}(f(x)) = x$, by the definition, so that $f^{-1} \circ f = i_X$. □

Theorem 2.2.1 is just one example of how the properties of functions, such as being one-to-one or onto, relate to compositions. We'll explore more in the exercises.

For a function $f : X \to Y$ and a subset $A \subset X$, we define the *restriction*of f to A, denoted by $f|_A : A \to Y$, by $f|_A(a) = f(a)$ for every $a \in A$. Intuitively, we ignore whatever elements of X lie outside A and just perform the function f on the elements of A. Another way to define this is to look at the composition $f \circ i$, where $i : A \hookrightarrow X$ is the inclusion of A into X, defined by $i(a) = a$ for every $a \in A$. (See Exercise 1.) This construction is quite useful for producing an invertible function, when the function you're dealing with fails to be one-to-one. For example, the function $\sin : \mathbb{R} \to [-1, 1]$ is very far from being one-to-one. Indeed, its periodic nature means that for each $y \in [-1, 1]$, there are infinitely many $x \in \mathbb{R}$ with $\sin x = y$. However, we *can* construct an inverse function, denoted $\sin^{-1}$ or arcsin, by restricting the domain of $\sin(-)$ to a portion of the real line that is small enough so that $\sin(-)$ is one-to-one (as done in most calculus or trigonometry courses). We have infinitely many choices for such a subset of $\mathbb{R}$, of course, but traditionally we restrict to the interval $[-\pi/2, \pi/2]$, and define

$$\sin^{-1}(-) : [-1, 1] \to [-\pi/2, \pi/2].$$

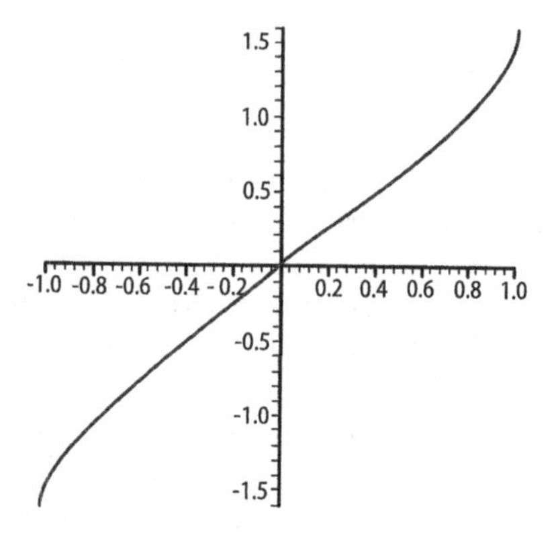

Similarly, given a function $g : A \to Y$ for $A \subset X$, we say that $\hat{g} : X \to Y$ is an *extension* of g if $\hat{g}|_A = g$. Another way to think of this is to say that the following diagram has to commute:

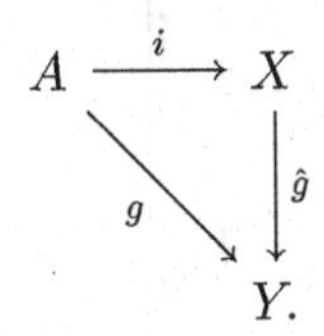

Note that a restriction of a function f is defined uniquely, by definition, but an extension of a function $g : A \to Y$ for $A \subset X$ over the set X is generally not unique – one usually has a choice of many such extensions.

Given a function $f : X \to Y$, we define the *image of a subset* $A \subset X$ under the function f as $f(A) = \{f(a) : a \in A\} \subset Y$. In this notation, then, $Im(f) = f(X)$. This construction respects the operation of union of sets quite nicely, but is a bit more problematic when applied to intersections, as the following theorem shows.

Theorem 2.2.2. *For a function $f : X \to Y$, and for A and B any subsets of X, we have*

$$f(A \cup B) = f(A) \cup f(B)$$

and

$$f(A \cap B) \subset f(A) \cap f(B).$$

Further, if f is one-to-one, then

$$f(A \cap B) = f(A) \cap f(B).$$

Any function which is not one-to-one provides a counterexample to equality for the second statement. For example, when $f(x) = x^2$ is considered as a function from $\mathbb{R}$ to $\mathbb{R}$, we see that $f([-2, -1] \cap [1, 2]) = f(\emptyset) = \emptyset$, but $f([-2, -1]) \cap f([1, 2]) = [1, 4] \neq \emptyset$, as the following graph shows clearly.

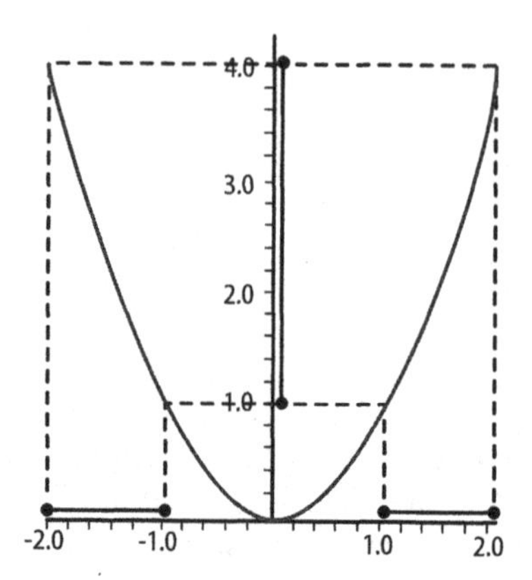

We'll prove the second and third statements, leaving the proof of the first statement to the exercises, along with some generalizations.

Proof: Let $A, B \subset X$, with $f : X \to Y$ any function. Let $y \in f(A \cap B)$. Then $y = f(x)$ for some $x \in A \cap B$, so that $y \in f(A)$ and $y \in f(B)$. Clearly, then, $y \in f(A) \cap f(B)$, so that we can conclude that $f(A \cap B) \subset f(A) \cap f(B)$, as we wished.

To see the other inclusion, we use the added hypothesis that f is one-to-one. Let $y \in f(A) \cap f(B)$, so that $y \in f(A)$ and $y \in f(B)$. Since $y \in f(A)$, we know that $y = f(x_1)$ for some $x_1 \in A$. Similarly, since $y \in f(B)$, we see that $y = f(x_2)$ for some $x_2 \in B$. Knowing that f is one-to-one allows us to conclude that $x_1 = x_2$, and hence, that this single point $x_1 = x_2 \in A \cap B$, so that $y \in f(A \cap B)$, as we wished. □

In addition to thinking about the image of a subset of the domain of a function, we can think about what is commonly known as the *inverse image* of a subset of the range of a function. Unfortunately, the notation for this idea is seriously misleading, but it has become quite standard throughout the mathematical world, so we'll use it (albeit grudgingly).

Definition 2.2.5. *For $C \subset Y$ and $f : X \to Y$ any function, the inverse image of C under f is the subset of X given by*

$$f^{-1}(C) = \{x \in X : f(x) \in C\}.$$

Note that we do *not* assume that f is one-to-one or onto here, so that the function f need not have an inverse function f^{-1}, despite the fact that the notation looks exactly the same as the *image* of C under the function f^{-1}. The notation is regrettable, but impossible to avoid in the mathematical literature, so we'll learn to deal with it.

Note that for an arbitrary function $f : X \to Y$ and a subset $C \subset Y$, $f^{-1}(C)$ need not be nonempty. For example, consider $f = \sin(-) : \mathbb{R} \to \mathbb{R}$. Then $f^{-1}(\{2\}) = \emptyset$.

The connection between images and inverse images is not as obvious as one might guess from the definitions.

Theorem 2.2.3. *For $C \subset Y$ and any function $f : X \to Y$, we have*

$$f(f^{-1}(C)) \subset C.$$

Proof: Let $y \in f(f^{-1}(C))$. Then $y = f(x)$ for some $x \in f^{-1}(C)$, so that $y = f(x) \in C$, by the definition of the inverse image of a set. □

Theorem 2.2.4. *For $C \subset Y$ and any **onto** function $f : X \to Y$, we have*

$$f(f^{-1}(C)) = C.$$

Proof: We see that $f(f^{-1}(C)) \subset C$ by the proof above. To see the other inclusion, let $y \in C$. Since f is onto, $y = f(x)$ for some $x \in X$, and that this particular x must be an element of $f^{-1}(C)$. Thus $y \in f(f^{-1}(C))$ as we wished. □

Theorem 2.2.5. *For $A \subset X$ and $f : X \to Y$ and function, we have*

$$A \subset f^{-1}(f(A)).$$

If, in addition, f is one-to-one, then

$$A = f^{-1}(f(A)).$$

The proof is deferred to the exercises.

Finally, we consider how the process of taking inverse images of subsets behaves with respect to the operations of unions and intersections. Again, we'll work with just two subsets here, leaving the more general results for the exercises.

Theorem 2.2.6. *Let $f : X \to Y$ be any function, and let $C, D \subset Y$. Then*

$$f^{-1}(C \cup D) = f^{-1}(C) \cup f^{-1}(D),$$

and

$$f^{-1}(C \cap D) = f^{-1}(C) \cap f^{-1}(D).$$

We'll prove the first statement, leaving the second for the exercises.

Proof: $\subset$) Let $x \in f^{-1}(C \cup D)$, so that $f(x) \in C \cup D$. So $f(x) \in C$ or $f(x) \in D$, which implies that $x \in f^{-1}(C)$ or $x \in f^{-1}(D)$. We conclude, then, that $x \in f^{-1}(C) \cup f^{-1}(D)$, as we wished.

$\supset$) Let $x \in f^{-1}(C) \cup f^{-1}(D)$, so that $x \in f^{-1}(C)$ or $x \in f^{-1}(D)$. Then $f(x) \in C$ or $f(x) \in D$, which implies that $x \in f^{-1}(C \cup D)$, as we wished. □

Exercises

1. For $A \subset X$ and $f : X \to Y$, prove that $f|_A = f \circ i$. Here i denotes the inclusion $i : A \hookrightarrow X$. (Recall that showing two functions to be equal requires showing that their domains are equal and that they take the same value on each element of the common domain.)

2. For $f : \mathbb{R} \to \mathbb{R}$ given by $f(x) = x^2$, find and sketch the following sets:
   (a) $f([1, 2])$
   (b) $f^{-1}([1, 2])$
   (c) $f(f^{-1}([3, 4])$
   (d) $f^{-1}(f(2, 3])$

3. Prove or disprove: for $B \subset Y$ and $f : X \to Y$, $f^{-1}(Y \setminus B) = X \setminus f^{-1}(B)$.

4. Prove the first part of Theorem 2.2.2: For a function $f : X \to Y$, and for A and B any subsets of X, we have $f(A \cup B) = f(A) \cup f(B)$.

5. Prove Theorem 2.2.5: For $A \subset X$ and $f : X \to Y$ any function, we have $A \subset f^{-1}(f(A))$. If, in addition, f is one-to-one, then $A = f^{-1}(f(A))$.

6. Prove the second part of Theorem 2.2.6: Let $f : X \to Y$ be any function, and let $C, D \subset Y$. Then $f^{-1}(C \cap D) = f^{-1}(C) \cap f^{-1}(D)$.

7. Prove the more general version of Theorem 2.2.6: Let $f : X \to Y$ be any function, and let $C_\alpha \subset Y$ for every $\alpha \in \Lambda$. Then $f^{-1}(\bigcup_{\alpha\in\Lambda} C_\alpha) = \bigcup_{\alpha\in\Lambda} f^{-1}(C_\alpha)$.

8. Let $f : X \to Y$ and $g : Y \to Z$ be any functions.
   (a) Prove that if f is one-to-one and g is one-to-one, then $g \circ f : X \to Z$ is one-to-one. Is the converse true?
   (b) If g is onto and f is onto, then is $g \circ f$ always onto? Is the converse true?
   (c) If g is onto and f is any function, then is $g \circ f$ always onto?

(d) Prove that for any subset $U \subset Z$, $(g \circ f)^{-1}(U) = f^{-1}(g^{-1}(U))$.

9. Given $f : X \to Y$ and $A \subset X$, prove that $f(f^{-1}(f(A))) = f(A)$.

10. Given $f : X \to Y$ and $B \subset Y$, prove that $f^{-1}(f(f^{-1}(B))) = f^{-1}(B)$.

2.3 EQUIVALENCE RELATIONS

The concept of a *relation* is, in some ways, even more fundamental than that of a function. In fact, we can define the term function in terms of relations, if we so desire.

Definition 2.3.1. *A relation R from a set X to a set Y is any subset $R \subset X \times Y$.*

As with functions, we use a piece of notation designed to make the idea clearer: we write xRy whenever $(x, y) \in R$, to reinforce the concept that "x is related to y." The simplest example, of course, is the empty relation $\emptyset \subset X \times Y$, where no element from X is related to any element from Y. More interesting examples include any function $f : X \to Y$, where xfy whenever $y = f(x)$. Indeed, a function is a relation from X to Y with the added property that xfy_1 and xfy_2 implies $y_1 = y_2$.

Most often, though, we're interested in a relation R from a set X to itself, usually regarded as a *relation on the set* X. Here are several examples, all on the set H, the set of all human beings.

- Let A be the relation on H given by xAy if x is an ancestor of y.
- Let D be the relation on H given by xDy if x is a descendant of y.
- Let S be the relation on H given by xSy if x is a (full) sibling of y.
- Let F be the relation on H given by xFy if x is in the immediate family of y.

The following properties of relations are very important and will show up a great deal in many contexts. A relation R on a set X is said to be

- *Reflexive* if xRx for every $x \in X$.
- *Symmetric* if x_1Rx_2 implies x_2Rx_1.
- *Transitive* if x_1Rx_2 and x_2Rx_3 implies x_1Rx_3.

Note that the relations A, D and S on the set H are not reflexive (by most peoples' definitions of the terms ancestor, descendant and sibling), while F is reflexive. Note that A and D are not symmetric, while S and F are. The relations A, D and S are all transitive. Perhaps surprisingly, F is not transitive, by most definitions, since one *is* related to one's spouse, and one's spouse is related to his/her siblings, but one is not in the same immediate family as one's brother-in-law, for example.

Definition 2.3.2. *An equivalence relation on a set X is a relation which is reflexive, symmetric and transitive.*

Traditionally, an equivalence relation on a set X is denoted by the symbol $\sim$ [with $x_1 \sim x_2$, our shorthand for $(x_1, x_2) \in \sim$, pronounced "x_1 is equivalent to x_2]. A nice example is quite familiar – two integers n_1 and n_2 are equivalent if $n_1 \equiv n_2$ modulo p, for some prime p (pronounced "n_1 is congruent to n_2 modulo p"); that is, $n_1 \sim n_2$ if $n_1 - n_2$ is a multiple of p. It's easy to check that this is an equivalence relation on the set $\mathbb{Z}$. An easier example holds for all sets X: $x_1 \sim x_2$ if and only if $x_1 = x_2$. It's trivial to check that this is an equivalence relation for any set X, which we'll refer to as the "trivial equivalence relation." We'll have more examples in the exercises.

Given an equivalence relation $\sim$ on a set X, for each $x \in X$, we use the notation $[x]$ to denote the *equivalence class* of x, $[x] = \{y \in X : y \sim x\}$. Keep in mind that for each $x \in X$, $[x]$ is a set, not an element of X. For example, with the equivalence relation congruence mod 3 on $\mathbb{Z}$, we see that $\cdots \sim -3 \sim 0 \sim 3 \sim 6 \sim \ldots$, that $\cdots \sim -2 \sim 1 \sim 4 \sim 7 \sim \ldots$, and that $\cdots \sim -1 \sim 2 \sim 5 \sim 8 \sim \ldots$. Thus

$$[-3] = [0] = [3] = [6] = \{\cdots - 3, 0, 3, 6, \ldots\},$$

and so on, so that there are exactly three equivalence classes: namely, $[0]$, $[1]$ and $[2]$. For the trivial equivalence relation on $\mathbb{R}$, say, $[0] = \{0\}$, $[\pi] = \{\pi\}$, so that each equivalence class has exactly one point in it. Note that these examples illustrate the following key point about equivalence relations.

Lemma 2.3.1. *For $\sim$ any equivalence relation on a set X, any two equivalence classes are either equal or disjoint.*

Proof: Let $x, y \in X$. If $[x] \cap [y] \neq \emptyset$, then there exists a $z \in X$ such that $z \sim x$ and $z \sim y$. Since $\sim$ is transitive, we see that $x \sim y$, so that $[x] = [y]$ (as we can confirm by a simple double inclusion, using transitivity). So we've shown that if two equivalence classes are not disjoint, then they're equal. To see the other implication, if $[x] \neq [y]$, then we can conclude that $[x] \cap [y] = \emptyset$, since any $w \in [x] \cap [y]$ would be simultaneously equivalent to both x and y, allowing one to use transitivity to show the two equivalence classes are equal. □

The upshot of this lemma is that an equivalence relation $\sim$ on a set X has the effect of partitioning the entire set X into disjoint subsets called the equivalence classes.

For example, the relation congruence modulo 3 on $\mathbb{Z}$ partitions all the integers into three disjoint subsets, which we might denote as $3\mathbb{Z}, 3\mathbb{Z}+1$ and $3\mathbb{Z}+2$. We'll use the notation

$$X_{/\sim} = \{[x] : x \in X\}$$

to denote this set of all equivalence classes.

If this all looks vaguely familiar, recall that in group theory, given a subgroup $K \leq G$, the set of left cosets of K forms a partition of the group G. We use the notation G/K to denote this set of cosets:

$$G/K = \{gK : g \in G\}.$$

In other words, we've put an equivalence relation on G: $g_1 \sim g_2$ if and only if $g_1^{-1}g_2 \in K$. (It's easy to verify that $g_1 \sim g_2$ if and only if the left cosets g_1K and g_2K are equal.) If we're lucky enough that K is a *normal* subgroup of G (i.e. $gKg^{-1} = K$ for every $g \in G$), then G/K inherits the group operation from G, forming a factor group or quotient group.

Exercises

1. Let $f : X \to Y$ be any function. Prove that the relation $x_1 \sim x_2$ if and only if $f(x_1) = f(x_2)$ is an equivalence relation on X.

2. Define two points (x_1, y_1) and (x_2, y_2) in $\mathbb{R}^2$ to be equivalent if $y_1 + x_1^2 = y_2 + x_2^2$. Check that this is an equivalence relation, then describe and sketch the equivalence classes.

3. Define two points (x_1, y_1) and (x_2, y_2) in $\mathbb{R}^2$ to be equivalent if $y_1^2 + x_1^2 = y_2^2 + x_2^2$. Check that this is an equivalence relation, then describe and sketch the equivalence classes.

4. Critique this "proof" that every relation R on a set X that is both symmetric and transitive must be reflexive; since R is symmetric, xRy implies yRx, which together imply xRx by transitivity.

5. Consider the relation on the set $\mathbb{R}$: $x \sim y$ if $x - y \in \mathbb{Q}$. Is this an equivalence relation? Why or why not?

6. Consider the relation on the set $\mathbb{R}$: $x \sim y$ if $x \leq y$. Is this relation reflexive? Symmetric? Transitive?

7. Consider the relation on the set $\mathbb{R}$: $x \sim y$ if both $x > y$ and $y > x$. Is this reflexive? Symmetric? Transitive?

8. Consider the relation $\sim$ on the set H of all human beings: $x \sim y$ if x and y have the same eye color. Is this an equivalence relation? Why or why not?

2.4 INDUCTION

The principle of mathematical inductionis a very handy tool for proving properties of the natural numbers (or subsets of $\mathbb{Z}$ which are bounded below). To be truly rigorous about this, we should construct the set $\mathbb{N}$ from some axioms (perhaps the Peano version of such axioms[5]) and prove the induction principle from there. Instead, we'll assume that the reader has dealt with these formalities at some point in life and present a terse account of how to use induction to make your life easier.

Theorem 2.4.1. *(Induction principle) If $S \subset \mathbb{N}$ has the following two properties:*

- $1 \in S$
- *if $n \in S$ then $n+1 \in S$*

then $S = \mathbb{N}$.

We'll use this tool, in general, by letting S denote the set of natural numbers having some property P. We first show that the number 1 has property P, then prove the following implication: if n has property P, then $n+1$ must also have P. If we can do this, we can conclude that every natural number has property P. A good analogy for understanding this principle is that the implication if n has property P, then $n+1$ must also have P can be viewed as putting together an infinite row of dominoes, each balanced on its edge, close enough together that if domino n were to topple, it would push domino $n+1$ down, which would push domino $n+2$, and so forth. The initial case, $n = 1$, is akin to toppling the very first domino in the row, with the implication toppling the rest from there. Of course, one need not start at the number 1. If one can prove the implication if n has property P, then $n+1$ must also have P, then any integer can serve as a starting point. Note that the implication is thought of as "**if** property P is true for the number n, then it is also true for $n+1$." In proving this implication, then, one assumes that property P holds for a natural number n (the so-called "inductive hypothesis"), and then uses this to show that the property holds for $n+1$. This does *not* mean that we're assuming what we want to prove: we're proving that *if* it holds for any one n, then it also holds for $n+1$. A simple example will make this clear.

[5]Guiseppe Peano (1858–1932) developed a formalized language that was intended to contain mathematical logic and all important branches of mathematics. His axioms give a formal construction of the natural numbers. See Ref. [4] for details and context.

EXAMPLE 2.1

We'll use induction to prove the following: for any $n \in \mathbb{N}, 1+2+3+\cdots+n=\frac{n(n+1)}{2}$.

Of course, one might prefer to express this as $\sum_{i=1}^{n} i=\frac{n(n+1)}{2}$. This equation can be interpreted as giving a simple closed calculational form for the nth *triangular number*, as the figures below indicate:

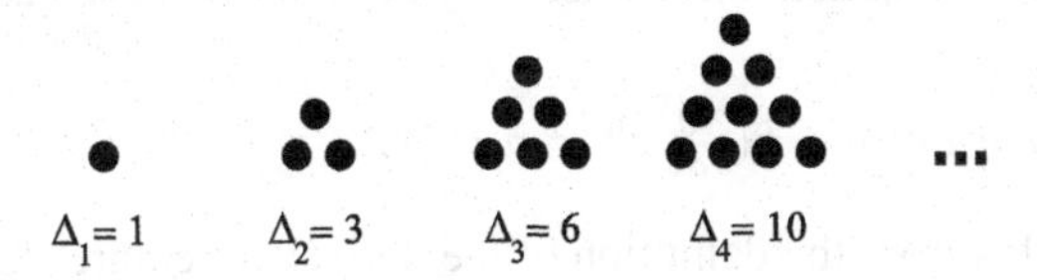

Proof: First, we observe that the initial case for $n=1$ is true: $1=\frac{1(1+1)}{2}=1$. Now we'll prove that if the formula holds for a natural number n, then it also holds for $n+1$. In other words, we wish to prove that if $1+2+\cdots+n=\frac{n(n+1)}{2}$, then $1+2+\cdots+(n+1)=\frac{(n+1)(n+2)}{2}$. To prove this, consider

$$\begin{aligned}
1+2+3+\cdots+(n+1) &= (1+2+3+\cdots+n)+(n+1) \\
&= \frac{n(n+1)}{2}+(n+1) \\
&= (n+1)\left(\frac{n}{2}+1\right) \\
&= (n+1)\left(\frac{n}{2}+\frac{2}{2}\right) \\
&= (n+1)\left(\frac{n+2}{2}\right) \\
&= \frac{(n+1)(n+2)}{2},
\end{aligned}$$

as we wished. By induction, then, we conclude that $1+2+3+\cdots+n=\frac{n(n+1)}{2}$ for every $n \in \mathbb{N}$. □

Similarly, one can provide easy inductive proofs of the various summation formulas that were so useful in working out Riemann integrals by definition in calculus:

$$\sum_{i=1}^{n} i^2=\frac{n(n+1)(2n+1)}{6}, \quad \sum_{i=1}^{n} i^3=\frac{n^2(n+1)^2}{4},$$

and so on. This last summation has a nice geometric explanation:[6]

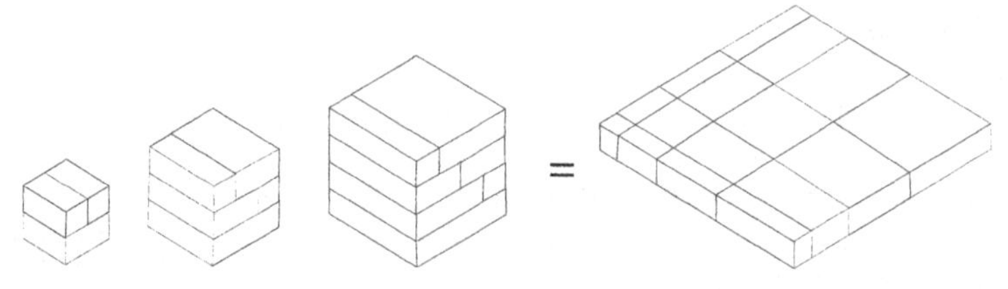

One can also use induction to establish very basic results, such as the following

Theorem 2.4.2. *For any* $n \in \mathbb{N}$, $\frac{d}{dx}x^n = nx^{n-1}$.

Proof: To prove this from the definition of the derivative requires knowledge of the Binomial Theorem. Here's an easy proof, which assumes knowledge of the product rule for derivatives. We first note that the initial $n+1$ case is easy: $\frac{d}{dx}x^1 = 1 = 1x^0$, as we expected. To prove the inductive implication, that $\frac{d}{dx}x^n = nx^{n-1}$ implies $\frac{d}{dx}x^{n+1} = (n+1)x^n$, we consider

$$\begin{aligned}\frac{d}{dx}x^{n+1} &= \frac{d}{dx}(x^n x) \\ &= \left(\frac{d}{dx}x^n\right)x + x^n\left(\frac{d}{dx}x\right) \quad \text{by the product rule} \\ &= nx^{n-1}x + x^n 1 \\ &= (n+1)x^n,\end{aligned}$$

as we had hoped. By induction, then, we see that $\frac{d}{dx}x^n = nx^{n-1}$ for all $n \in \mathbb{N}$. □

One can also use induction to prove things that are true for all natural numbers greater than or equal to some m, but perhaps false for the whole set $\mathbb{N}$. For example, we can use this slightly more general form of induction to prove that the inequality

$$2^n \geq 100n$$

is valid for all integers $n \geq 10$. Note that the inequality is false for $n = 1, 2, \ldots, 9$, but that $2^{10} = 1024 > 1000$, so that we can begin our induction at $n = 10$. Now we'll prove that if $2^n \geq 100n$, then $2^{n+1} \geq 100(n+1)$. To see why this implication

[6]A version of this picture appeared in the *American Mathematical Monthly* with no explanation, merely a title of *Behold!* and a signature, Michael Hirschorn. Neat, eh? The practice of labeling a geometric proof that requires no words with "Behold!" apparently started with the Hindu mathematician Bhaskara (c1150), whose dissection proof of the Pythagorean theorem was presented in this manner. Just to be completely clear, the picture shows that $1^3 + 2^3 + 3^3 + 4^3 = (1+2+3+4)^2$.

is true, consider

$$\begin{aligned} 2^{n+1} &= 2^n \cdot 2 \\ &\geq 100n \cdot 2, \text{ by the inductive hypothesis} \\ &= 100 \cdot (2n) \\ &> 100 \cdot (n+1), \text{ since } n > 1, \end{aligned}$$

as we wished to show. We conclude that the inequality holds for all $n \in \{10, 11, 12, \ldots\}$, by induction.

One must exercise care in using induction, however, as the following discussion shows. The "False Theorem" below is clearly not true (as the equestrian examples of Trigger and Black Beauty show), but the "proof" by induction is quite simple.

False Theorem: *All horses are the same color.*

"Proof": We proceed by induction. For each natural number n, we will prove that all sets of n horses consist only of horses of a single color. The $n = 1$ case is obvious, since any set with exactly one horse in it consists only of horses of one color. Now we'll prove the inductive implication: If all sets of n horses are single-colored, then all sets of $n + 1$ horses are single-colored. To see why this is true, consider any set H with $n + 1$ horses. Remove any horse h_1 from H, so that $H \setminus \{h_1\}$ has exactly n horses and must be single-colored by the inductive hypothesis. Now replace h_1 in H and remove any other horse h_2. Then $H \setminus \{h_2\}$ is single-colored, also. We conclude that all the horses in H are of one color, proving the inductive implication. By induction, we conclude that the False Theorem is valid. □

Exercises

1. Find the mistake in the "proof" of the False Theorem.

2. Prove the summation formulas:

$$\sum_{i=1}^{n} i^2 = \frac{n(n+1)(2n+1)}{6}, \quad \sum_{i=1}^{n} i^3 = \frac{n^2(n+1)^2}{4}.$$

3. Prove the generalized distributive law for $\mathbb{R}$:

$$b(a_1 + a_2 + \cdots + a_n) = ba_1 + ba_2 + \cdots + ba_n$$

for any $n \in \mathbb{N}$, where b and all the a_is are real numbers. [Of course, you already know the usual distributive law: $d(e + f) = de + df$.]

4. Prove that $\begin{bmatrix} 1 & 1 \\ 0 & 1 \end{bmatrix}^n = \begin{bmatrix} 1 & n \\ 0 & 1 \end{bmatrix}$ for all $n \in \mathbb{N}$.

5. Prove that $\begin{bmatrix} 2 & 0 \\ 0 & -1 \end{bmatrix}^n = \begin{bmatrix} 2^n & 0 \\ 0 & (-1)^n \end{bmatrix}$ for all $n \in \mathbb{N}$.

6. Find a suitable integer m and prove that $3^n > 200n$ for all $n \geq m$.

2.5 CARDINAL NUMBERS

The goal of this section is to provide a brief introduction to some of the profound ideas involving infinite sets. Along the way, we'll encounter the concept of cardinal numbers and their arithmetic. We'll be a bit light on rigor at times, but there will be enough details provided to make the ideas clear.

Definition 2.5.1. *Two sets A and B are equivalent (denoted $A \sim B$) if there exists a bijection $f : A \to B$.*

For example, the sets $\{1, 2, 3, \ldots, 26\}$ and $\{a, b, c, \ldots, z\}$ are equivalent, with 26! possible bijections (one-to-one and onto functions) from one set to the other. This "relation" of equivalence does indeed satisfy the reflexive, symmetric and transitive properties. However, we can't meaningfully discuss a "set of all sets,"[7] so this isn't really an equivalence relation in the precise sense when we think of sets in general. If we restrict our attention to just the subsets of X for some set X, then $\sim$ is indeed an equivalence relation on $\mathcal{P}(X)$, the power set of X.

Intuitively, the *cardinal number* of a set A [hereafter denoted $Card(A)$] is just the number of elements in the set. This would work reasonably well as a definition, if we were to restrict our attention to finite sets, since two sets that are equivalent would have cardinal numbers that are equal. However, what we commonly think of as infinite sets behave rather differently, as we see by the following examples.

■ EXAMPLE 2.2

The set $\mathbb{N}$ is equivalent to its proper subset $A = \{2, 3, 4, \ldots\}$, with the equivalence $f : \mathbb{N} \to A$ defined by $f(n) = n + 1$. We see that f is a bijection since its inverse function is $f^{-1}(m) = m - 1$.

This example seems very counterintuitive if one is used to dealing only with finite sets. We tend to think of a proper subset $R \subsetneq S$ as meaning that R is strictly "smaller" than S. However, we see that $A \sim \mathbb{N}$, despite the fact that $A \subsetneq \mathbb{N}$. One might be

[7]Beware of Russell's Paradox lurking in the bushes.

inclined to dismiss this example, since the set A is constructed by removing a single element from the set $\mathbb{N}$, but the following example shows that our intuition needs to be overhauled.

■ **EXAMPLE 2.3**

The set $\mathbb{N}$ is equivalent to its proper subset $\mathbb{E} = \{2, 4, 6, 8, \dots\}$, the set of even natural numbers, with the bijection $f : \mathbb{N} \to \mathbb{E}$ defined $f(n) = 2n$ (with inverse function $f^{-1}(m) = \frac{m}{2}$).

This sort of odd behavior prompts the following definition.

Definition 2.5.2. *A set A is infiniteif A is equivalent to some proper subset of itself.*

A set B is finite if and only if it is not infinite.

So our examples above show that the set $\mathbb{N}$ is infinite (but do not show that $A = \{2, 3, 4 \dots\}$ or $\mathbb{E}$ are infinite, at least directly by the definition). One can also show easily that the closed interval $[0, 1]$ is infinite, by this definition, by showing that $f : [0, \frac{1}{2}] \to [0, 1]$ given by $f(x) = 2x$ is a bijection. Similarly, one can show that $\mathbb{Z}$, $\mathbb{Q}$ and $\mathbb{R}$ are infinite (see below for details), although at least one of these requires a great deal of thought (see Example 2.5). Luckily, we can prove the following theorem, which shows that our definition of infinite agrees with our intuitive ideas.

Theorem 2.5.1. *A set A is infinite if and only if there exists a one-to-one function*

$f : \mathbb{N} \to A$.

Note that this implies that any infinite set contains a subset which is equivalent to $\mathbb{N}$, which fits our intuition well.

Sketch of proof: $\Longrightarrow$) Assume that A is infinite, so that $A \sim C$ for some $C \subsetneq A$. Since A is nonempty and $A \sim C$, there exists an element $a_1 \in C$. We will start our definition of the inclusion $f : \mathbb{N} \to A$ by letting $f(1) = a_1$. Now $C \setminus \{a_1\} \neq \emptyset$ (otherwise we can't get a bijection $A \to C$) so there exists an element $a_2 \in C \setminus \{a_1\}$, and we'll define $f(2) = a_2$. Similarly, $C \setminus \{a_1, a_2\} \neq \emptyset$ (else the function $A \to C$ wouldn't be a bijection), so we have $a_3 \in C \setminus \{a_1, a_2\}$ and define $f(3) = a_3$. We proceed in this way, choosing at each stage $a_n \in C \setminus \{a_1, a_2, \dots, a_{n-1}\}$. By induction, we've defined the desired function $f : \mathbb{N} \to A$.

$\Longleftarrow$) Let $f : \mathbb{N} \to A$ be the given one-to-one function, and let $D = Im(f)$. Then $D = \{d_1, d_2, d_3, \dots\}$, where d_i denotes $f(i)$. Then we have the following bijection $g : A \to A \setminus \{d_1\}$: $g(a) = a$ for every $a \in A \setminus D$ (i.e. $g|_{A \setminus D} = i_{A \setminus D}$ and $g(d_i) = d_{i+1}$. It's easy to check that g is a bijection, so that A is infinite, by Definition 2.5.2. □

We call this a "sketch of proof" rather than a proof, because some tedious details of the construction of the function f are suppressed. By the way, the proof of this theorem uses the "Axiom of Choice," which states that given a nonempty collection of nonempty sets, one may choose one element from each set. This innocuous sounding axiom has some interesting consequences, which we'll see later. Theorem 2.5.1 allows us to show that our intuitive ideas about finite sets are exactly correct, as well.

Theorem 2.5.2. *A set B is finite if and only if B is empty or $B \sim \{1, 2, \ldots, n\}$ for some $n \in \mathbb{N}$.*

Proof: $\Longrightarrow$) Let B be finite in the sense of Definition 2.5.2. So B is not equivalent to any proper subset of itself. Then, by Theorem 2.5.2, there is no one-to-one function $f : \mathbb{N} \to B$. If $B = \emptyset$, we're done, so assume that B is nonempty and there exists an element $b_1 \in B$. If $B \setminus \{b_1\}$ is empty, we're done, so assume $B \setminus \{b_1\}$ is nonempty. Thus we have $b_2 \in B \setminus \{b_1\}$. If $B \setminus \{b_1, b_2\}$ is empty, we're finished ($B \sim \{1, 2\}$), so assume there exists $b_3 \in B \setminus \{b_1, b_2\}$. Continue in this manner. Since there is no injection $f : \mathbb{N} \to B$, this process must terminate for some $n \in \mathbb{N}$, meaning that $B \setminus \{b_1, b_2, \ldots, n\} = \emptyset$, so that $B \sim \{1, 2, \ldots, n\}$.

$\Longleftarrow$) If $B = \emptyset$, then B is not equivalent to any proper subset of itself and we're done. Assume $B \sim \{1, 2, \ldots, n\}$ for some $n \in \mathbb{N}$. We show that $B \not\sim C$ for any proper subset $C \subsetneq B$ by induction. We let $S \subset \mathbb{N}$ consist of the set of natural numbers k such that any set equivalent to $\{1, 2, \ldots, k\}$ is not equivalent to any proper subset of itself. Observe that $1 \in S$, since the only proper subset of a singleton set $\{a\}$ is the empty set. Assume that $n \in S$. Consider a set $R = \{r_1, r_2, \ldots, r_{n+1}\}$, and let $T \subsetneq R$. Without loss of generality, assume $r_1 \in T$. Then

$$T \setminus \{r_1\} \subsetneq R \setminus \{r_1\} \sim \{1, 2, \ldots, n\},$$

so that $T \setminus \{r_1\} \not\sim R \setminus \{r_1\}$ by the inductive hypothesis. Thus $T \not\sim R$ and we see that $n + 1 \in S$. By induction, $S = \mathbb{N}$ and the theorem is proved. □

So we have a useful criterion to determine whether a set is finite or infinite. We'll delve a bit further into the intricacies of infinite sets.

Definition 2.5.3. *A set A is said to be countable if there exists an onto function $f : \mathbb{N} \to A$. Otherwise, we say that A is uncountable.*

Note that any finite set is countable by this definition. (Why? Think carefully.) We'll refer to sets like $\mathbb{N}$ and $\mathbb{E}$ as *countably infinite*. It's not too hard to prove that a set that is countably infinite is equivalent to $\mathbb{N}$, using Theorem 2.5.1. Note that a bijection $f : \mathbb{N} \to A$ is a "counting" of the elements of A: $f(1) = a_1, f(2) = a_2$, and so forth.

EXAMPLE 2.4

The set $\mathbb{Z}$ of integers is countably infinite.

On the surface, this seems unlikely, since $\mathbb{Z} = \{\ldots, -2, -1, 0, 1, 2, \ldots\}$ seems twice as large as $\mathbb{N}$. However, it's quite easy to count the integers, though, by starting at 0 and then alternating between positive and negatives:

$$\begin{array}{ccccccccccc} \mathbb{Z} & \cdots & -3 & -2 & -1 & 0 & 1 & 2 & 3 & \cdots \\ \uparrow & \cdots & \uparrow & \uparrow & \uparrow & \uparrow & \uparrow & \uparrow & \uparrow & \cdots \\ \mathbb{N} & \cdots & 7 & 5 & 3 & 1 & 2 & 4 & 6 & \cdots \end{array}$$

This counting is given by the function $f : \mathbb{N} \to \mathbb{Z}$ defined by

$$f(n) = \begin{cases} 0 & \text{if } n = 1 \\ \frac{-n+1}{2}, & \text{for } n \text{ odd} \\ \frac{n}{2} & \text{for } n \text{ even.} \end{cases}$$

In essence, we're "hopping back and forth" between the positives and negatives. Note that this verifies that the set $\mathbb{Z}$ is infinite, by our definition.

In fact, it's easy to prove the following result.

Theorem 2.5.3.

1. *The finite union or finite product of finite sets is finite.*

2. *The finite union or finite product of countable sets is countable.*

The proof is deferred to the exercises.

Our intuition is about to be strained further. Consider the set of rational numbers, $\mathbb{Q} = \{\frac{m}{n} : m \in \mathbb{Z}, n \in \mathbb{N}\}$. On the surface, this looks like it's equivalent to the union of infinitely many copies of $\mathbb{Z}$, because the positive rationals contain the following elements:

$$\begin{array}{cccccc} \frac{1}{1} & \frac{1}{2} & \frac{1}{3} & \frac{1}{4} & \frac{1}{5} & \cdots \\ \frac{2}{1} & \frac{2}{2} & \frac{2}{3} & \frac{2}{4} & \frac{2}{5} & \cdots \\ \frac{3}{1} & \frac{3}{2} & \frac{3}{3} & \frac{3}{4} & \frac{3}{5} & \cdots \\ \frac{4}{1} & \frac{4}{2} & \frac{4}{3} & \frac{4}{4} & \frac{4}{5} & \cdots \\ \frac{5}{1} & \frac{5}{2} & \frac{5}{3} & \frac{5}{4} & \frac{5}{5} & \cdots \\ \vdots & \vdots & \vdots & \vdots & \vdots & \cdots \end{array}$$

This table of elements is quite redundant, since every rational number appears infinitely many times (with $\frac{1}{3} = \frac{2}{6} = \frac{3}{9} = \ldots$, etc). However, it's clear from this presentation that the "hopping" procedure used to show that $\mathbb{Z}$ is countable is doomed to failure here. If we try counting row by row, we run out of natural numbers before we finish the first row. If we try counting column by column, we run into the same difficulty. If we try to do the first two rows first (just like counting $\mathbb{Z}$), we run out of naturals before getting to row 3, and so forth. However, we can count the fractions in this table by counting "down the diagonals" of the table – count in this order : $\frac{1}{1}, \frac{1}{2}, \frac{2}{1}, \frac{1}{3}, \frac{2}{2}, \frac{3}{1}, \frac{1}{4}, \frac{2}{3}, \frac{3}{2}, \frac{4}{1}, \frac{5}{1}, \cdots$. In other words, we define a function $f : Table \to \mathbb{N}$ by $f(\frac{m}{n}) = \frac{1}{2}(m + n - 2)(m + n - 1) + m$. For example, $f(\frac{3}{2}) = \frac{1}{2}(3 + 2 - 2)(3 + 2 - 1) + 3 = 9$, and we saw that $\frac{3}{2}$ was the ninth fraction counted. This doesn't show that the set $\mathbb{Q}$ is countable, because of the lack of uniqueness of representations of fractions ($\frac{1}{3} = \frac{2}{6}$, etc.) but it does give an function from the set $\mathbb{N}$ *onto* the set of positive rationals (by looking at f^{-1}). Combining this with the obvious assignments for 0 and the negative rationals, we have a function $g : \mathbb{Z} \to \mathbb{Q}$ which is onto, verifying the following example.

■ EXAMPLE 2.5

The set $\mathbb{Q}$ of rational numbers is countably infinite.

The following example is due to Georg Cantor and caused an enormous uproar in mathematics when it first appeared.[8]

■ EXAMPLE 2.6

The set $\mathbb{R}$ of all real numbers is uncountable; that is, there is no onto function $f : \mathbb{N} \to \mathbb{R}$.

Rather than proving this for all real numbers, we'll prove an even more surprising result, as follows.

[8]Georg Cantor (1845–1918) was the first mathematician to consider the problems raised by infinite sets in a systematic fashion. His counterintuitive results caused a backlash in the mathematical community, bringing some degree of doubt about any result using infinite sets or infinite processes (such as nearly all of calculus). The most extreme view was voiced by the "constructivists," who demanded that proofs eschew the Axiom of Choice in order to be valid. This school of thought seems to have faded from mainstream thought in modern mathematics. See Ref. [4] for details. By the way, Cantor spent a considerable amount of time in mental institutions, so you should be warned about the perils of pondering infinite sets.

Theorem 2.5.4. *The closed interval* $[0, 1]$ *is uncountable.*

This result may seem quite hard to believe, given that the set $\mathbb{Q}$ of rational numbers *is* countable, and given how few irrational numbers we know. Of course, we're aware that π, $\sqrt{2}$, e, and so on are all irrational (as are all of their rational multiples), but this set seems unlikely to be larger in any essential way than $\mathbb{Q}$. However, there are indeed *more* irrational numbers than rationals in the set $\mathbb{R}$.

We present a sketch of the proof, rather than present all the details necessary for complete rigor. We'll indicate where we've "fudged" a bit in the proof afterward. This proof is originally due to Cantor. A nice presentation of it appears in Ref. [9].
Sketch of proof: Each real number x in the interval $[0, 1]$ can be expressed as a decimal $x = 0.x_1x_2x_3 \dots x_n \dots$, where each x_i is a digit $0, 1, 2, \dots, 9$. (Yes, we mean to include 1 in the interval, since $1 = 0.99999\dots$.) Assume that we have a counting $a : \mathbb{N} \to [0, 1]$. Then we can, in principle, "list" every element of $[0, 1]$ as follows:

$$\begin{aligned} a_1 &= 0.a_{11}a_{12}a_{13}a_{14} \dots a_{1n} \dots \\ a_2 &= 0.a_{21}a_{22}a_{23}a_{24} \dots a_{2n} \dots \\ a_3 &= 0.a_{31}a_{32}a_{33}a_{34} \dots a_{3n} \dots \\ a_4 &= 0.a_{41}a_{42}a_{43}a_{44} \dots a_{4n} \dots \\ &\vdots \\ a_m &= 0.a_{m1}a_{m2}a_{m3}a_{m4} \dots a_{mn} \dots \\ &\vdots \end{aligned}$$

Since a is a bijection, *every* element of $[0, 1]$ shows up on the list as some $a(i) = a_i$. However, we can construct the following element:

$$b = 0.b_1b_2b_3b_4 \dots b_n \dots,$$

where b_1 is any digit $0, 1, 2, \dots, 9$ except that we require that $b_1 \neq a_{11}$. Similarly, b_2 is any digit, except $b_2 \neq a_{22}$, and so forth: $b_3 \neq a_{33}, \dots, b_n \neq a_{nn}, \dots$. So the element $b \neq a_1$, since b and a_1 differ in the first decimal place. Similarly, $b \neq a_2$ since b and a_2 differ in the second decimal place. Indeed, b cannot be any of the a_ns on the list, since b and a_n differ in the nth decimal place. We conclude that the function f cannot be onto, since $b \in [0, 1]$ is not in the image of f. This contradicts our assumption that $[0, 1]$ is countable, so that we conclude that $[0, 1]$ is uncountable. □

Some comments on the proof are in order.

1. We used a proof by contradiction, but it's easy to see that our proof shows that no function $f : \mathbb{N} \to [0, 1]$ can be onto, so that there's a simple direct proof hiding inside ours (for those who dislike indirect proofs). This proof seems clearer the first time through, however.

2. The proof constructs not just one $b \notin Im(f)$, but rather infinitely many such bs, since one has 9 choices for each decimal place b_i that differ from a_{ii}.

3. The proof uses the Axiom of Choice in a very essential way, which caused some of the controversy surrounding this axiom.[9]

4. The proof is not precisely correct as stated, since it fails to account for the nonunique decimal representation of real numbers. Because, for example, $1.0000\cdots = 0.99999\ldots$, we would need to specify a unique way which of these representations of the number 1 we use (to avoid the number b actually showing up in the list as an alternate representation of some a_i). This is not too hard to do: specify that whatever digit one chooses for b_i (out of the nine that are different from a_{ii}), we will *never* choose b_i to be 9 itself. We then make a rule that our decimal representations of numbers will never end with an infinite sequence of 9s.

Now we have enough background to begin our exploration of cardinal numbers and their arithmetic. We'll use an axiomatic approach. The word "number" here should be interpreted rather loosely, of course, since we're interested in infinite sets as well as finite ones.
The following are our *axioms/definitions for cardinal numbers*:

1. For every set A, there exists a cardinal number $Card(A)$.

2. If $A \sim B$, then $Card(A) = Card(B)$.

3. A is equivalent to a subset of B if and only if $Card(A) \leq Card(B)$.

4. If $A \subset B$ and $A \not\sim B$, then $Card(A) < Card(B)$.

Lots of "cardinal numbering systems" could fit these axioms. We'll specify one to be used from here on: $Card(\emptyset) = 0$ and $Card(\{1, 2, 3, \ldots, n\}) = n$. Thus $Card(\{a, b, c, \ldots, z\}) = 26$. We have a striking inequality, due to Theorem 2.5.4:

$$Card(\mathbb{N}) < Card(\mathbb{R}).$$

Intuitively, this means that there are numbers larger than what we commonly refer to as "infinity": $Card(\mathbb{N})$. This sort of counterintuitive result is what caused the

[9]Most of the controversy was no doubt due to the Well-ordering Theorem, which states that any set can be given an ordering in which every nonempty subset has a least element. This result seems *extremely* counterintuitive when one thinks about an uncountable set like $\mathbb{R}$, which is very far from well-ordered in its *usual* ordering. The Well-ordering Theorem is, in fact, equivalent to the Axiom of Choice. See Ref. [9] for details.

mathematical community to be slow to accept Cantor's results, although the results have become widely accepted.

Note that we have no names yet for cardinal numbers like $Card(\mathbb{N}) = Card(\mathbb{Z}) = Card(\mathbb{Q}) < Card(\mathbb{R})$. We can call them whatever we want. Over the years, challenged with naming these numbers, my students have suggested, for example, calling $Card(\mathbb{N})$ "Henry," with the resulting inequality that Henry $> n$ for any $n \in \mathbb{N}$. Cantor's notation has become pretty standard: $Card(\mathbb{N}) = \aleph_0$ (pronounced "aleph-nought") and $Card(\mathbb{R}) = c$ (for "continuum"). So we have the following sequence of cardinal numbers and sets representing them:

Card	0	1	2	3	...	$\aleph_0$	...	c	...
Set	$\emptyset$	$\{a\}$	$\{e, f\}$	$\{1, 2, 3\}$	...	$\mathbb{N}$	...	$\mathbb{R}$	...

Note that extra space was left in the list beyond $c = Card(\mathbb{R})$. Are there sets with cardinal number larger than that of $\mathbb{R}$? To see what the answer is to this question, we first back up to some simple finite sets.

Theorem 2.5.5. *If* $Card(A) = n$ *for any* $n \in \mathbb{N}$*, then* $Card(\mathcal{P}(A)) = 2^n$.

The proof is an exercise.

In light of this, the following theorem (called Cantor's Theorem) is not surprising.

Theorem 2.5.6. *For any set* A*, we have*

$$Card(\mathcal{P}(A)) > Card(A).$$

Note especially that this applies to infinite sets as well as finite.

Proof: The proof is quite similar to Cantor's proof of Theorem 2.5.4. First, note that A is equivalent to a proper subset of $\mathcal{P}(A)$, by a function $s : A \to \mathcal{P}(A)$ given by $s(a) = \{a\}$. In other words, just the singleton subsets of A are in one-to-one correspondence with the elements of A, with possibly lots of other subsets of A in $\mathcal{P}(A)$ unaccounted for. This shows $Card(\mathcal{P}(A)) \geq Card(A)$. We now show that this inequality is strict. Let $f : A \to \mathcal{P}(A)$ be any one-to-one function. So f assigns to each element a a subset $f(a) \subset A$. We'll show that f cannot be onto. Let $Z = \{a \in A : a \notin f(a)\}$. Then $Z \in \mathcal{P}(A)$, but Z is not in the image of the function f, because if $Z = f(x)$ for some $x \in A$, then $x \notin Z$, by the definition of Z. But if $x \notin f(x) = Z$, then $x \in Z$. This contradicts our assumption that $Z \in Im(f)$. Since no one-to-one function $f : A \to \mathcal{P}(A)$ can be onto, we see that $Card(\mathcal{P}(A)) > Card(A)$. □

Thus $Card(\mathcal{P}(\mathbb{R})) > Card(\mathbb{R})$, and $Card(\mathcal{P}(\mathcal{P}(\mathbb{R}))) > Card(\mathcal{P}(\mathbb{R}))$, and so on, so that the chain of cardinals goes on forever.

Note also the dots between $\aleph_0 = Card(\mathbb{N})$ and $c = Card(\mathbb{R})$ in our listing of the cardinal numbers. Are there any cardinal numbers between these two [i.e. are there any sets W such that $\mathbb{N} \sim Y$ with $Y \subset W$ and $W \sim Z$ for some $Z \subset \mathbb{R}$, with $\aleph_0 < Card(W) < Card(\mathbb{R})$]? This question is extremely deep. Cantor conjectured that no such set W exists and referred to this conjecture as the continuum hypothesis. Kurt Godel proved in 1940 that the continuum hypothesis is consistent with the axioms of set theory. Paul Cohen proved in 1960 that it's independent of the axioms of set theory. Thus, the continuum hypothesis is, according to the axioms of set theory, completely undecidable: it's neither true nor false. Most modern set theorists assume the continuum hypothesis as an added axiom, tacked onto the usual Zermelo-Frankel axioms. The generalized continuum hypothesis asserts the nonexistence of cardinals between $Card(\mathcal{P}(\mathbb{R}))$ and $Card(\mathbb{R})$, between $Card(\mathcal{P}(\mathcal{P}(\mathbb{R})))$ and $Card(\mathcal{P}(\mathbb{R}))$, and so forth.

Exercises

1. Prove that the "relation" of $\sim$ on sets is reflexive, symmetric and transitive.

2. Show that the set $\mathbb{N}$ is equivalent to the set $\{1, 4, 7, 10, \ldots\}$, the set of natural numbers congruent to 1 modulo 3. Note that this verifies that the set $\mathbb{N}$ is infinite, by our definition, but not that $\{1, 4, 7, 10, \ldots\}$ is.

3. Verify that the set $\{1, 4, 7, 10, \ldots\}$ is infinite, by Definition 2.5.2

4. Prove that the finite union or finite product of finite sets is finite.

5. Prove that the finite union of countable sets is countable. ((*Hint: think of the counting of $\mathbb{Z}$ in Example 2.4. Also, to show that any finite "thing" has a certain property, induction is a very useful tool. Show that the desired result works for the case $n = 2$, then use this to do the inductive implication.*)

6. *Prove that the countable union of countable sets is countable. (Hint: think of the table of positive rationals in Example 2.5.)*

7. *Prove that the finite product of countable sets is countable. (Hint: think of the counting of the table of positive rationals in Example 2.5.)*

8. *Verify the formula counting the table of positive rationals for the case $\frac{3}{5}$.*

9. *Prove (directly from the definition) that the set $\mathbb{R}$ of real numbers is infinite. (Hint: Can you think of a 1-1, continuous function f from some interval to the whole real line?)*

10. *Prove that if c is irrational then rc is irrational for any nonzero $r \in \mathbb{Q}$.*

11. *Prove that if $Card(A) = n$ for any $n \in \mathbb{N}$, then $Card(\mathcal{P}(A)) = 2^n$. [Hint: any subset B of A can be specified by designating which elements of A are in B and which are not. Use this idea to associate to each $B \subset A$ a unique function $\chi_B : A \to \{0, 1\}$. (Such a function is called the "characteristic function of B.") Now show there are 2^n such functions.]*

2.6 GROUPS

This section sketches out the basics of groups and homomorphisms, sufficient to take the student through the work on fundamental groups later. The ideas here will also help clarify the use of the group analogy in Chapter 1 and the questions about groups in Section 2.1.

The idea of a group is one of the most fundamental concepts in all of mathematics. It occurs very naturally in nearly all areas of math (and in many parts of physics). It is appropriate, then, that the idea seems to have arisen as a result of separate streams of development in several areas of mathematics, roughly at the same time. (See Ref. [18] and especially Ref. [20] for more details on how the concept of an abstract group came into being.) First, the study of geometry underwent enormous change in the beginning of the nineteenth century, becoming less dependent on measurement and more on "incidence," as seen in projective or other noneuclidean geometries. In particular, the work of A. F. Moebius[10] in the 1820s began a rather fundamental reassessment of what is meant by a particular geometry, thinking of one as the study of properties invariant under certain transformations. Although Moebius seems never to have thought about the idea of a group as an interesting object in its own right, his work anticipated Felix Klein's Erlangen program remarkably closely. The idea of a group also seems to be implicit in Julius Plucker's work on "line geometries."

[10] Yes, the same Moebius as the Moebius band! August Ferdinand Moebius (1790–1868) studied under C. F. Gauss, but spent much of his later career rather isolated from the mathematical community. Because of this,

Second, the area of number theory advanced dramatically in the late eighteenth and early nineteenth centuries, largely because of the work of Leonard Euler and Carl Friedrich Gauss. Euler studied the remainders modulo n of powers of a number, with the idea of a group being implicitly used. Gauss went further with Euler's ideas, and in the context of number theory proves several results about subgroups of cyclic groups.

Third, a very active area of mathematics in the late eighteenth century was the theory of algebraic equations. In particular, many mathematicians looked at the question of which polynomial equations over the rationals could be solved by radicals. Ruffini claimed to have proved in 1799 that quintic polynomials were not solvable, using groups of permutations, although there were gaps in his proof. Niels Abel gave the first generally accepted proof of this result in 1824, using ideas of permutations of roots but apparently not pushing the notion of an abstract group any further. Evariste Galois' work in the early 1830s made the fundamental link between the algebraic solution of an equation and the group of permutations related to it. He worked out the idea now known as *normal subgroups* and showed that the smallest simple group (with no proper nontrivial normal subgroups) has 60 elements. Galois' work was not published until 1846, long after his tragic death and about the time that Cauchy's major work on permutations appeared.

With the idea of a group having shown up in so many areas roughly contemporaneously, it had to have been clear that the idea was of importance in its own right. Jordan's work with permutation groups in the 1860s, Klein's group-theoretic description of geometry in the 1870s and Cayley's work on permutations and matrices all point to how central this idea had become. However, it seems to have taken Cayley's series of papers on "The theory of groups" in the late 1870s for the area to take root on its own.

So what *is* a group, anyway?

Definition 2.6.1. *A group is a set G with a binary operation $\bullet : G \times G \to G$ such that*

1. *The operation $\bullet$ is associative; that is, for every $g, g_2, g_3 \in G$,*

$$g_1 \bullet (g_2 \bullet g_3) = (g_1 \bullet g_2) \bullet g_3.$$

his early work on the idea of looking at geometries in terms of properties invariant under transformations was largely ignored at the time. Later, his ideas won a great deal of attention as others came to understand this rather abstract notion of a geometry.

2. *G contains an identity for the operation $\bullet$; that is, there exists an element $e \in G$ such that for every $g \in G$*

$$g \bullet e = g = e \bullet g.$$

3. *Each element of G has an inverse; that is, for every $g \in G$, there exists and element $g^{-1} \in G$ such that*

$$g \bullet g^{-1} = e = g^{-1} \bullet g.$$

To make sense of this, we'll first look at some very familiar examples of sets and operations to see which of them form groups. Then we'll look a bit more closely at *why* these properties are what we want to group to have, as opposed to other nice properties of sets and operations, like commutativity. When the operation in G is clear, we'll often denote it by juxtaposition: $g_1 g_2 = g_1 \bullet g_2$.

First, consider the set of natural numbers, $\mathbb{N} = \{1, 2, 3, \dots\}$, under the operation of usual addition, +. This is a very nice binary operation on the set $\mathbb{N}$, and you probably remember from very early on that usual addition is associative, so $\mathbb{N}$ under + has at least a chance of being a group. However, there is no identity for addition in $\mathbb{N}$, since, for example, $3 + x \neq 3$ for every $x \in \mathbb{N}$, so that $\mathbb{N}$ under + is *not* a group. Clearly, we'd "like" the identity to be 0, but zero isn't in the set of naturals. The obvious way to "fix" this is to look at the set of nonnegative integers, $\mathbb{N}^+ = \{0, 1, 2, 3, \dots\}$, under addition so that we've somewhat artificially put an identity for the operation into the set. Even $\mathbb{N}^+$ fails to be a group under addition, however, since it fails to contain inverses for its nonzero elements. For example, there is no number x in $\mathbb{N}^+$ that yields the identity 0 when added to 3. In other words, we have no inverse for 3 in $\mathbb{N}^+$.

This suggests what we need to add in to the natural numbers in order to form a group under addition; consider the set of integers $\mathbb{Z} = \{\dots, -2, -1, 0, 1, 2, \dots\}$ under addition. The operation is associative, has an identity, namely, 0, and has inverses for every element: $x^{-1} = -x$. Further, the group $\mathbb{Z}$ under addition has a property that many other groups do not: $\mathbb{Z}, +$ is commutative. That is $x + y = y + x$ for every $x \in \mathbb{Z}$. (It is customary to use the term *abelian* in place of commutative for groups, in honor of N. H. Abel.)

One might reasonably ask about whether the integers form a group under other operations, such as multiplication. Note that the operation $\bullet$ on $\mathbb{Z}$ is associative, and there is a multiplicative identity, namely, 1, but that we have problems finding inverses: $3 \bullet x = 1$ has no solution in $\mathbb{Z}$. This suggests again what we might add in to the integers to get a group under multiplication; consider the set of rational numbers $\mathbb{Q} = \{\frac{m}{n} : m \in \mathbb{Z}, n \in \mathbb{N}\}$. This set *almost* forms a group under multiplication: $\bullet$ is indeed associative in $\mathbb{Q}$, $1 \in \mathbb{Q}$ is the identity, and it looks, at first glance, as if inverses are indeed present. For example, $3^{-1} = \frac{1}{3}$, since $3 \bullet \frac{1}{3} = 1$. In fact, it appears that $(\frac{a}{b})^{-1} = \frac{b}{a}$, until one thinks a bit more carefully: 0 has no multiplicative inverse, so that $\mathbb{Q}$ under $\bullet$ is not a group. One can circumvent this problem by looking instead at $\mathbb{Q}^x$, the set of nonzero rational numbers, which does indeed form an abelian group under multiplication.

Looking just at groups formed of familiar sets of numbers is inappropriate, though, because it hides just how natural this structure is and how it pervades mathematics. As a better example, consider

$$M_2(\mathbb{R}) = \left\{ \begin{bmatrix} a & b \\ c & d \end{bmatrix} : a, b, c, d \in \mathbb{R} \right\},$$

the set of 2×2 matrices with real entries. Under addition of matrices, $M_2(\mathbb{R})$ is a group, with the zero matrix as the identity and the inverse of a matrix A being $-A$. Under matrix multiplication, the situation is more interesting in several ways. First, the identity matrix is

$$I = \begin{bmatrix} 1 & 0 \\ 0 & 1 \end{bmatrix},$$

since $AI = A = IA$ for every $A \in M_2(\mathbb{R})$. However, many two by two matrices have no inverse. For example, the equation

$$\begin{bmatrix} 1 & 0 \\ 0 & 0 \end{bmatrix} X = I = \begin{bmatrix} 1 & 0 \\ 0 & 1 \end{bmatrix}$$

has no solution X in $M_2(\mathbb{R})$. One must restrict attention to the set of invertible 2×2 matrices:

$$Gl_2(\mathbb{R}) = \{ \begin{bmatrix} a & b \\ c & d \end{bmatrix} : a, b, c, d \in \mathbb{R}, ac - bd \neq 0\} \subset M_2(\mathbb{R}),$$

the *general linear group* of 2×2 invertible real matrices. This group is nonabelian, as one can check by computing

$$\begin{bmatrix} 1 & 1 \\ 1 & 1 \end{bmatrix} \begin{bmatrix} 1 & 1 \\ 1 & 0 \end{bmatrix} = \begin{bmatrix} 2 & 1 \\ 2 & 1 \end{bmatrix} \neq \begin{bmatrix} 2 & 2 \\ 1 & 1 \end{bmatrix} = \begin{bmatrix} 1 & 1 \\ 1 & 0 \end{bmatrix} \begin{bmatrix} 1 & 1 \\ 1 & 1 \end{bmatrix}.$$

Among the first examples of groups studied by geometers and algebraists were the symmetry groups. Perhaps the best way to understand this family of groups is to look carefully at a particular example, the group of symmetries of an equilateral triangle. A *symmetry* of such a triangle is a transformation of the triangle that preserves its shape. For example, we denote by r the symmetry that rotates the triangle by 120°:

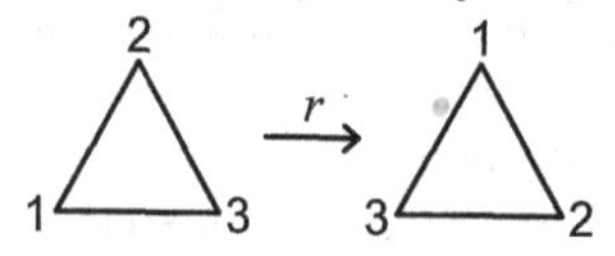

The symmetry r is one of three rotations, along with e and s, specified below. In addition, we have three reflections (or "flips"), such as

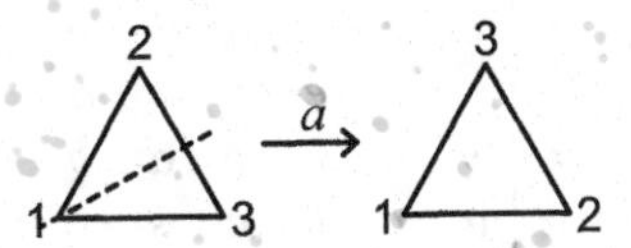

Here is a "glossary" of the 6 symmetries of a triangle:

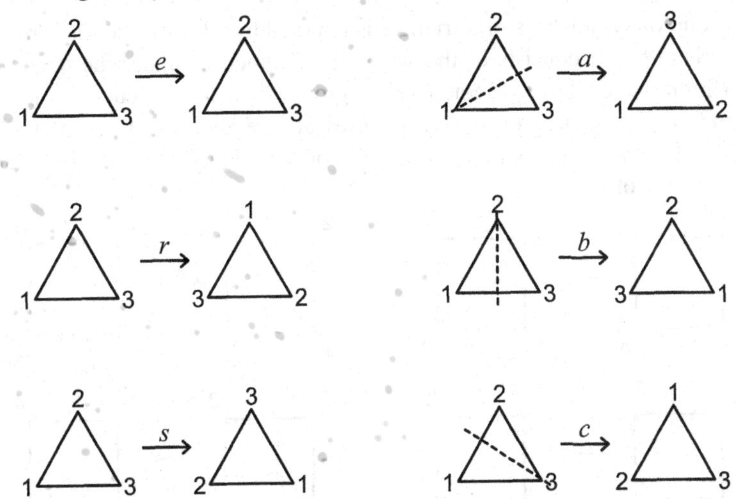

These six symmetries are the elements of the group that we want. For the operation, we compose the transformations, although it's better to think of the operation as *concatenation* – simply performing the symmetries one after the other. For example, here's how one determines which of the six symmetries one gets from doing first r, then a:

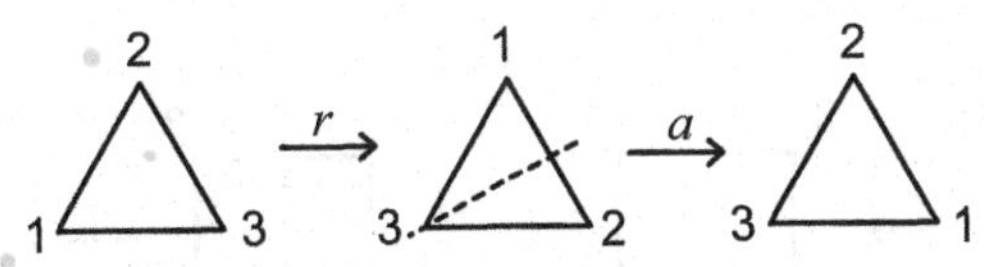

At this point, it would be instructive to construct a physical manipulative to help sort out how this operation on symmetries works. In other words, cut out a triangle from a piece of cardstock and carefully label each of the three vertices as above. Then physically perform the rotations and reflections to verify the following table, which gives the concatenations of all possible pairs of symmetries of the triangle. Note that the table is read as follows: the row specifies the first element in the "product," and the column the second – thus $ra = b$ but $ar = c$.

Concatenation of symmetries of a triangle

	e	r	s	a	b	c
e	e	r	s	a	b	c
r	r	s	e	b	c	a
s	s	e	r	c	a	b
a	a	c	b	e	s	r
b	b	a	c	r	e	s
c	c	b	a	s	r	e

The group of symmetries of a triangle is denoted by D_3, for the *dihedral group* on three "letters." It consists of the set of six symmetries under the operation of concatenation. Note that D_3 is a nonabelian group, since $ra = b$ but $ar = c$.

In general, D_n is the group of symmetries of a regular n-gon, consisting of n rotations and n reflections, with the operation of concatenation. Here, for example, are the elements of D_4:

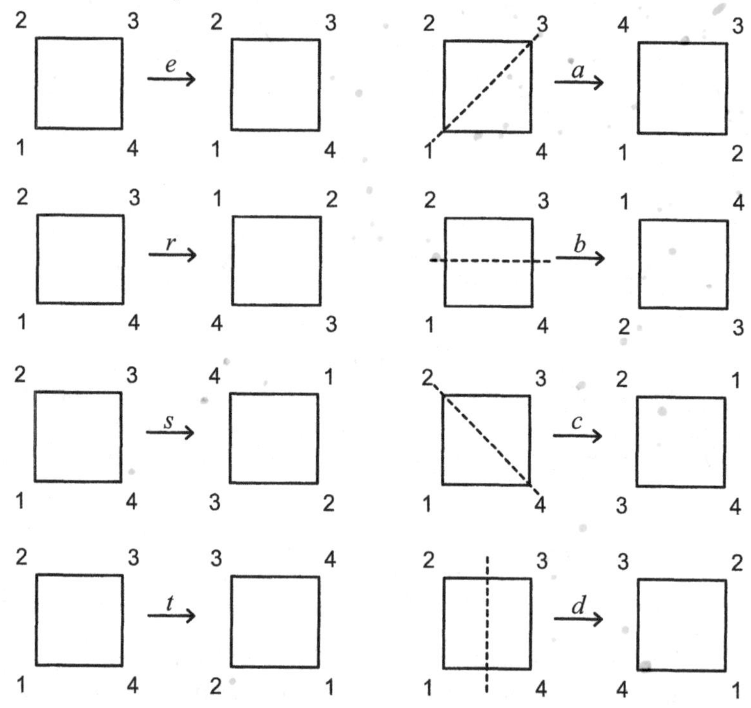

One can see that the dihedral groups are nonabelian (for $n \geq 3$) by looking at the concatenation of any nontrivial rotation with any reflection.

At this point, it's appropriate to look more closely at the *definition* of a group and try to see why the properties listed are so important. The key reason is that the properties of associativity, identity and inverse are the minimum that one needs to be able to solve equations. For example, in the group D_3, one can find the solution to the equation $aX = s$ by looking at the table and seeing that $ab = s$, so that the unknown X must equal b. It's much more enlightening to solve the equation explicitly, to see where the properties that define the group structure come into play. Considering the equation $aX = b$, we'll try to solve it as we solved equations in elementary algebra – by isolating the variable X. To do so, we'll "multiply" both sides of the equation by a^{-1} (on the left!), reasoning that if two elements are equal, then their products with a^{-1} must be equal. Note that $a^{-1} = a$, since $aa = e$. So, since $b = aX$, we see that

$$a^{-1}b = a^{-1}(aX) = (a^{-1}a)X = eX = X,$$

where the second equality holds because of associativity, the third by the definition of inverse, and the fourth because e is the identity in the group. Thus, $X = a^{-1}b = ab = s$, verifying what we found from the table. More importantly, we've seen where each of the the three properties of associativity, identity and inverse is necessary in the process.

This idea of solving equations in groups yields a pleasant consequence: an easy way to check whether a table for an operation on a finite set forms a group. Consider for example, the following table which describes an operation on the set $A = \{m, n, o\}$:

	m	n	o
m	m	n	o
n	n	o	m
o	o	m	n

Since each element shows up in each row and each column of the table, the equations $gX = h$ and $Xg = h$ all have solutions in A for each g and h in A. Thus the set A under this operation is potentially a group. (In fact, it is, as one can easily see, with m acting as the identity.) If any row failed to contain, say, the element n, then the equation $gX = n$ would have no solution in A for some $g \in A$. Similarly, if any column failed to contain n, then the equation $Xg = n$ would have no solution for some g.

Another family of groups that show up in many areas of mathematics are the cyclic groups. One can form these groups based on the integers by using "modular arithmetic." We say that two integers a and b are *congruent modulo* n (for any natural number n) if $a - b$ is divisible by n. It's trivial to check that this is an equivalence relation on $\mathbb{Z}$, partitioning $\mathbb{Z}$ into n distinct equivalence classes. This set of n equivalence classes is usually denoted by

$$\mathbb{Z}_{/n} = \{[0], [1], \ldots, [n-1]\}.$$

Under the operation of usual addition "reduced" modulo n, $\mathbb{Z}_{/n}$ is an abelian group of order n. To see why these groups are called "cyclic," one need only look at the table for $\mathbb{Z}_{/6}$ under addition modulo 6, for example:

$\mathbb{Z}_{/6}$ under addition modulo 6

	0	1	2	3	4	5
0	0	1	2	3	4	5
1	1	2	3	4	5	0
2	2	3	4	5	0	1
3	3	4	5	0	1	2
4	4	5	0	1	2	3
5	5	0	1	2	3	5

Here we use k as a shorthand for the equivalence class $[k]$. One can see easily that this group is abelian (that $xy = yx$ for all x and y in $\mathbb{Z}_{/6}$) by observing the symmetry across the line of slope -1. The identity is

In essentially all areas of mathematics, discussing any kind of "object" leads naturally to looking at "subobjects." Groups are no different in this regard.

Definition 2.6.2. *Let $G, \bullet$ be a group. A subset H of G is said to be a subgroup of G if H forms a group under the operation $\bullet$.*

The notation $A \leq G$ is often used as a shorthand notation for the statement "A is a subgroup of G." Note that the subset $\{e, r, s\}$ of the dihedral group D_3 is a subgroup, as the following table shows:

	e	r	s
e	e	r	s
r	r	s	e
s	s	e	r

Similarly, the set $\{0, 3\}$ is a subgroup of $\mathbb{Z}_{/6}$ under addition modulo 6, as one can easily confirm.

A simple criterion for checking whether a subset of a group is a subgroup is the following.

Theorem 2.6.1. *Let G be a group. A subset $A \subset G$ is a subgroup of G if and only if A is nonempty and whenever $a, b \in A$, then $ab^{-1} \in A$.*

Proof: $\Longrightarrow$) Let A be a subgroup of G. If $b \in A$, then its inverse b^{-1} must also be in A by the definition of a group. So if a and b are in A, then $ab^{-1} \in A$ since A is closed under the operation.

$\Longleftarrow$) Let $A \subset G$ satisfy the criterion of the theorem. First, A must contain the identity of G, which we'll denote by e, because A is nonempty: There must be some

$a \in A$, so that $aa^{-1} = e \in A$ by the criterion. Next, having $e \in A$ shows that A must contain an inverse for each of its elements, since if $a \in A$ then $1a^{-1} = a^{-1} \in A$. Note also that A is closed under the operation: If a and b are in A, then $ab = a(b^{-1})^{-1} \in A$. Finally, since the operation is associative in all of G, it must be associative when restricted to A. Thus A is a group under the operation inherited from G. □

One can "build" subgroups of a group G using any given set of elements of G. For example, if $a \in G$, the associativity shows that a^n is unambiguously defined for any $n \in \mathbb{N}$. (Why?) So one can consider, a finite group G and any $a \in G$, the set

$$\langle a \rangle := \{a, a^2, a^3, \ldots\}.$$

If $a^n = e$ for any $n \in \mathbb{N}$, then $\langle a \rangle$ is a finite subset of G which forms a subgroup called the subgroup *generated* by a. Note that we've borrowed a piece of notation from the computer scientists here, using the symbol := as a shorthand for "defined to be."

In any area of mathematics, one needs a clear notion of what an "equivalence of objects" means in that area. To a group theorist, two groups are equivalent if they are *isomorphic*. Intuitively, this means that the elements of one group are just a "relabeling" of the other group, in a way compatible with the operation, with this relabeling correspondence called an *isomorphism*.For example, the group A defined above and the subgroup $\{e, r, s\}$ of D_3 are isomorphic, with the most obvious correspondence being $m \mapsto e$, $n \mapsto r$ and $o \mapsto s$. (There is another correspondence that works, as well!) One clearly gets the "same" operation table after relabeling by this correspondence. To be more precise, we need to define a more general concept.

Definition 2.6.3. *Let $G, \bullet$ and $H, \star$ be groups. A homomorphism from $G, \bullet$ to $H, \star$ is a function $f : G \to H$ from one set to the other such that for every g_1 and g_2 in G, $f(g_1 \bullet g_2) = f(g_1) \star f(g_2)$.*

The requirement that $f(g_1 \bullet g_2) = f(g_1) \star f(g_2)$ is a precise way of saying that a homomorphism must respect the group operations. An *isomorphism* of groups is a homomorphism that is both 1–1 and onto.

For example, the function $f : \mathbb{Z}_{/3} \to D_3$ given by $0 \mapsto e$, $1 \mapsto r$ and $2 \mapsto s$ is a homomorphism whose image is the subgroup $\{e, r, s\} \subset D_3$. Thus the subgroup $\{e, r, s\}$ is isomorphic to $\mathbb{Z}_{/3}$. Similarly, as an exercise, one should check that the dihedral group D_4 contains a subgroup isomorphic to $\mathbb{Z}_{/4}$.

For another example, let $a \in G$ and consider the subgroup generated by a, $\langle a \rangle \leq G$, defined above. Note that if n is the least natural number with $a^n = e$, then $\langle a \rangle$ is a group of order n which is isomorphic to $\mathbb{Z}_{/n}$ To see why, one should check that the function $f : \mathbb{Z}_{/n} \to \langle a \rangle$ given by $f([k]) = a^k$ is a homomorphism.

Note that two groups which are isomorphic must have the same order, since an isomorphism is both a one-to-one and onto function. However, the converse of this statement is easily seen to be false, since $\mathbb{Z}_{/6}$ and D_3 are both groups of order 6. However, they are not isomorphic, since $\mathbb{Z}_{/6}$ is abelian and D_3 is not. Alternatively, $\mathbb{Z}_{/6}$ is generated by a single element ([1]), whereas D_3 requires at least two generators. (You can check this easily: look at all the powers of the elements of D_3. All the elements of D_3 generate subgroups of order 1, 2 or 3. So no function f from D_3 onto $\mathbb{Z}_{/6}$ can be a homomorphism, since homomorphisms send a^k to $(f(a))^k$ for all $k \in \mathbb{N}$.)

Since homomorphisms must respect the operations of the groups involved, we get the following result:

Theorem 2.6.2. *Let $f : G \to H$ be a homomorphism of groups. Then f sends the identity of G to the identity of H and $f(g^{-1}) = (f(g))^{-1}$ for every $g \in G$.*

The proof is left as an exercise.

One might expect that subgroups and homomorphisms are closely related.

Theorem 2.6.3. *If $f : G \to H$ is a homomorphism of groups, then $Im(f)$ is a subgroup of H.*

The proof is also left as an exercise.

A nice example of a homomorphism that points out several interesting properties is "reduction" mod n,

$$\rho_n : \mathbb{Z} \to \mathbb{Z}_{/n}$$

given by $\rho_n(k) = [k]$. Note that In fact, one can define this sort of reduction from $\mathbb{Z}_{/mn} \to \mathbb{Z}_{/n}$ as well. In the case of $\rho_3 : \mathbb{Z}_{/6} \to \mathbb{Z}_3$, we see that ρ_3 is onto, but not 1-1. Note that ρ_3 sends both 0 and 3 to the identity of the range, and that $\{0, 3\}$ is a subgroup of the domain $\mathbb{Z}_{/6}$. This is a universal phenomenon.

Definition 2.6.4. *For any homomorphism of groups $f : G \to H$, the kernel of f is*

$$Ker(f) := \{g \in G : f(g) = e_H\},$$

the set of elements mapping to the identity of H.

Theorem 2.6.4. *For any homomorphism of groups $f : G \to H$, $Ker(f)$ is a subgroup of G.*

The proof is an (easy) exercise.

Given a subgroup A of G, one might hope to form a "quotient group" by "modding out by A." This very vague idea can be made very precise using the idea of *cosets*. For $A \leq G$, a left coset of A is simply

$$gA := \{ga : a \in A\}$$

for any particular element $g \in G$. It's quite easy to show that two cosets of a subgroup A are either disjoint or equal, so that the set of cosets partitions the group G. In other words, we've put an equivalence relation on G: $g_1 \sim g_2$ if and only if $g_1^{-1}g_2 \in A$. (The proofs of these statements are left to the exercises. Yes, we're being rather terse with our discussion of cosets, but this topic will be used only very briefly in the sequel. Almost any introductory book on abstract algebra, such as Ref. [10] will give full details on this material.) To form a "quotient group," one hopes to make a group out of the set of cosets of A, which is usually denoted by G/A, by using the operation from the group G. In other words, we define the operation on G/A by $g_1A \bullet g_2A := (g_1g_2)A$. Unfortunately, this operation is not well-defined for an arbitrary subgroup $A \leq G$. If we're lucky enough that A has an additional property, that of being a *normal* subgroup of G (i.e. $gAg^{-1} = A$ for every $g \in G$) then G/A is a group under the operation it inherits from G, forming a factor group or quotient group. The cyclic groups $\mathbb{Z}_{/n}$ are very nice examples of factor groups, where one mods out by the subgroup $n\mathbb{Z} = \{\dots, -n, 0, n, 2n, \dots\}$ of the additive group of the integers $\mathbb{Z}$.

Exercises

1. The following table specifies an operation on the set $A = \{a, b, c, d\}$. Is the set A with this operation a group?

	a	b	c	d
a	b	a	d	c
b	a	b	c	d
c	d	c	a	b
d	c	d	b	a

2. Fill in the following table for the group D_4 of symmetries of a square.

Symmetries of the square

	e	r	s	t	a	b	c	d
e								
r								
s								
t								
a								
b								
c								
d								

3. Find a subgroup of D_4 that is isomorphic to the cyclic group of order 4, showing the isomorphism explicitly.

4. Prove that if G is a group, then there is a *unique* identity element in G.. (*Hint*: If there are two identity elements in G, their product should help you see that the two are equal.)

5. Prove that each element in a group G has a unique inverse. (*Hint*: If there are two inverses for a particular element g, then one of these times g times the other is an interesting element to think about.)

6. Prove that for any homomorphism $f : G \to H$, f sends the identity of G to the identity of H and $f(g^{-1}) = (f(g))^{-1}$ for every $g \in G$.

7. Prove that for any homomorphism $f : G \to H$, f sends a^k to $(f(a))^k$ for all $k \in \mathbb{N}$.)

8. Prove that the image of a homomorphism is a subgroup (of the target group).

9. Prove that for any homomorphism of groups $f : G \to H, Ker(f)$ is a subgroup of G.

10. Let $h \in G$ be any element of a group. Then the function $c_h : G \to G$ defined by $c_h(g) = hgh^{-1}$ is an isomorphism of G, generally called *conjugation* by h.

11. Prove that for a subgroup $A \leq G$, any two left cosets are either disjoint or empty. That is, for any $g_1, g_2 \in G$, either $g_1A \cap g_2A = \emptyset$ or $g_1A = g_2A$.

12. Prove that for a subgroup $A \leq G$, two cosets g_1A and g_2 are equal if and only if $g_1^{-1}g_2 \in A$.

CHAPTER 3

TOPOLOGICAL SPACES

3.1 INTRODUCTION

As in many other areas of mathematics, the precise definition of a topological space is that it is a set X, plus some additional structure, in this case known as a *topology* on X, which we'll define precisely in the next section. The topology on the set X will encode in some sense how "close" points are to each other in the resulting space. This formal approach to the subject makes it a bit difficult to see exactly how one can interpret "closeness," especially in some of the more esoteric examples. To avoid this level of abstractness, many texts look first at the topology of subsets of $\mathbb{R}$ or $\mathbb{R}^n$ (regarded as "Euclidean spaces"), which can be worked out rather easily, then move to more general topological spaces later. The danger of this approach is that while the student will develop intuition about what works for subsets of $\mathbb{R}$ or $\mathbb{R}^n$ (a good thing), these intuitive ideas will not always be true in more general settings. Given how often one needs to work in settings that are too complicated to be subsets of $\mathbb{R}^n$ (such as spaces of functions, etc.), it seems more prudent to work in full generality immediately. For this reason, we'll start with the formal, abstract definition of a topological space, looking at $\mathbb{R}$ or $\mathbb{R}^n$ as particular examples.

3.2 DEFINITIONS AND EXAMPLES

One begins with an arbitrary set X, then adds this additional structure that somehow determines the geometric form that X can take. For example, we usually think of the set $\mathbb{R}$ as forming a line, but this interpretation uses additional structure above and beyond the *set* of real numbers (namely, the order). Our approach will be to add an additional structure, a *topology*, to a set X to give it some sort of geometric form. Here's the precise definition.[1]

Definition 3.2.1. *A topology on a set X is a collection τ of subsets of X, satisfying the following properties:*

1. $\emptyset, X \in \tau$.
2. *τ is closed under union; that is, if $U_\alpha \in \tau$ for every $\alpha \in \Lambda$, then $U = \bigcup_{\alpha \in \Lambda} U_\alpha \in \tau$.*
3. *τ is closed under finite intersection, that is, if $U_i \in \tau$ for $i = 1, 2, \ldots, n$, then $U = \bigcap_{i=1}^{n} U_i \in \tau$.*

The pair (X, τ) is called a *topological space*, and we'll generally use the notation X_τ to denote a set X endowed with the particular topology τ. A set $U \in \tau$ will be referred to as a *τ-open set* (although we will use the term *open set* if there is no ambiguity about the topology).

Here are some examples of topological spaces.

■ **EXAMPLE 3.1**

Let X be any set. Let $\mathcal{D} = \mathcal{P}(X)$, the *power set* of X.

[1]This axiomatic approach to topology is due to Felix Hausdorff (1868–1942), who will show up eponymously a lot in later chapters. His 1914 book *Grundzuge der Mengenlehre* was the first to present this idea and proved enormously influential. Hausdorff actually added one more axiom to those presented here, as we'll see in Chapter 8.

Then $\mathcal{D}$ is a topology on X, as one can see easily:

- $\emptyset, X \in \mathcal{D}$ since they're both subsets of X.
- $\mathcal{D}$ is closed under union, since the union of subsets of X is a subset of X.
- $\mathcal{D}$ is closed under finite intersection, since any intersection of subsets of X is a subset of X.

The topology $\mathcal{D}$ is referred to as the *discrete topology* on X, and the space $X_\mathcal{D}$ is often called a *discrete space*. Note that this is a pretty simple example, since we can define $\mathcal{D}$ for any set (including the empty set!). Note also that $\mathcal{D}$ is the largest topology one can define on a set X, since any topology τ on X obviously satisfies $\tau \subset \mathcal{D}$.

The following terminology for comparing topologies on a set X has become standard in recent years: if τ, τ' and τ'' are topologies on X, and if $\tau \subset \tau''$, then we say that τ'' is *finer* than τ. If $\tau' \subset \tau$, then we say τ' is *coarser* than τ. Of course, the terms *strictly finer* and *strictly coarser* refer to proper inclusions of the topologies.

■ **EXAMPLE 3.2**

Let X be any set. Let $\mathcal{I} = \{\emptyset, X\}$.

Then $\mathcal{I}$ is a topology on X, as one can see easily:

- $\emptyset, X \in \mathcal{I}$ by definition.
- $\mathcal{I}$ is closed under union, since the only subsets in $\mathcal{I}$ are $\emptyset$ and X, whose union is X.
- $\mathcal{I}$ is closed under finite intersection, since the only subsets in $\mathcal{I}$ are $\emptyset$ and X, whose intersection is $\emptyset$.

The topology $\mathcal{I}$ is referred to as the *indiscrete topology* on X, and the space $X_\mathcal{I}$ is often called an *indiscrete space*. Note that this is also a pretty simple example, since we can define $\mathcal{I}$ for any set (including the empty set!). The indiscrete topology is the coarsest topology one can have on a set X.

■ **EXAMPLE 3.3**

Let $X = \{a, b, c, d\}$. Let $\sigma = \{\emptyset, X, \{a, b\}, \{b, c, d\}\}$ and $\tau = \{\emptyset, X, \{a\}, \{b\}, \{a, b\}\}$. Then σ is not a topology, since $\{a, b\} \cap \{b, c, d\} \notin \sigma$, but τ is a topology, as one can check easily.

■ **EXAMPLE 3.4**

Let $\mathcal{U} = \{V \subset \mathbb{R} : \text{ if } x \in V, \text{ then there exists an open interval } (a, b) \text{ such that } x \in (a, b) \subset V\}$. The collection $\mathcal{U}$ is referred to as the *usual topology* on $\mathbb{R}$.

Note that the usual topology $\mathcal{U}$ on $\mathbb{R}$ has the nice property that intervals of the form (a, b) are indeed $\mathcal{U}$-open (i.e., "open" intervals are $\mathcal{U}$-open), although one should keep in mind that lots of other sets are $\mathcal{U}$-open as well. This is not the case, for example, with the indiscrete topology on $\mathbb{R}$, where "open" intervals are not $\mathcal{I}$-open. While the set $\mathbb{R}$ is often thought of as the "real line," this description is misleading: the geometric structure is completely determined by the topology on $\mathbb{R}$. For example, we shall see that $\mathbb{R}_{\mathcal{D}}$ is a discrete "cloud" of points, without the ordered structure that a line has. Also, $\mathbb{R}_{\mathcal{I}}$ will be seen to be "one large point," in some appropriate sense. It will turn out that the topology $\mathcal{U}$ is the correct one to put on the set $\mathbb{R}$ to give it the structure of a line.

We can easily prove that $\mathcal{U}$ is a topology on $\mathbb{R}$ as follows:

- The set $\mathbb{R} \in \mathcal{U}$, since for any $x \in \mathbb{R}$, we have $x \in (x - 10^{50}, x + 10^{50}) \subset \mathbb{R}$. The empty set is in $\mathcal{U}$ vacuously,[2] since there is no x in $\emptyset$ to violate the condition that defines $\mathcal{U}$.

- The collection $\mathcal{U}$ is closed under union; let $V_\alpha \in \mathcal{U}$ for every $\alpha \in \Lambda$. Consider $V = \bigcup_{\alpha \in \Lambda} V_\alpha$. If $x \in V$, then $x \in V_\beta$ for some $\beta \in \Lambda$. Since $V_\beta \in \mathcal{U}$, we know there exists an "open" interval (a, b) such that $x \in (a, b) \subset V_\beta$. So we see that $x \in (a, b) \subset V_\beta \subset V = \bigcup_{\alpha \in \Lambda} V_\alpha$, as we wished.

- To show that the collection $\mathcal{U}$ is closed under finite intersection, let V_1 and V_2 belong to $\mathcal{U}$. If $V_1 \cap V_2 = \emptyset$, we're done. If not, let $x \in V_1 \cap V_2$. Since $x \in V_1 \in \mathcal{U}$, there exists an "open" interval (a, b) with $x \in (a, b) \subset V_1$. Since $x \in V_2 \in \mathcal{U}$, there exists an "open" interval (c, d) with $x \in (c, d) \subset V_2$. Let $e = \max\{a, c\}$ and $f = \min\{b, d\}$. Then $x \in (e, f) \subset V_1 \cap V_2$, so that $V_1 \cap V_2 \in \mathcal{U}$. A simple induction shows that $\mathcal{U}$ is closed under finite intersections.

[2] Many students have some difficulty with the concept of an implication being true "vacuously." To be precise about this concept, recall that the implication $p \implies q$, alias "If p then q," is generally thought of as "If p is true then q must also be true." But, what if the hypothesis p is false? The easiest way to analyze this is by a simple example: consider the implication "If you give Dr. Kolesar one million dollars, then you will get an A for this course." To see that this implication is true, let's think about the possible cases. Say that Nick has given Dr. K. \$$10^6$ and Nick receives an A. In his case, the implication is certainly true. What if Steve does not give Dr. K. \$$10^6$, and then receives a final grade of C? Again the implication is true. How about if Molly gives Dr. K. \$$10^6$ but receives a B? In this case the implication is false. What if Noura does not give Dr. K. \$$10^6$ and receives an A? In this case, the implication is still true: remember that the promise is "**If** you give Dr. K. \$$10^6$, then you will receive an A." It promises nothing if you don't give Dr. K. \$$10^6$. So the only way to see if the implication is a lie is to give Dr. K. the money and wait for

This example illustrates several interesting points about the definition of a topology.

1. For the first three examples, we specified the topology by listing all of the open sets. For the topology $\mathcal{U}$ on the set $\mathbb{R}$, though, we specified the topology "locally"; we gave a condition on a set V that could be checked at each point in V, rather than specifying some "global" condition.

2. This example points out the danger of using the term "open" indiscriminately: "open" intervals like $(2, 4)$ are $\mathcal{U}$-open, but are not $\mathcal{I}$-open.

3. To prove that a potential topology τ is closed under finite intersection, one need only show that $V_1 \cap V_2 \in \tau$ whenever V_1 and V_2 are in τ, then use induction to see that *any* finite intersection of sets from τ is back in τ.

4. Defining the topology $\mathcal{U}$ locally, as we did, runs the risk of misleading readers about exactly which sets are $\mathcal{U}$-open: there are many more $\mathcal{U}$-open sets than just intervals of the form (a, b). (Think of a few, before you continue on.)

■ EXAMPLE 3.5

Let $\mathcal{L} = \{V \subset \mathbb{R} : \text{ if } x \in V, \text{ then there exists } a, b \in V,\ a < b, \text{ such that } x \in [a, b) \subset V\}$. The collection $\mathcal{L}$ is referred to as the *left-hand topology* on $\mathbb{R}$.

We defer the proof that $\mathcal{L}$ is a topology on $\mathbb{R}$ to the exercises. Note that any set V which is $\mathcal{U}$-open is also $\mathcal{L}$-open; for any $x \in V$, a $\mathcal{U}$-open set, there exists an "open" interval (a, b) with $x \in (a, b) \subset V$, since $V \in \mathcal{U}$. Then $x \in [x, b) \subset (a, b) \subset V$, so that V is also $\mathcal{L}$-open. However, the set $[0, 1)$ is obviously $\mathcal{L}$-open, but is not $\mathcal{U}$-open, since there is no "open" interval around 0 that fits inside $[0, 1)$. So we have the chain of *proper* inclusions of topologies on $\mathbb{R}$:

$$\mathcal{I} \underset{\neq}{\subset} \mathcal{U} \underset{\neq}{\subset} \mathcal{L} \underset{\neq}{\subset} \mathcal{D}.$$

■ EXAMPLE 3.6

Consider the collection

$$\mathcal{RR} = \{V \subset \mathbb{R} : \text{ if } x \in V, \text{ then there exists a ray } (a, +\infty) \text{ for some } a \in \mathbb{R} \text{ with } x \in (a, +\infty) \subset V\}.$$

Then $\mathcal{RR}$ is a topology on $\mathbb{R}$, called the *right ray topology.*

your final grade. By the way, the trap students often fall into is confusing the implication $p \implies q$ with its *converse*, $q \implies p$, a favorite tactic of defense attorneys everywhere.

Again, the proof that $\mathcal{RR}$ is a topology on $\mathbb{R}$ is deferred to the exercises. We note that $\mathcal{RR}$ is coarser than the usual topology on $\mathbb{R}$. Note also that, even though $\mathcal{RR}$ is defined locally, only the sets used to define it are in fact $\mathcal{RR}$-open, in addition to $\emptyset$ and $\mathbb{R}$, in marked contrast to the usual topology. (Why?)

■ **EXAMPLE 3.7**

Let X be any set. The collection

$$\mathcal{FC} = \{V \subset X : X \setminus V \text{ is finite, or } V = \emptyset.\}$$

is the finite complement topology on X.

Before we prove that $\mathcal{FC}$ is a topology on X, a few comments are in order.

1. If X is finite, then $\mathcal{FC} = \mathcal{D}$, since any subset of a finite set will have a finite complement.

2. If X is infinite, then $\mathcal{FC} \underset{\neq}{\subset} \mathcal{D}$, since no nonempty finite subset of X will be $\mathcal{FC}$-open.

3. If we look at $\mathbb{R}_{\mathcal{FC}}$, we see that a typical $\mathcal{FC}$-open set is $\mathbb{R} \setminus \{0, 1\} = (-\infty, 0) \cup (0, 1) \cup (1, +\infty)$, which is also $\mathcal{U}$-open (and hence $\mathcal{L}$-open). So it's easy to see that $\mathcal{U}$ is finer that $\mathcal{FC}$ as topologies on $\mathbb{R}$. Be careful, though, to remember that $\mathcal{U}$ is defined only on $\mathbb{R}$, (because it depends on the ordering of $\mathbb{R}$) whereas $\mathcal{FC}$ is defined for any set. Although the idea of the proof works for any set X, we'll prove that $\mathcal{FC}$ is a topology only for the special case $X = \mathbb{R}$. The general case is left to the exercises.

To prove that $\mathcal{FC} = \{V \subset \mathbb{R} : \mathbb{R} \setminus V \text{ is finite, or } V = \emptyset\}$ is a topology on $\mathbb{R}$, first observe that $\emptyset \in \mathcal{FC}$ by definition. The set $\mathbb{R} \in \mathcal{FC}$ since $\mathbb{R} \setminus \mathbb{R} = \emptyset$, which is awfully finite. To verify that $\mathcal{FC}$ is closed under union, let $V_\alpha \in \mathcal{FC}$ for every $\alpha \in \Lambda$. If all of the the V_αs are empty, then the union is empty and we're done. If at least one V_β is nonempty, then DeMorgan's Law allows us to see that $V = \bigcup_{\alpha \in \Lambda} V_\alpha$ is in $\mathcal{FC}$ since

$$\mathbb{R} \setminus V = \mathbb{R} \setminus \bigcup_{\alpha \in \Lambda} V_\alpha = \bigcap_{\alpha \in \Lambda} (\mathbb{R} \setminus V_\alpha) \subset \mathbb{R} \setminus V_\gamma$$

for any $\gamma \in \Lambda$, so $\mathbb{R} \setminus V$ is finite, since $\emptyset \neq V_\beta \in \mathcal{FC}$. To confirm that $\mathcal{FC}$ is closed under finite intersections, let V_1 and V_2 be in $\mathcal{FC}$. If either is empty, so is $V_1 \cap V_2$, and we're done. If both V_1 and V_2 are nonempty, then $\mathbb{R} \setminus (V_1 \cap V_2) = (\mathbb{R} \setminus V_1) \cup (\mathbb{R} \setminus V_2)$, which is finite (as the union of two finite sets), so that $V_1 \cap V_2 \in \mathcal{FC}$. Use induction to see that $\mathcal{FC}$ is closed under all finite intersections.

Here's one more interesting topology on the set $\mathbb{R}$.

EXAMPLE 3.8

Let x_0 be any real number. The *distinguished point* topology on $\mathbb{R}$ is given by

$$\tau_{x_0} := \{U \subset \mathbb{R} : x_0 \in U \text{ or } U = \emptyset\}.$$

When $x_0 = 1$, for example, the sets $[0,1]$, $(-2,3)$ and $\{1\}$ are all τ_1-open, but the set $(0,1)$ is not. We leave it as an exercise to show that τ_{x_0} is a topology on $\mathbb{R}$. Note that we've borrowed a piece of notation from the computer scientists here, using the symbol := as a shorthand for "defined to be."

Exercises

1. Is $\{\emptyset, X, \{a\}, \{b\}, \{c\}\}$ a topology on $X = \{a, b, c\}$? Why or why not?

2. Is $\{\{a\}, \{b\}, \{c\}, \{a,b\}, \{a,c\}, \{b,c\}\}$ a topology on $X = \{a, b, c\}$? Why or why not?

3. Is $\{\emptyset, X, \{a\}, \{c\}, \{a,c\}, \{a,b\}\}$ a topology on $X = \{a, b, c\}$? Why or why not?

4. Find all the topologies on the set $X = \{a, b, c\}$. There are 29 of them. (*Hint*: be extremely organized in how you write them down – a pattern will help you sort things out.)

5. Work out the following question for $n \leq 5$: If a set X has n elements, how many distinct topologies are there on X? [*Hint*: First work out the case where $X = \{a, b\}$. The exercise above tells you the answer for a set with three elements. Then see if you can see a pattern for determining how to include singletons, doubletons, etc. This is not a trivial exercise.] The answers for this exercise are 1,4,29,355 and 6942. These are the first five terms in a very interesting sequence of integers (number A000798 in the Online Encyclopedia of Integer Sequences, `http://www.research.att.com/cgi-bin/access.cgi/as/njas/eis.`) There appears to be no known formula for the number of topologies on a set with n elements.

6. Prove that if $\tau \subset \mathcal{P}(X)$ such that if $V_1, V_2 \in \tau$ then $V_1 \cap V_2 \in \tau$, then τ is closed under finite intersections. (Yes, actually write out the induction proof!)

7. Prove that $\mathcal{L}$ is a topology on $\mathbb{R}$.

8. Prove that $\mathcal{RR}$ is a topology on $\mathbb{R}$.

9. Find an example which shows that $\mathcal{U}$ is *not* closed under arbitrary intersections.

10. Is $\mathcal{RR}$ coarser than $\mathcal{FC}$? Finer than $\mathcal{FC}$? (If neither, then they're noncomparable.)

11. Prove that for any $x_0 \in \mathbb{R}$, the distinguished point topology of Example 3.8 is indeed a topology for $\mathbb{R}$.

12. Prove that the collection $\mathcal{FC}$ is a topology on X for any set X, the finite-complement topology.

3.3 BASICS ON OPEN AND CLOSED SETS

Given a set X with a topology τ, we'll continue to use the notation X_τ for the resulting topological space. We recall that a subset $V \subset X_\tau$ is τ-*open* if $V \in \tau$.

Definition 3.3.1. *A subset $C \subset X_\tau$ is τ-closed if $X \setminus C$ is τ-open.*

As with the term τ-open, we'll occasionally use the term *closed* as a shorthand for τ-closed, when there's no ambiguity about the topology. In any topological space X_τ, we see that $\emptyset$ and X are both τ-closed, since their complements are X and $\emptyset$, respectively, which are guaranteed to be τ-open.

In $\mathbb{R}_\mathcal{U}$, we see easily that any singleton subset is $\mathcal{U}$-closed, since $\mathbb{R} \setminus \{a\} = (-\infty, a) \cup (a, +\infty)$, which is $\mathcal{U}$-open. Similarly, we see that intervals of the form $[a, b]$ are $\mathcal{U}$-closed, because the complement $\mathbb{R} \setminus [a, b] = (-\infty, a) \cup (b, +\infty)$ is $\mathcal{U}$-open. Note that intervals of the form $[a, b]$ are not $\mathcal{U}$-open, since we can't find appropriate "open" intervals around the endpoints a and b that are contained in $[a, b]$. In fact, it will turn out that the only subsets of $\mathbb{R}$ which are simultaneously $\mathcal{U}$-open and $\mathcal{U}$-closed are $\emptyset$ and $\mathbb{R}$. This property has to do with the "connectedness" of $\mathbb{R}_\mathcal{U}$, as we'll see in Chapter 6.

However, in a general topological space X_τ, one can indeed have subsets that are both τ-open and τ-closed, and subsets that are neither τ-open nor τ-closed. For example, if X is any set, then in the discrete space $X_\mathcal{D}$, *any* subset W of X is both $\mathcal{D}$-open and $\mathcal{D}$-closed, since W and $X \setminus W$ are both subsets of X and hence elements of $\mathcal{D} = \mathcal{P}(X)$. In the indiscrete space $X_\mathcal{I}$, any proper, nonempty subset $Z \subset X$ is neither $\mathcal{I}$-open nor $\mathcal{I}$-closed, since $\mathcal{I} = \{\emptyset, X\}$. So don't fall into the trap of assuming that "closed" means "not open" and vice–versa.

One might assume that we get such odd behavior for open and closed sets in $X_\mathcal{D}$ and $X_\mathcal{I}$ because these are such extreme examples of topologies, being the largest and smallest topologies, respectively, on a set X. However, even for a nice space like the left-hand topology on $\mathbb{R}$, one can get seemingly counterintuitive phenomena involving open and closed sets. Consider, for example, the set $[0, 1) \subset \mathbb{R}_\mathcal{L}$. Clearly $[0, 1)$ is $\mathcal{L}$-open, since it contains an interval of the form $[a, b)$ around each of its points (viz., the entire interval $[0, 1)$). However, $[0, 1)$ is also $\mathcal{L}$-closed, since $\mathbb{R} \setminus [0, 1) = (-\infty, 0) \cup [1, +\infty)$, which is easily shown to be open in $\mathbb{R}_\mathcal{L}$. (Yes, you should explicitly show that the complement is $\mathcal{L}$-open, just to be certain.)

The following theorem is often useful for showing a given set is closed.

Theorem 3.3.1. *For $X_\mathcal{T}$ any topological space,*

1. *The intersection of $\mathcal{T}$-closed sets is $\mathcal{T}$-closed.*

2. *The finite union of $\mathcal{T}$-closed sets is $\mathcal{T}$-closed.*

Before we prove the theorem, we'll note that there's no guarantee that an infinite union of $\mathcal{T}$-closed sets is $\mathcal{T}$-closed. For example, in $\mathbb{R}_\mathcal{U}$, we know that singleton subsets are $\mathcal{U}$-closed. However, $(0,1) = \bigcup_{x\in(0,1)}\{x\}$ is not $\mathcal{U}$-closed. The proof of the theorem follows directly from DeMorgan's Laws.

Proof. We'll prove the first result, leaving the second for the exercises. Let C_α be $\mathcal{T}$-closed for every $\alpha \in \Lambda$. Consider $C = \bigcap_{\alpha\in\Lambda} C_\alpha$. Since each C_α is $\mathcal{T}$-closed, we see that for every $\alpha \in \Lambda$, $X \smallsetminus C_\alpha$ must be $\mathcal{T}$-open. Then one of DeMorgan's Laws tells us that

$$X \smallsetminus C = X \smallsetminus \bigcap_{\alpha\in\Lambda} C_\alpha = \bigcup_{\alpha\in\Lambda}(X \smallsetminus C_\alpha),$$

which is $\mathcal{T}$-open since the arbitrary union of open sets is open. Hence $C = \bigcap_{\alpha\in\Lambda} C_\alpha$ is $\mathcal{T}$-closed, as we wished. □

Often, for a given point $x \in X_\mathcal{T}$, we'll wish to talk about an open set containing x. This occurs so frequently that we adopt the following shorthand notation: A *neighborhood* of x is any $\mathcal{T}$-open set U containing x. This notation is not universal, however, since some texts allow for neighborhoods which are not $\mathcal{T}$-open sets.[3] Here, we'll *always* use the word "neighborhood" to refer to an open set. An example of the usefulness of this concept is the following theorem, which provides a useful characterization of $\mathcal{T}$-open sets.

Theorem 3.3.2. *For X any topological space, a subset $U \subset X_\mathcal{T}$ is $\mathcal{T}$-open if and only if U contains a neighborhood of each of its points. That is, U is $\mathcal{T}$-open if and only if for every $x \in U$, there exists a $\mathcal{T}$-open set N with $x \in N \subset U$.*

[3] The definition used as an alternative to ours is this: A set U is a $\mathcal{T}$*-neighborhood* of x if there exists a $\mathcal{T}$-open set V such that $x \in V \subset U$. One advantage of our approach is that an *open* neighborhood is a neighborhood of each of its points. This isn't true for the nonopen neighborhoods allowed by the other definition. Again, we reserve the word "neighborhood" for $\mathcal{T}$-open sets.

Proof: $\Longrightarrow$) Let U be any τ-open set, with $x \in U$. Then U is a neighborhood of x, with $x \in U \subset U$.
$\Longleftarrow$) Let U be a subset of X_τ which contains a neighborhood of each of its points. So for each $x \in U$, there exists a τ-open set N_x with $x \in N_x \subset U$. Then $U = \bigcup_{x \in U} N_x$, so that U is a union of τ-open sets. □

The following concept will be very useful as we proceed more deeply into topological spaces.

Definition 3.3.2. *For X_τ any topological space and $A \subset X$, the closure of A, denoted $Cl(A)$, is the smallest τ-closed subset of X that contains A.*

The closure of A is also often denoted by $\overline{A}$. As an example, in $\mathbb{R}_\mathcal{U}$, the set of reals in the usual topology, $Cl((0,1)) = [0,1]$, since any $\mathcal{U}$-closed set containing $(0,1)$ must contain $[0,1]$, while any smaller $\mathcal{U}$-closed set would fail to contain some elements in $(0,1)$. In any indiscrete space $X_\mathcal{I}$, if A is any nonempty subset of X, then $Cl(A) = X$, since the only $\mathcal{I}$-closed subsets of X are $\emptyset$ and X. Similarly, in any discrete space $X_\mathcal{D}$, any subset A is its own closure. In fact, it's easy to see that for any $A \subset X_\tau$, $Cl(A) = A$ if and only if A is τ-closed.

The following characterization of closures is handy

Theorem 3.3.3. *If X_τ is any topological space and $A \subset X$, then $Cl(A)$ is the intersection of all the τ-closed subsets of X that contain A.*

Proof: Given $A \subset X_\tau$. Let

$$C = \bigcap_{\substack{A \subset K \\ K\ \tau\text{-closed}}} K.$$

We need to show $Cl(A) = C$, so we'll use a double inclusion:
$\subset$) Note that

$$C = \bigcap_{\substack{A \subset K \\ K\ \tau\text{-closed}}} K$$

is an intersection of τ-closed sets, so C is τ-closed. Further, each set K in the intersection that defines C contains A as a subset, so that $A \subset C$. Hence $Cl(A) \subset C$, since $Cl(A)$ is the smallest τ-closed set containing A.
$\supset$) Since $Cl(A)$ is τ-closed and contains A, $Cl(A)$ is one set that goes into the intersection

$$C = \bigcap_{\substack{A \subset K \\ K\ \tau\text{-closed}}} K.$$

Since $C \subset K$ whenever K is a term in the intersection, we see that $C \subset Cl(A)$. □

A sort of "dual" construction to $Cl(A)$ is given in the following definition.

Definition 3.3.3. *For X_τ any topological space and $A \subset X$, the interior of A, denoted $Int(A)$, is the largest τ-open subset of X that is contained in A.*

The terminology "x is an interior point of A" will often be used as an alternative to saying $x \in Int(A)$. As an example, in $\mathbb{R}_\mathcal{U}$, $Int((0,1)) = (0,1)$ and $Int([0,1]) = (0,1)$. Also in $\mathbb{R}_\mathcal{U}$, $Int(\{0\}) = \emptyset$. In the set $X = \{a,b,c,d\}$, endowed with the topology $\tau = \{\emptyset, X, \{a\}, \{b\}, \{a,b\}, \{a,b,c\}\}$, then the subset $B = \{a,c\}$ has $Int(B) = \{a\}$ and $Cl(B) = \{a,c,d\}$.

Dual to Theorem 3.3.3 is the following:

Theorem 3.3.4. *If X_τ is any topological space and $A \subset X$, then $Int(A)$ is the union of all τ-open sets contained in A.*

The proof is deferred to the exercises. The construction of the closure of a subset works well with the operations of union and intersection.

Theorem 3.3.5. *For $A, B \subset X_\tau$,*

1. $Cl(A \cup B) = Cl(A) \cup Cl(B)$;

2. $Cl(A \cap B) \subset Cl(A) \cap Cl(B)$.

To see that inclusion is the best one can do with closure and intersection, consider the following example. In $\mathbb{R}_\mathcal{U}$, let $A = (0,1)$ and $B = (1,2)$. Then $A \cap B = \emptyset$ so that $Cl(A \cap B) = \emptyset$. However, $Cl(A) = [0,1]$ and $Cl(B) = [1,2]$, so that $Cl(A) \cap Cl(B) = \{1\}$.

Proof: To prove 1, we'll use our usual double-inclusion approach.

$\subset$) Note that $A \cup B \subset Cl(A) \cup Cl(B)$, so that $Cl(A \cup B) \subset Cl(A) \cup Cl(B)$, by definition.

$\supset$) Note that $Cl(A) \subset Cl(A \cup B)$, since $A \subset A \cup B$. Similarly, $Cl(B) \subset Cl(A \cup B)$. Hence, $Cl(A) \cup Cl(B) \subset Cl(A \cup B)$.

To prove 2, let $x \in Cl(A \cap B)$. Let K_1 be any closed set containing A and K_2 any closed set containing B. Then $K_1 \cap K_2$ is a closed set containing $A \cap B$, so $x \in K_1 \cap K_2$. We conclude that x is an element of any closed set containing A and any closed set containing B, so that $x \in Cl(A) \cap Cl(B)$, as we wished. □

The following term is widely used in topology and analysis.

Definition 3.3.4. *For X_τ any topological space, a subset $A \subset X_\tau$ is said to be dense in X_τ if $Cl(A) = X$.*

For example, any nonempty subset of an indiscrete space is dense. (Why?) No proper subset of a discrete space is dense. (Again, why? Be sure to think through this before you move on.) In $\mathbb{R}_{\mathcal{U}}$, the set $\mathbb{Q}$ is dense, as one can see by considering a neighborhood of any point $x \in \mathbb{R}$: if $x \notin Cl(\mathbb{Q})$, then there has to be an entire neighborhood of x that misses $Cl(\mathbb{Q})$, since $\mathbb{R} \setminus Cl(\mathbb{Q})$ is $\mathcal{U}$-open. Since every interval contains both irrationals and rationals, we see that this can't occur for any $x \in \mathbb{R}$. Thus the space $\mathbb{R}_{\mathcal{U}}$ has a countable dense subset (a property known as being *separable*).

Exercises

1. Prove that no proper subset of $\mathbb{R}_{\mathcal{RR}}$ is simultaneously open and closed (in the $\mathcal{RR}$ topology).

2. Prove that no proper subset of $\mathbb{R}_{\mathcal{FC}}$ is simultaneously open and closed.

3. Prove that in any topological space $X_{\mathcal{T}}$, a finite union of $\mathcal{T}$-closed sets is $\mathcal{T}$-closed.

4. Consider the set $X = \{a, b, c\}$ with the topology $\mathcal{T} = \{\emptyset, X, \{a\}, \{b\}, \{a, b\}\}$.

   (a) Find $Cl(\{a\})$.

   (b) Find $Cl(\{c\})$.

   (c) Find $Cl(\{b, c\})$.

   (d) List all the closed subsets of $X_{\mathcal{T}}$.

5. Consider the set $X = \{a, b, c\}$ with the topology $\mathcal{T} = \{\emptyset, X, \{a\}, \{b\}, \{a, b\}\}$.

   (a) Find $Int(\{a\})$.

   (b) Find $Int(\{c\})$.

   (c) Find $Int(\{b, c\})$.

6. Is any infinite set aside from $\mathbb{R}$ closed in $\mathbb{R}_{\mathcal{FC}}$?

7. Is any finite set aside from $\emptyset$ open in $\mathbb{R}_{\mathcal{U}}$?

8. Are the following subsets of $\mathbb{R}$ $\mathcal{L}$-open? $\mathcal{L}$-closed? Neither? Both? $A = (0, 1]$; $B = \{0\} \cup [2, 3)$; $C = (-\infty, 0) \cup [1, +\infty)$; $D = \{0, 1, 2, 3\}$.

9. Prove Theorem 3.3.4; that is, prove that for $A \subset X_{\mathcal{T}}$,

$$Int(A) = \bigcup_{\substack{U \subset A \\ U\ \mathcal{T}\text{-open}}} U$$

10. Give an example of a topology on a finite set, with two subsets A and B, showing that $Cl(A \cap B) \underset{\neq}{\subset} Cl(A) \cap Cl(B)$.

11. Give an example of two subsets A and B of $\mathbb{R}_{\mathcal{U}}$ such that $Cl(A \cap B) = \emptyset$ and $Cl(A) \cap Cl(B) = \mathbb{R}$. Does your example work in $\mathbb{R}_{\mathcal{L}}$?

3.4 THE SUBSPACE TOPOLOGY

When X_τ is any topological space, any subset $A \subset X_\tau$ "inherits" a topology from X_τ as follows:

Definition 3.4.1. *If X_τ is any topological space, for $A \subset X_\tau$, the subspace topology on A, denoted by τ_A, is given by*

$$\tau_A = \{U \subset A : U = V \cap A \text{ for some } V \in \tau\}.$$

Note that this is equivalent to defining τ_A to be $\{V \cap A : V \in \tau\}$. In either description, the τ_A-open subsets of A are exactly those obtained by the intersection of τ-open subsets of X with A. For example, consider $[0, 1] \subset \mathbb{R}_{\mathcal{U}}$. Then $[0, \frac{1}{2})$ is $\mathcal{U}_{[0,1]}$-open, since $[0, \frac{1}{2}) = (-\frac{1}{2}, \frac{1}{2}) \cap [0, 1]$. So τ_A-open sets need *not* be τ-open, in general.

Theorem 3.4.1. *If X_τ is any topological space and $A \subset X$, then τ_A is a topology on A.*

We'll refer to the topological space A_{τ_A} as a *subspace* of X_τ.

Proof: We observe that $\emptyset = \emptyset \cap A$ and $A = X \cap A$, so that $\emptyset$ and A belong to τ_A, since $\emptyset, X \in \tau$. To see that τ_A is closed under union, let $U_\alpha \in \tau_A$ for every $\alpha \in \Lambda$. Then each $U_\alpha = V_\alpha \cap A$ for some τ-open V_α. So

$$U = \bigcup_{\alpha \in \Lambda} U_\alpha = \bigcup_{\alpha \in \Lambda} (V_\alpha \cap A) = \left(\bigcup_{\alpha \in \Lambda} V_\alpha \right) \cap A,$$

which is back in τ_A since $\bigcup_{\alpha \in \Lambda} V_\alpha \in \tau$, because τ is a topology for X. Similarly, if $U_1, U_2 \in \tau_A$, then $U_i = V_i \cap A$ for some τ-open V_i, for $i = 1, 2$. Hence $U_1 \cap U_2 = (V_1 \cap A) \cap (V_2 \cap A) = (V_1 \cap V_2) \cap A$, which is back in τ_A since $V_1 \cap V_2 \in \tau$. □

Of course, by definition, for $A \subset X_\tau$, a subset $C \subset A$ is τ_A-closed if $A \setminus C$ is τ_A-open. For example, $[\frac{1}{2}, 1]$ is $\mathcal{U}_{[0,1]}$-closed, precisely because $[0, \frac{1}{2})$ is $\mathcal{U}_{[0,1]}$-open. We have an easier criterion to check whether such a subset is closed in the subspace topology.

Theorem 3.4.2. *Let X_τ be any topological space and $A \subset X$. Then a subset $C \subset A$ is τ_A-closed if and only if $C = K \cap A$ for some $K \subset X$, with K τ-closed.*

Proof: $\Longrightarrow$) Let $C \subset A$ be τ_A-closed. Then $A \setminus C \in \tau_A$, so that $A \setminus C = U \cap A$ for some $U \in \tau$. Then $C = (X \setminus U) \cap A = K \cap A$, where $K = X \setminus U$ is τ-closed, so that C is τ_A-closed by definition.
$\Longleftarrow$) Let $C = K \cap A$ for K any τ-closed set. Then $K = X \setminus U$, where $U \in \tau$, by definition. So $A \setminus C = (X \setminus K) \cap A = U \cap A$, so that $A \setminus C$ is τ_A-open, and hence C is τ_A-closed. □

Exercises

1. Prove that any subspace of a discrete space is discrete.

2. Prove that any subspace of an indiscrete space is indiscrete.

3. Prove that if $A \subset X$ is τ-open, then any τ_A-open set is also τ-open.

4. Which of the following subsets of $\mathbb{R}$ are open in $\mathcal{U}_{[1,2]}$? $\mathcal{U}_{[1,2]}$-closed? Neither? Both? $A = [1, 2]$; $B = \{1\}$; $C = [\frac{3}{2}, 2]$; $D = (\frac{3}{2}, 2]$.

5. Which of the following subsets of $\mathbb{R}$ are open in $\mathcal{L}_{[1,2]}$? $\mathcal{L}_{[1,2]}$-closed? Neither? Both? $A = [1, 2]$; $B = \{1\}$; $C = [\frac{3}{2}, 2]$; $D = (\frac{3}{2}, 2]$.

6. Consider the set $X = \{a, b, c\}$ with the topology $\tau = \{\emptyset, X, \{a\}, \{b\}, \{a, b\}\}$. Let $A = \{a, c\}$ and let $B = \{a, b\}$.

   (a) List all the sets in τ_A.

   (b) List all the sets in τ_B.

   (c) List all of the closed subsets of A_{τ_A}.

   (d) List all of the closed subsets of B_{τ_B}.

3.5 CONTINUOUS FUNCTIONS

We have a good intuitive idea about what we mean by a continuous function $f : \mathbb{R} \to \mathbb{R}$, at least from the perspective of calculus. We recall this definition.

Definition 3.5.1. *A function $f : \mathbb{R} \to \mathbb{R}$ is "calc-continuous" if for every $a \in \mathbb{R}$ and for every $\epsilon > 0$, there exists $\delta > 0$ such that if $0 < |x-a| < \delta$, then $|f(x)-f(a)| < \epsilon$.*

In other words, f is continuous in the sense of calculus if $\lim_{x \to a} f(x) = f(a)$ for every $a \in \mathbb{R}$. Intuitively, then, a function $f : \mathbb{R} \to \mathbb{R}$ is continuous in the sense of calculus if points which are close together in the domain are mapped by f to points that are close together in the range. The following graph illustrates this.

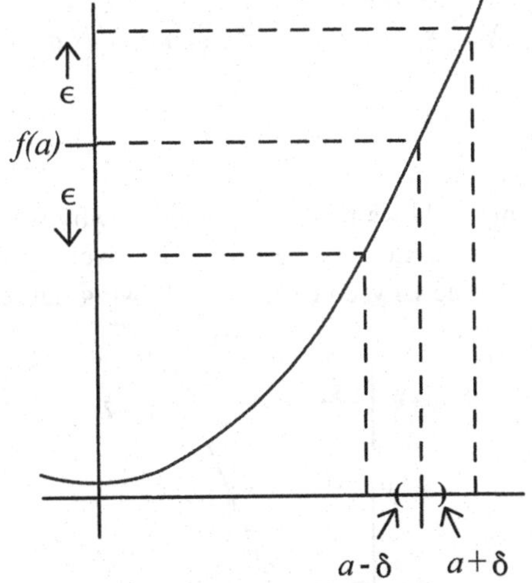

To generalize this idea to functions between arbitrary topological spaces, we need to get a feel for what "close together" means in a typical space X_τ. The idea (intentionally rather vague) is that two points x_1 and x_2 are "close together" in X_τ if "most" of the τ-open sets containing one point also contain the other. With this in mind, we have the following definition:

Definition 3.5.2. *A function $f : X_\tau \to Y_\nu$ of topological spaces is $(\tau - \nu)$-continuous if for every ν-open subset $V \subset Y$, we have $f^{-1}(V)$ τ-open as a subset of X.*

As before, we'll be a bit less precise and write that f is continuous if there's no ambiguity about the topologies on X and Y. If there's any doubt at all about what topologies are being discussed, we'll use the full terminology: "f is $(\tau - \nu)$-continuous."

As a very simple example, let X_τ be any topological space and let $Y_\mathcal{I}$ be an indiscrete space. Then any function $f : X_\tau \to Y_\mathcal{I}$ is $(\tau - \mathcal{I})$-continuous. To see why, note that the only $\mathcal{I}$-open subsets of Y are $\emptyset$ and Y, with $f^{-1}(\emptyset) = \emptyset$ and $f^{-1}(Y) = X$, both open subsets of X_τ.

Similarly, let Y_ν be any topological space and let $X_\mathcal{D}$ be a discrete space. Then any function $f : X_\mathcal{D} \to Y_\nu$ is $(\mathcal{D} - \nu)$-continuous. The proof is left as an exercise.

As a less simple example, consider the identity function $i_\mathbb{R} : \mathbb{R}_\mathcal{L} \to \mathbb{R}_\mathcal{U}$. Then $i_\mathbb{R}$ is $(\mathcal{L} - \mathcal{U})$-continuous, since if V is any $\mathcal{U}$-open set, then $i_\mathbb{R}^{-1}(V) = V$ is also $\mathcal{L}$-open. Note that $i_\mathbb{R} : \mathbb{R}_\mathcal{U} \to \mathbb{R}_\mathcal{L}$ is *not* $(\mathcal{U} - \mathcal{L})$-continuous, because $i_\mathbb{R}^{-1}([0,1)) = [0,1)$ is not $\mathcal{U}$-open. In fact, we have the following observation:

Theorem 3.5.1. *Let X be any set, with τ and τ' two topologies on X. Then the identity function $i_X : X_\tau \to X_{\tau'}$ is $(\tau - \tau')$-continuous if and only if τ' is coarser than τ.*

The proof is an exercise.

The following example shows much more clearly why we discuss continuity in terms of closeness of points. Let $f : \mathbb{R}_\mathcal{U} \to \mathbb{R}_\mathcal{U}$ be given by $f(x) = 7x$. Then f is $(\mathcal{U} - \mathcal{U})$-continuous. To see why, consider the following figure:

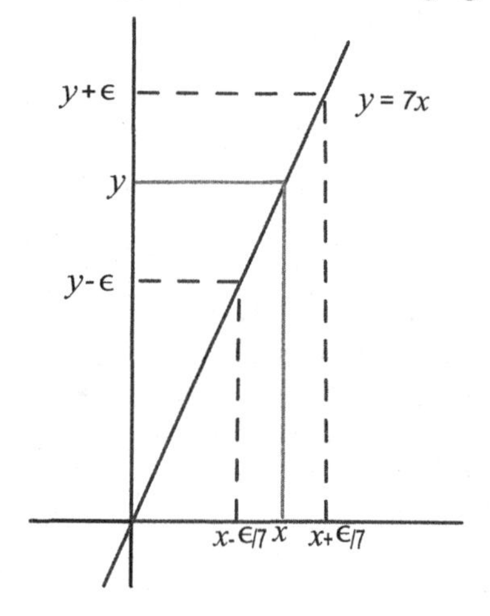

Let V be any $\mathcal{U}$-open set in the range of f. Then for each $y \in V$, there exists a neighborhood of y completely contained in the set V. In other words, there exists an $\epsilon > 0$ such that $(y - \epsilon, y + \epsilon) \subset V$. We need to show that $f^{-1}(V)$ is $\mathcal{U}$-open in the domain of f. Let $a \in f^{-1}(V)$. Then, according to the discussion above, there exists a neighborhood $(f(a) - \epsilon, f(a) + \epsilon) \subset V$. Note, then, that for $\delta = \frac{\epsilon}{7}$, we have $(a - \delta, a + \delta) \subset f^{-1}(V)$, since for any $x \in (a - \delta, a + \delta)$, we get $f(x) \in (f(a) - \epsilon, f(a) + \epsilon) \subset V$. So for each point $a \in f^{-1}(V)$, we have an open neighborhood around a whose image is completely contained in V, implying that $f^{-1}(V)$ is $\mathcal{U}$-open, as we wished.

This example is very nice for seeing the connection between "calc-continuity" and our precise Definition 3.5.2. Fortunately, our definition is much more interesting than just the calculus concept, as the following example shows.

■ EXAMPLE 3.9

Let $g : \mathbb{R}_{\mathcal{L}} \to \mathbb{R}_{\mathcal{U}}$ be defined by

$$g(x) = \begin{cases} x+1, & \text{for } x \geq 1 \\ x^3, & \text{for } x < 1. \end{cases}$$

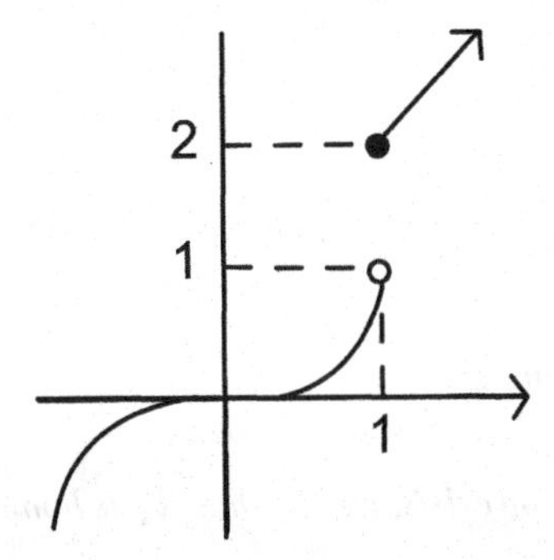

Then g is $(\mathcal{L}-\mathcal{U})$-continuous! To see why, informally, note that the one point that has us worried is $x = 1$ in the domain. Take a typical $\mathcal{U}$-open set containing $f(x) = 2$, say, $(0,3)$. Then $g^{-1}((0,3)) = (0,1) \cup [1,2)$ which is indeed $\mathcal{L}$-open. Note that g is not $(\mathcal{U}-\mathcal{U})$-continuous, since it's easy to find a $\mathcal{U}$-open subset W of the range containing 2 with $g^{-1}(W)$ not $\mathcal{U}$-open. (You should explicitly find a $\mathcal{U}$-open set whose inverse image under g is not $\mathcal{U}$-open, in order to make this precise.) We leave the details of why g is *not* $(\mathcal{U}-\mathcal{L})$-continuous or $(\mathcal{U}-\mathcal{U})$-continuous to the exercises.

This rather abstract concept of continuity has the sorts of properties that one might hope. For example, the composition of continuous functions is continuous:

Theorem 3.5.2. *Let $f : X_\tau \to Y_\nu$ and $g : Y\nu \to Z_\sigma$ be continuous functions. Then the composition $g \circ f : X_\tau \to Z_\sigma$ is continuous.*

Proof: Let W be any σ-open subset of Z. Then the inverse image $g^{-1}(W)$ is a ν-open subset of Y, since g is $(\nu-\sigma)$-continuous. Hence $(g \circ f)^{-1}(W) = f^{-1}(g^{-1}(W))$ is indeed τ-open, since it's f^{-1} of a ν-open set and we know f is continuous. By definition, then, $g \circ f$ is continuous. □

Continuous functions also work well with subspaces:

Theorem 3.5.3. *Let $f : X_\tau \to Y_\nu$ be continuous. For any subset $A \subset X_\tau$, the restriction map $f|_A : A_{\tau_A} \to Y_\nu$ is continuous.*

The proof is left as an exercise.

The concept of continuity is also part of our definition of an "equivalence" of spaces:

Definition 3.5.3. *A function $f : X_\tau \to Y_\nu$ of topological spaces is a homeomorphism if it satisfies all of the following conditions:*

- *f is one-to-one.*
- *f is onto.*
- *f is $(\tau - \nu)$-continuous.*
- *f^{-1} is $(\nu - \tau)$-continuous.*

When such a homeomorphism exists, we say that X_τ is homeomorphic to Y_ν, denoted $X_\tau \cong Y_\nu$.

Note that we could just as well have said that a homeomorphism is a continuous function $f : X_\tau \to Y_\nu$ which has a continuous inverse. In this sense, a homeomorphism is exactly what we mean by an equivalence of topological spaces. An alternative definition uses the following idea.

Definition 3.5.4. *A function $f : X_\tau \to Y_\nu$ of topological spaces is open if $f(U)$ is ν-open in Y whenever U is a τ-open subset of X.*

So a homeomorphism is a function that is 1-1, onto, continuous and open.

Of course, for any space X_τ, the identity function i_X is a homeomorphism. In fact, given two topologies τ and τ' on X, $i_X : X_\tau \to X_{\tau'}$ is a homeomorphism if and only if $\tau = \tau'$. (Why?) However, there are lots of examples of different topologies on a space that yield homeomorphic topological spaces. For example, consider the "left ray topology" on $\mathbb{R}$:

$$\mathcal{LR} = \{V \subset \mathbb{R} : \text{ if } x \in V, \text{ then there exists a ray } (-\infty, a) \text{ for some } a \in \mathbb{R} \text{ with } x \in (-\infty, a) \subset V\}.$$

Then $\mathcal{LR}$ is a topology very different from the right ray topology $\mathcal{RR}$, since $(-\infty, 0)$ is $\mathcal{LR}$-open but not $\mathcal{RR}$-open. The identity function $i : \mathbb{R} \to \mathbb{R}$ is not $(\mathcal{LR} - \mathcal{RR})$-continuous. However, the resulting spaces $\mathbb{R}_{\mathcal{LR}}$ and $\mathbb{R}_{\mathcal{RR}}$ *are* homeomorphic; the function $n : \mathbb{R}_{\mathcal{LR}} \to \mathbb{R}_{\mathcal{RR}}$ is a homeomorphism, where $n(x) = -x$. (Check!)

Exercises

1. Prove that if Y_ν is any topological space and $X_\mathcal{D}$ is a discrete space, then any function $f : X_\mathcal{D} \to Y_\nu$ is $(\mathcal{D} - \nu)$-continuous.

2. Let τ and τ' be any two topologies on X. Then $i_X : X_\tau \to X_{\tau'}$ is $(\tau - \tau')$-continuous if and only if τ' is coarser than τ.

3. Show precisely that the function g of Example 3.9 is *not* $(\mathcal{U} - \mathcal{L})$-continuous.

4. Given two topologies τ and τ' on X, $i_X : X\mathcal{T} \to X_{\tau'}$ is a homeomorphism if and only if $\tau = \tau'$.

5. Prove Theorem 3.5.3: If $f : X_\tau \to Y_\nu$ be continuous, then for any subset $A \subset X_\tau$, the restriction map $f|_A : A_{\tau_A} \to Y_\nu$ is continuous. (*Hint*: you can view a restriction map as a composition.)

6. Consider the "right-hand topology" $\mathcal{R} = \{V \subset \mathbb{R} :$ if $x \in V$, there exists $a, b \in V, a < b,$ such that $x \in (a, b] \subset V\}$.

   (a) a. Show $\mathcal{R}$ is a topology on $\mathbb{R}$.

   (b) b. Show that $\mathbb{R}_\mathcal{R} \cong \mathbb{R}_\mathcal{L}$.

7. Consider the function $f : \mathbb{R} \to \mathbb{R}$ defined by

$$f(x) = \begin{cases} x + 1, & \text{if } x \geq 1, \\ x, & \text{if } x < 1. \end{cases}$$

   (a) Is f $\mathcal{U} - \mathcal{U}$ continuous?

   (b) Is f $\mathcal{L} - \mathcal{U}$ continuous?

   (c) Is f $\mathcal{L} - \mathcal{L}$ continuous?

   (d) Is f $\mathcal{U} - \mathcal{L}$ continuous?

   (e) Is f $\mathcal{R} - \mathcal{U}$ continuous?

   (f) Is f $\mathcal{R} - \mathcal{R}$ continuous?

   (g) Is f $\mathcal{U} - \mathcal{R}$ continuous?

8. Consider the function $g : \mathbb{R} \to \mathbb{R}$ defined by

$$g(x) = \begin{cases} x + 1, & \text{if } x > 1, \\ x, & \text{if } x \leq 1. \end{cases}$$

(a) Is $g\ \mathcal{U} - \mathcal{U}$ continuous?

(b) Is $g\ \mathcal{L} - \mathcal{U}$ continuous?

(c) Is $g\ \mathcal{L} - \mathcal{L}$ continuous?

(d) Is $g\ \mathcal{U} - \mathcal{L}$ continuous?

(e) Is $g\ \mathcal{R} - \mathcal{U}$ continuous?

(f) Is $g\ \mathcal{R} - \mathcal{R}$ continuous?

(g) Is $g\ \mathcal{U} - \mathcal{R}$ continuous?

CHAPTER 4

MORE ON OPEN AND CLOSED SETS AND CONTINUOUS FUNCTIONS

4.1 INTRODUCTION

In this chapter, we'll look more closely at what it means for a subset $A \subset X_\tau$ to be τ-open or τ-closed. Along the way, we'll find a new way to define a topology on a set X, working "locally" rather than "globally." We'll define the concept of a *limit point* of a subset $A \subset X_\tau$, and use this to help us determine whether A is open or closed. For a set $A \subset X_\tau$, we'll define the *boundary* of A and the *interior* of A, and see how these relate to our previous concept, the closure of A. Finally, we'll come up with some easier ways to determine whether a function $f : X_\tau \to Y_\nu$ is continuous.

4.2 BASIS FOR A TOPOLOGY

In this section, we set up some machinery that will enable us to define a topology on a set X "locally," much as we did in defining the usual topology on $\mathbb{R}$. This machinery is called a *basis* for a topology on X, defined below. The definition may seem a bit abstract at first, but it will soon become clear just why we need a basis to have certain properties.

Definition 4.2.1. *For a set X, a basis for (a topology on) X is a collection $\mathcal{B}$ of subsets of X satisfying the following properties:*

- *$\mathcal{B}$ "covers" X:*

$$\bigcup \mathcal{B} = \bigcup_{B \in \mathcal{B}} B = X.$$

- *If $x \in B_1 \cap B_2$ for $B_1, B_2 \in \mathcal{B}$, then there exists a set $B_3 \in \mathcal{B}$ such that $x \in B_3 \subset B_1 \cap B_2$.*

Any set $B \in \mathcal{B}$ will be called a basis set.

The second property, in particular, may seem a bit strange. Here's what it means: in general, the intersection of two basis sets $B_1 \cap B_2$ need *not* be a basis set; however, $B_1 \cap B_2$ must *contain* a basis set around each of its points. To make sense of this, we need some examples. Here are two examples of bases for the set $\mathbb{R}$:

$$\begin{aligned} \text{Let } \mathcal{B}_1 &= \{(a,b) : a, b \in \mathbb{R},\ a < b\}, \\ &\text{and} \\ \text{Let } \mathcal{B}_2 &= \{[a,b) : a, b \in \mathbb{R},\ a < b\}. \end{aligned}$$

We'll show that $\mathcal{B}_1$ is indeed a basis for $\mathbb{R}$. To see that $\bigcup_{B \in \mathcal{B}_1} B = \mathbb{R}$, note that every real number x lives in $(x-1, x+1) \in \mathcal{B}_1$. To check that $\mathcal{B}_1$ satisfies the intersection property, let $B_1 = (a_1, b_1)$ and $B_2 = (a_2, b_2)$ be any two basis sets. If $B_1 \cap B_2 = \emptyset$, then the intersection property is satisfied vacuously. If $x \in B_1 \cap B_2$, let $a = max\{a_1, a_2\}$ and $b = min\{b_1, b_2\}$. Then $a < x < b$ (else $x \notin B_1 \cap B_2 \neq \emptyset$) and we choose our required basis set $B_3 = (a, b)$. We leave it as an exercise to show that $\mathcal{B}_2$ is a basis.

We're interested in bases because they provide a convenient language for specifying a topology, as the following theorem shows.

Theorem 4.2.1. *If $\mathcal{B}$ is a basis for a set X, then*

$$\tau_{\mathcal{B}} := \{U \subset X : \textit{ if } x \in U \textit{ then there exists a basis set } B \in \mathcal{B} \textit{ such that } x \in B \subset U\}$$

is a topology on X.

We refer to $\tau_{\mathcal{B}}$ as the *topology generated by the basis $\mathcal{B}$.*

So, in the topology generated by a basis, a set is open if it contains a basis set around each of its points. We note that the bases $\mathcal{B}_1$ and $\mathcal{B}_2$ above generate the topologies $\mathcal{U}$ and $\mathcal{L}$, respectively, on $\mathbb{R}$. Note also that different bases can generate the same topology. For example,

$$\mathcal{B}_3 = \{(r_1, r_1) : r_1, r_2 \in \mathbb{Q},\ r_1 < r_2\} \subset \mathcal{P}(\mathbb{R})$$

is another basis for $\mathbb{R}$ that also generates $\mathcal{U}$. The proof of Theorem 4.2.1 will show clearly why our definition of basis included the unusual intersection condition.

Proof: Let $\mathcal{B}$ be a basis for X and let $\tau_{\mathcal{B}}$ be defined as in the theorem. Then $\emptyset \in \tau_{\mathcal{B}}$ vacuously, since there is no element in it to violate the definition. At the other extreme, $X \in \tau_{\mathcal{B}}$ since $\mathcal{B}$ must cover $X : \bigcup \mathcal{B} = X$, so that if $x \in X$, then there exists a basis set B such that $x \in B \subset X$. To see that $\tau_{\mathcal{B}}$ is closed under arbitrary unions, let $\{U_\alpha : \alpha \in \Lambda\} \subset \tau_{\mathcal{B}}$. If $U = \bigcup_{\alpha\in\Lambda} U_\alpha$ is empty, the $U \in \tau_{calB}$. If $U \neq \emptyset$, let $x \in U = \bigcup_{\alpha\in\Lambda} U_\alpha$. Then $x \in U_\beta$ for some $\beta \in \Lambda$, and since U_β is open, there exists a basis set B such that $x \in B \subset U_\beta$. Thus

$$x \in B \subset U_\beta \subset U = \bigcup_{\alpha\in\Lambda} U_\alpha,$$

as we hoped. Finally, to see that $\tau_{\mathcal{B}}$ is closed under finite intersections, let U_1 and U_2 belong to $\tau_{\mathcal{B}}$. If $x \in U_1 \cap U_2$, then $x \in U_1$ and $x \in U_2$, which are both $\tau_{\mathcal{B}}$-open, so there exists basis sets B_1 and B_2 such that $x \in B_1 \subset U_1$ and $x \in B_2 \subset U_2$. Thus $x \in B_1 \cap B_2$, so that the second property of the definition of basis implies there exists $B_3 \in \mathcal{B}$ such that $x \in B_3 \subset B_1 \cap B_2$, which is in turn a subset of $U_1 \cap U_2$. So $U_1, U_2 \in \tau_{\mathcal{B}}$ implies $U_1 \cap U_2$ is back in $\tau_{\mathcal{B}}$, as we wished to prove. □

For some interesting examples of bases, we consider the set $\mathbb{R}^2 = \{(x, y) : x, y \in \mathbb{R}\}$, which we intuitively regard as the real plane. However, this set must be endowed with an appropriate topology in order to have this geometric structure. For example, the discrete space $\mathbb{R}^2_{\mathcal{D}}$ can be interpreted as a "cloud" of points, with no two points "close" to each other. Here are the building blocks of three bases for the set $\mathbb{R}^2$: for each $\epsilon > 0$ and each point $(x_0, y_0) \in \mathbb{R}^2$, we define

$$\begin{aligned}
B_\epsilon((x_0, y_0)) &:= \{(x, y) \in \mathbb{R}^2 : \sqrt{(x - x_0)^2 + (y - y_0)^2} < \epsilon\} \text{ ("open } \epsilon\text{-ball")} \\
D_\epsilon((x_0, y_0) &:= \{(x, y) \in \mathbb{R}^2 : |x - x_0| + |y - y_0| < \epsilon\} \text{ ("open } \epsilon\text{-diamond")} \\
S_\epsilon((x_0, y_0) &:= \{(x, y) \in \mathbb{R}^2 : \max\{|x - x_0|, |y - y_0|\} < \epsilon\} \text{ ("open } \epsilon\text{-square").}
\end{aligned}$$

The reader should graph some examples of each set to see why descriptions are apt. The bases are given by

$$\begin{aligned}\mathcal{B}_b &:= \{B_\epsilon((x,y)) : (x,y) \in \mathbb{R}^2, \epsilon > 0\};\\ \mathcal{B}_d &:= \{D_\epsilon((x,y)) : (x,y) \in \mathbb{R}^2, \epsilon > 0\};\\ \mathcal{B}_s &:= \{S_\epsilon((x,y)) : (x,y) \in \mathbb{R}^2, \epsilon > 0\}.\end{aligned}$$

Then each set $\mathcal{B}_*$ is a basis for $\mathbb{R}^2$, for $* = b, d, s$. We'll prove that $\mathcal{B}_b$ is such a basis, leaving the details of the other two to the exercises. First, any $(x,y) \in \mathbb{R}^2$ belongs to an open ϵ-ball [viz., $B_{1000}((x,y))$, among many others]. Thus $\mathcal{B}_b$ covers $\mathbb{R}^2$. Next, let (x,y) lie in the intersection of two open ϵ-balls, as shown below:

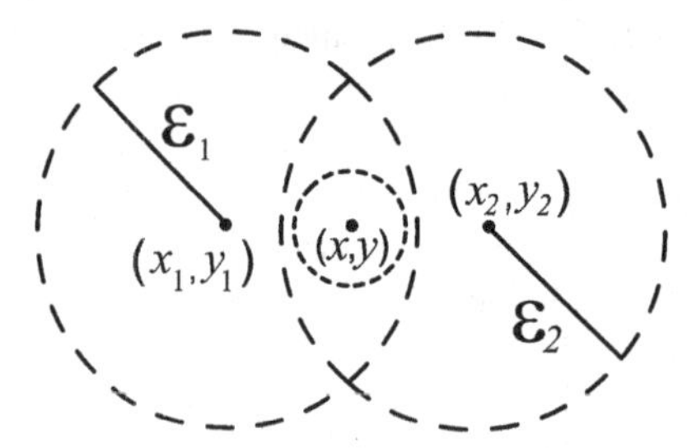

Precisely, let $(x,y) \in B_{\epsilon_1}((x_1,y_1)) \cap B_{\epsilon_2}((x_2,y_2))$. We need to find some positive ϵ_3 and a point at which to center an open ϵ_3-ball, contained in the intersection $B_{\epsilon_1}((x_1,y_1)) \cap B_{\epsilon_2}((x_2,y_2))$. The picture makes it clear; take $\epsilon_3 = min\{\epsilon_1 - \sqrt{(x-x_1)^2+(y-y_1)^2}, \epsilon_2 - \sqrt{(x-x_2)^2+(y-y_2)^2}\}$. Then the open ϵ_3-ball $B_{\epsilon_3}((x,y))$ fits neatly into the intersection of the first two balls, containing the point (x,y), as we wished.

Let us examine the topologies generated by these three bases. For easier notation, we'll denote $\tau_{\mathcal{B}_b}$ by τ_b, and so on.

$$\begin{aligned}\tau_b &:= \{U \subset \mathbb{R}^2 : \text{if } (x,y) \in U, \text{ then there exists } \epsilon > 0 \text{ and } (x_0,y_0) \in U\\ &\qquad \text{such that } (x,y) \in B_\epsilon((x_0,y_0)) \subset U\}.\\ \tau_d &:= \{U \subset \mathbb{R}^2 : \text{if } (x,y) \in U, \text{ then there exists } \epsilon > 0 \text{ and } (x_0,y_0) \in U\\ &\qquad \text{such that } (x,y) \in D_\epsilon((x_0,y_0)) \subset U\}.\\ \tau_s &:= \{U \subset \mathbb{R}^2 : \text{if } (x,y) \in U, \text{ then there exists } \epsilon > 0 \text{ and } (x_0,y_0) \in U\\ &\qquad \text{such that } (x,y) \in S_\epsilon((x_0,y_0)) \subset U\}.\end{aligned}$$

Then the following sets are easily seen to be τ_b-open:

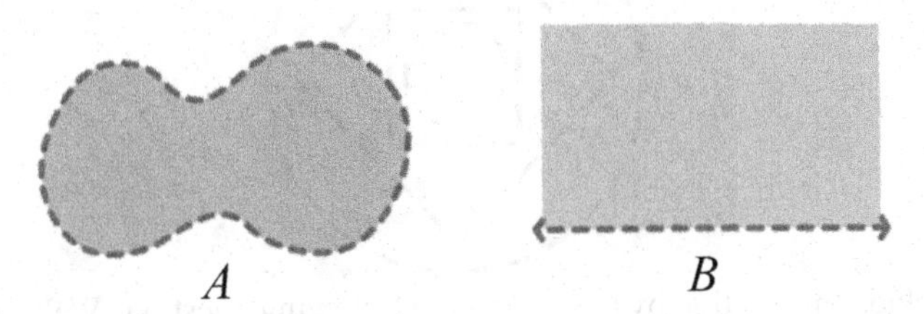

and the following sets are τ_b-closed:

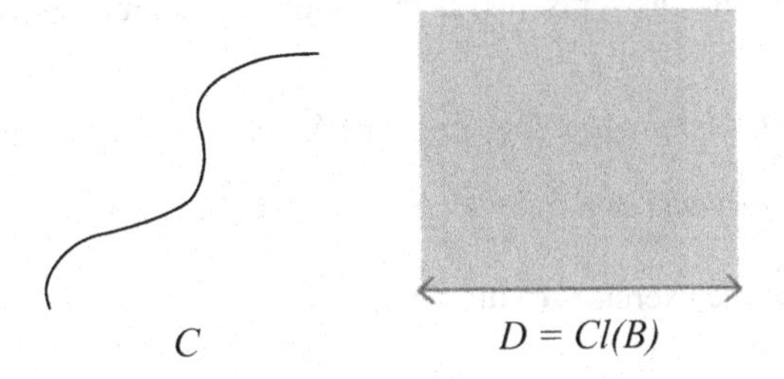

Here the set B is the "open" upper half-plane, $\{(x, y) : y > 0\}$, and the set D is the "closed" upper half-plane, $\{(x, y) : y \geq 0\}$. In fact, we have the following (possibly surprising) relationship between the topologies generated by these three bases:

$$\tau_b = \tau_d = \tau_s.$$

To prove this, we show the following chain of inclusions: $\tau_b \subset \tau_d \subset \tau_s \subset \tau_b$. Let U be τ_b-open and let $(x, y) \in U$. Then the following picture makes clear how to get an open ϵ-diamond around (x, y) inside U:

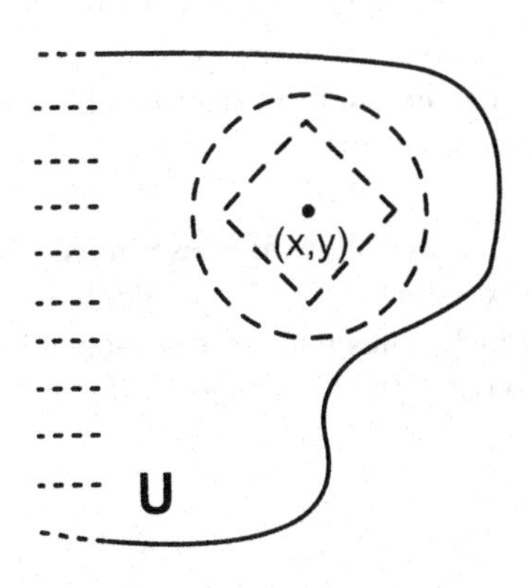

Hence, any set which is τ_b-open is also τ_d-open, showing our first inclusion. Similarly, the following picture shows how to show the other inclusions:

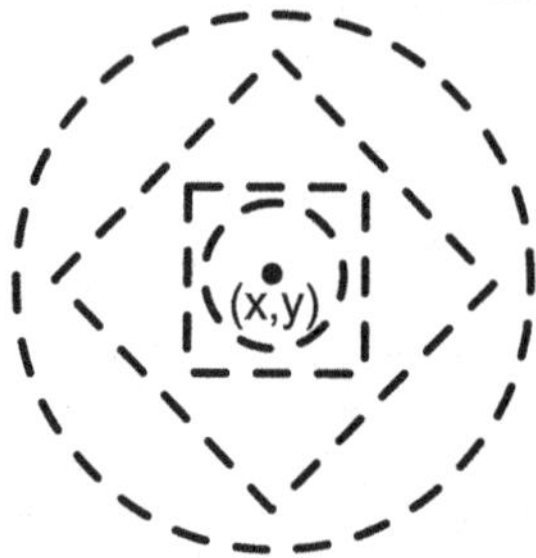

We conclude this section by answering the following question. We've shown that a basis on X generates a topology on X; given a topology τ on X, is there a basis $\mathcal{B}$ such that $\tau_{\mathcal{B}} = \tau$? The answer is yes, as the following theorem shows.

Theorem 4.2.2. *If τ is any topology on a set X, then τ is also a basis for X and, further, the topology generated by τ is τ back again; that is, $\tau_\tau = \tau$.*

The proof is left to the exercises (with hints).

Exercises

1. Show that $\mathcal{B}_2 = \{[a, b) : a, b \in \mathbb{R},\ a < b\}$ is a basis for $\mathbb{R}$.

2. Show that for $\mathcal{B}_3 = \{(r_1, r_2) : r_1, r_2 \in \mathbb{Q},\ r_1 < r_2\}$, we have $\tau_{\mathcal{B}_3} = \mathcal{U}$.

3. Prove that $\mathcal{B}_d$ and $\mathcal{B}_s$ are bases for $\mathbb{R}^2$.

4. Fill in the details, showing that $\tau_b = \tau_d$ as topologies on $\mathbb{R}^2$. Precisely, given U any τ_b-open set, around any $(x, y) \in U$ there exists $B_\epsilon(x, y)$ for some $\epsilon > 0$. Show that U is τ_d-open by giving a precise formula for an open ϵ'-diamond around any (x, y) that fits inside the open ϵ-ball. Then show the other inclusion (which takes a bit more work).

5. Prove Theorem 4.2.2 by following these hints: First, show τ is a basis by the definition. Next, show $\tau_\tau = \tau$ by a double inclusion. To do this, recall Theorem 3.3.2, which states that a set is τ-open if and only if it contains an open neighborhood of each of its points.

4.3 LIMIT POINTS

The concept of "closeness" that a topology is intended to convey becomes much clearer when we consider the *limit points* of a subset $A \subset X_{\mathcal{T}}$:

Definition 4.3.1. *Let A be any subset of a topological space $X_{\mathcal{T}}$. A point $x \in X$ is a limit point of A if for every open neighborhood N_x of x, $(N_x \smallsetminus \{x\}) \cap A \neq \emptyset$. The set of limit points of A is denoted by A'.*

A limit point of a set A is often called an *accumulation point* of A or a *cluster point* of A. The set of limit points of A, A' is (more rarely) referred to as the "derived set" of A. The definition of limit points may seem a bit strange. For a point x to be a limit point of A, we require not just that every open neighborhood N_x of x intersect A nontrivially but also that $N_x \cap A$ contain point(s) *other than x*. For example, consider the subsets $B = (0,1), C = \mathbb{Q}$, and $D = [0,1) \cup (2,3) \cup \{4\}$ of $\mathbb{R}_{\mathcal{U}}$. Then $B' = [0,1]$, since every $\mathcal{U}$-open neighborhood of any point in $[0,1]$ hits $B = (0,1)$ in infinitely many points, whereas, if $x > 1$ or $x < 0$, then we can find a $\mathcal{U}$-open neighborhood $N_x = (x-\epsilon, x+\epsilon)$, which misses B entirely. Similarly, $C' = \mathbb{Q}' = \mathbb{R}$. In both these cases, the set A is contained in its set of limit points A'. However, for $D = [0,1) \cup (2,3) \cup \{4\}$, $D' = [0,1] \cup [2,3]$, since there are many $\mathcal{U}$-open neighborhoods of 4 which intersect D only in the set $\{4\}$, such as $(\frac{7}{2}, 500)$. In this case, then, $D \not\subset D'$: when this occurs ($a \in A \smallsetminus A'$) we say a is an *isolated point* of A.

These same sets, considered as subsets of $\mathbb{R}_{\mathcal{L}}$, produce very different limit points: $B' = [0,1) = B$, since lots of $\mathcal{L}$-neighborhoods of 1 miss B entirely. $C' = \mathbb{R}$ and $D' = [0,1) \cup [2,3)$, as one can verify easily.

When considered as subsets of $\mathbb{R}_{\mathcal{FC}}$, the limit points are quite interesting: $B' = C' = D' = \mathbb{R}$, since for any $x \in \mathbb{R}$, an $\mathcal{FC}$-open neighborhood of x has to contain all but finitely many real numbers, and hence intersects the sets B, C and D in infinitely many points.

As a less intuitive example, let $X = \{a, b, c, d\}$ with topology $\mathcal{T} = \{\emptyset, X, \{a\}, \{c\}, \{a, c\}\}$. Then for $A = \{a, b\}$, we see that $A' = \{b, d\}$ since the only neighborhood of b or d is X itself, which intersects A nicely, whereas a and c have neighborhoods that hit A only in $\{a\}$ and $\{c\}$.

Intuitively, the limit points of A are as "close to" the set A as they can possibly be. This fits nicely into our intuition about what subsets of a space are open and closed:

Theorem 4.3.1. *Let A be any subset of a topological space $X_{\mathcal{T}}$. Then A is $\mathcal{T}$-closed if and only if A contains all of its limit points.*

Note how this explains several of our examples of limit points above.

Proof: $\Longrightarrow$) Let A be any τ-closed subset of X. Since A is closed, its complement $X \setminus A$ is open. Let $x \notin A$, so that $x \in X \setminus A$. Since $X \setminus A$ is open, it is a τ-open neighborhood of x that is disjoint from A. Hence x cannot be a limit point of A. So we've shown that $x \notin A \Longrightarrow x \notin A'$, which implies, by contrapositive, $A' \subset A$, as we wished.

$\Longleftarrow$) Let A be any subset of X_τ with $A' \subset A$. We hope to show that A is τ-closed, or, equivalently, $X \setminus A$ is τ-open. Let $x \notin A$. Since $A' \subset A$, we know that x is not a limit point of A, so there must exist some τ-open neighborhood N_x of x such that $N_x \cap A$ is either empty or just $\{x\}$. Since $x \notin A$, $N_x \cap A$ can't be $\{x\}$, so the only possibility is that $N_x \cap A = \emptyset$. Thus we have a neighborhood N_x of x that misses A entirely: $N_x \subset X \setminus A$. So $X \setminus A$ contains a neighborhood of each of its points, implying that $X \setminus A$ is indeed τ-open, as we hoped to prove. □

We recall that for $A \subset X_\tau$, the *closure* of A, $Cl(A)$ is the smallest τ-closed subset of X containing A. In light of the above theorem, the following is not surprising.

Theorem 4.3.2. *Let A be any nonempty subset of a topological space X_τ. Then $x \in Cl(A)$ if and only if every open neighborhood of x intersects A nontrivially.*

Proof: Lets prove the "only if" part first, since it's easier:
$\Longleftarrow$) Let $x \in X$ be such that every open neighborhood N_x of x intersects A nontrivially. If $x \notin Cl(A)$, then $x \in X \setminus Cl(A)$, an open set that fails to intersect A. (Easy, eh?)
$\Longrightarrow$) Let $x \in Cl(A)$. Let N_x be any open neighborhood of x. Assume that $N_x \cap A = \emptyset$, so then $X \setminus N_x$ is a closed set, with $A \subset X \setminus N_x$. Hence $Cl(A) \subset (X \setminus N_x)$, since $Cl(A)$ is the smallest closed set containing A. So since $x \notin X \setminus N_x$, we see that $x \notin Cl(A)$, contradicting our starting point. So our assumption (that some neighborhood N_x of x misses A) must be false, and the theorem is proved. □

We'll use this result to help prove the following characterization of closure.

Theorem 4.3.3. *For A any subset of a topological space X_τ, $Cl(A) = A \cup A'$.*

Proof: We proceed by double inclusion:
$\subset$) If A is empty, the inclusion is obvious. If $A \neq \emptyset$, let $x \in Cl(A)$, the smallest closed set containing A and let N_x be any open neighborhood of x. Then $N_x \cap A \neq \emptyset$, by Theorem 4.3.2. If $N_x \cap A = \{x\}$, then $x \in A$. If $x \notin A$, then $N_x \cap A$ contains points other than x, since the intersection is nonempty, so that $x \in A'$. Thus, $x \in A \cup A'$.
$\supset$) If A is empty, we're done. If not, let $x \in A \cup A'$. If $x \in A$, then $x \in Cl(A)$, since $A \subset Cl(A)$. If $x \notin A$, then $x \in A'$, which says that every open neighborhood of x hits A (in point other than x, but all we need is that the intersection is nonempty). By Theorem 4.3.2, x must be in $Cl(A)$. □

Exercises

1. For each subset of $\mathbb{R}_{\mathcal{U}}$ below, find the set of limit points. Give reasons where appropriate:

    - $\mathbb{I} = \{x \in \mathbb{R} : x \notin \mathbb{Q}\}$
    - $B = \{1, 2, 3, 4\}$
    - $C = (0, 1]$
    - $D = (0, 1) \cup [2, 3) \cup (4, 5] \cup [6, 7] \cup \{8\}$

2. Find the set of limit points for each of the sets in Exercise 1, considered now as subsets of $\mathbb{R}_{\mathcal{L}}$.

3. Find the set of limit points for each of the sets in Exercise 1, considered now as subsets of $\mathbb{R}_{\mathcal{RR}}$.

4. Find the set of limit points for each of the sets in Exercise 1, considered now as subsets of $\mathbb{R}_{\mathcal{FC}}$.

5. Find the set of limit points for each of the sets in Exercise 1, considered now as subsets of $\mathbb{R}_{\mathcal{I}}$.

6. Prove that if A is any subset of any discrete space $X_{\mathcal{D}}$, then $A' = \emptyset$.

7. Prove that if A is any subset of the indiscrete space $X_{\mathcal{I}}$, with $Card(A) \geq 2$, then $A' = X$.

4.4 INTERIOR, BOUNDARY AND CLOSURE

We recall the definition of the interior of a set, from Definition 3.3.3: For a subset $A \subset X_{\mathcal{T}}$, the *interior of* A (denoted $Int(A)$) is the largest $\mathcal{T}$-open subset of X contained in A.

So for every $A \subset X_{\mathcal{T}}$, we have the chain of inclusions

$$Int(A) \subset A \subset Cl(A).$$

To recall an easy example, for $A = [0, 1) \subset \mathbb{R}_{\mathcal{U}}$, we see that $Int(A) = (0, 1)$. Also in $\mathbb{R}_{\mathcal{U}}$, $Int(\mathbb{Q}) = \emptyset$, since any $\mathcal{U}$-open set must contain an entire neighborhood around each of its points, and any $\mathcal{U}$-neighborhood must contain irrationals as well as rationals. In $\mathbb{R}_{\mathcal{FC}}$, however, $Int([0, 1)) = \emptyset$, because nonempty $\mathcal{FC}$-open sets contain all but finitely many real numbers. Clearly, for any $A \subset X_{\mathcal{T}}$, A is $\mathcal{T}$-open if and only if $A = Int(A)$, as one can check very easily from the definition. (Yes, you should do so. Now.)

The following theorem gives an easy characterization of $Int(A)$.

Theorem 4.4.1. *For A any subset of a topological space $X_{\mathcal{T}}$, $x \in Int(A)$ if and only if there exists a $\mathcal{T}$-open neighborhood N_x of x, such that $N_x \subset A$.*

The proof is an exercise.

Related to the concept of the interior of a set is its boundary:

Definition 4.4.1. *For A any subset of a topological space $X_{\mathcal{T}}$, a point $x \in X$ is a boundary point of A if for every neighborhood N_x of x, both $N_x \cap A$ and $N_x \cap (X \setminus A)$ are nonempty. The set of all boundary points of A is denoted $Bdy(A)$.*

In other texts, you might see $Bdy(A)$ denoted by ∂A. For some simple examples, consider the subsets of $\mathbb{R}_{\mathcal{U}}$ given above: $Bdy([0,1)) = \{0,1\}$ and $Bdy(\mathbb{Q}) = \mathbb{R}$, as one can check easily by looking at neighborhoods of points in each set. For some more geometric intuition, consider the subsets $A, B \subset \mathbb{R}^2_{\mathcal{T}_b}$ given below:

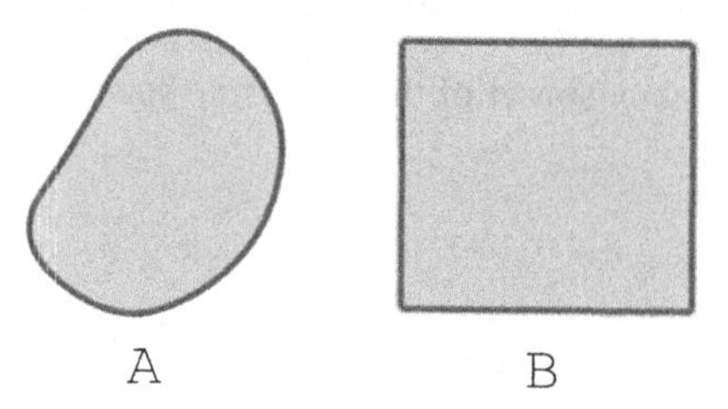

Then $Int(A), Bdy(A), Int(B)$ and $Bd(B)$ are:

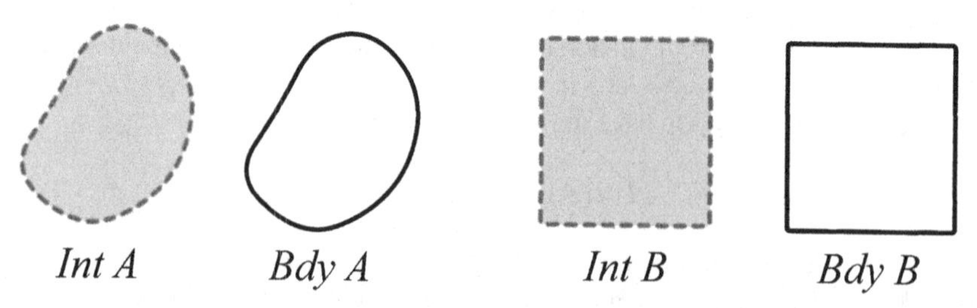

Remember, though, that we need to have much more "general" intuition than what we acquire if we look only at subsets of the real line or plane. For example, in $\mathbb{R}_{\mathcal{FC}}$, we see that $Int([0,1)) = \emptyset$ and $Bdy([0,1)) = \mathbb{R}$, since $\mathcal{FC}$-neighborhoods contain all but finitely many reals. We can use pictures like those of the subsets of the plane in the "ball" topology to get some intuition, but there are far too many interesting spaces in the world that don't act like the plane to let such examples dominate our thinking.

The concepts of the interior, boundary and closure of a set $A \subset X_\tau$ fit together beautifully, as the following theorems show.

Theorem 4.4.2. *A set $A \subset X_\tau$ is τ-closed if and only if A contains its boundary.*

Proof: $\Longrightarrow$) Let A be τ-closed. Let $x \in X \setminus A$, which is τ-open since A is closed. Since $x \in X \setminus A$, an open set, there exists an open neighborhood N_x of x, such that $x \in N_x \subset (X \setminus A)$. Hence $x \notin Bdy(A)$, because we have a neighborhood of x that fails to intersect A. By contrapositive, $Bdy(A) \subset A$.
$\Longleftarrow$) Let $Bdy(A) \subset A$. We'll show A is closed by showing $X \setminus A$ is open. Let $x \in X \setminus A$. Since $Bdy(A) \subset A$, we know $x \notin Bdy(A)$. So there exists some open neighborhood N_x of x with $N_x \cap A = \emptyset$ or $N_x \cap (X \setminus A) = \emptyset$. But $x \in X \setminus A$, so we know $N_x \cap (X \setminus A) \neq \emptyset$, so that $N_x \cap A$ must be empty: i.e. $N_x \subset (X \setminus A)$. Thus every point of $X \setminus A$ has a neighborhood completely contained in $X \setminus A$, so that $X \setminus A$ is open, as we wished to show. □

Theorem 4.4.3. *For any subset A of a topological space X_τ, $Cl(A) = Int(A) \cup Bdy(A)$.*

Proof: This is left as an exercise, with the following hints:

1. be sure to deal with the case $A = \emptyset$ separately;
2. showing $Int(A) \cup Bdy(A) \subset Cl(A)$ is pretty straightforward – the interior part is free (why?) and the boundary part follows from relating boundary points to limit points (See Theorem 4.3.3.); and
3. For the other inclusion, work by contrapositive. Note that, in general, $Int(A) \cap Bdy(A) = \emptyset$, so that we can denote $Int(A) \cup Bdy(A)$ by the more informative $Int(A) \coprod Bdy(A)$, the usual notation for "disjoint union."

Exercises

1. Find the interiors of the following subsets of the set $\mathbb{R}_\mathcal{U}$:
   - $A = \{1, 2, 3\}$

- $B = (0,1) \cap \mathbb{Q}$
- $C = [0,1] \cup (2,3] \cup \{4\}$

2. Do Exercise 1 with the sets considered as subsets of $\mathbb{R}_{\mathcal{RR}}$.

3. Do Exercise 1 with the sets considered as subsets of $\mathbb{R}_{\tau_2}$, the distinguished point topology (a nonempty set is τ_2-open if it contains 2).

4. Prove Theorem 4.4.1: For A any subset of a topological space X_τ, $x \in Int(A)$ if and only if there exists a τ-open neighborhood N_x of x, such that $N_x \subset A$.

5. For the sets and spaces of Exercises 1-3, find the boundaries of the sets.

6. For $A \subset X_\tau$ define the *exterior of A* (denoted $Ext(A)$) as $Ext(A) := Int(X \setminus A)$.

   (a) For the sets and spaces of Exercises 1-3, find the exteriors of the sets.

   (b) For any $A \subset X_\tau$, prove that the three sets $Int(A)$, $Bdy(A)$ and $Ext(A)$ are pairwise disjoint, and show that

   $$X = Int(A) \coprod Bdy(A) \coprod Ext(A).$$

   [Hint: Prove first that $Bdy(A) = X \setminus (Int(A) \cup Ext(A))$.]

7. Prove Theorem 4.4.3.

4.5 MORE ON CONTINUITY

We recall that our definition of continuity is that a function $f : X_\tau \to Y_\nu$ is $(\tau - \nu)$-continuous if the inverse image if every ν-open subset of Y is τ-open. Unfortunately, this appears, at least on the surface, to be a very difficult definition to check, since one has to examine *every* open set in the range. The following theorem gives a number of properties that are equivalent to a function being continuous, some of which are much easier to check.

Theorem 4.5.1. *For a function $f : X_\tau \to Y_\nu$ of topological spaces, the following are equivalent:*

1. *f is $(\tau - \nu)$-continuous.*

2. *If $\mathcal{B}$ is a basis for the topology ν on Y, then for each basis set $B \in \mathcal{B}$, $f^{-1}(B) \in \tau$.*

3. *For every $x \in X$ and every ν-neighborhood $M_{f(x)}$ of $f(x)$, $f^{-1}(M_{f(x)})$ contains a τ-neighborhood of x.*

4. *For every $x \in X$ and for every ν-neighborhood $M_{f(x)}$ of $f(x)$, there exists a τ-neighborhood N_x of x with $f(N_x) \subset M_{f(x)}$.*

5. *For every subset $A \subset X$, $f(Cl(A)) \subset Cl(f(A))$.*

6. *For every ν-closed subset $C \subset Y$, $f^{-1}(C)$ is τ-closed in X.*

Note that properties 3 and 4 are "local" properties, in that they can be checked at each point, rather than having to deal with all possible open sets.

Proof: We will show that all six statements are equivalent by proving that

$$(1) \Longrightarrow (2) \Longrightarrow (3) \Longrightarrow (4) \Longrightarrow (5) \Longrightarrow (6) \Longrightarrow (1).$$

$(1) \Longrightarrow (2)$: Let $\mathcal{B}$ be a basis for ν. Then each set $B \in \mathcal{B}$ is ν-open (since B contains a basis set (viz., itself) around each of its points. Since f is continuous, $f^{-1}(B)$ is τ-open for every $B \in \mathcal{B}$.
$(2) \Longrightarrow (3)$: Let $x \in X$ and let $M = M_{f(x)}$ be any ν-neighborhood of $f(x)$. Since M is ν-open, M must contain a basis set around each of its points, so there exists $B \in \mathcal{B}$ with $f(x) \in B \subset M$. Then $x \in f^{-1}(B) \subset f^{-1}(M)$, where $f^{-1}(B)$ is τ-open by (2), so $f^{-1}(B)$ is the desired τ-neighborhood of x contained in $f^{-1}(M)$.
$(3) \Longrightarrow (4)$: Let $x \in X$ and let $M = M_{f(x)}$ be any ν-neighborhood of $f(x)$. By (3), we see that $N = f^{-1}(M)$ is a τ-neighborhood of x, since $f^{-1}(M)$ is τ-open and contains x. Hence $f(N) \subset M$, as we wished to show.
$(4) \Longrightarrow (5)$: Let A be any subset of X. Let $y \in f(Cl(A))$, so there exists at least one $x \in Cl(A)$ with $y = f(x)$. We wish to show $y \in Cl(f(A))$, so we need to show every ν-neighborhood of y intersects $f(A)$ (using Theorem 4.3.2). Let $M = M_y$ be any ν-neighborhood of y. By (4), we know there exists a τ-neighborhood N_x of x with $f(N_x) \subset M$. Now $x \in Cl(A)$, so that $N_x \cap A \neq \emptyset$. In particular, let $x_1 \in N_x \cap A$. Then $f(x_1) \in f(N_x \cap A) \subset f(N_x) \cap f(A) \subset M \cap f(A)$, so that $M \cap f(A) \neq \emptyset$, as we wished to show.
$(5) \Longrightarrow (6)$: Let C be any closed subset of Y_ν. We need to show $f^{-1}(C)$ is τ-closed. By (5), we know

$$f(Cl(f^{-1}(C))) \subset Cl(f(f^{-1}(C))) \subset Cl(C) = C,$$

so that $Cl(f^{-1}(C)) \subset f^{-1}(C)$. Since any set is contained in its closure, we see that $f^{-1}(C) = Cl(f^{-1}(C))$, so that $f^{-1}(C)$ is τ-closed, as we had hoped.
(6) $\Longrightarrow$ (1) : Let U be any ν-open subset of Y. We need to show that $f^{-1}(U)$ is open in X. Note that $Y \setminus U$ is ν-closed, so that (6) tells us that $f^{-1}(Y \setminus U)$ is τ-closed. However, we know that $f^{-1}(Y \setminus U) = X \setminus f^{-1}(U)$ (by Exercise 3 of Section 2.2. We conclude that $f^{-1}(U)$ is indeed τ-open, as we wished to prove. □

Now we're in a position to show whether certain functions are continuous with much less efforts. For example, we can now show the following:

Theorem 4.5.2. *A function $f : \mathbb{R} \to \mathbb{R}$ is $(\mathcal{U} - \mathcal{U})$-continuous if and only if f is "calc-continuous" (in the sense of Definition 3.5.1).*

The proof is an exercise.

Exercises

1. Prove Theorem 4.5.2, relating $(\mathcal{U} - \mathcal{U})$-continuity with the calculus definition. (*Hint*: see the discussion following Theorem 3.5.1 for some ideas on where to start.)

2. Consider the function $f : \mathbb{R} \to \mathbb{R}$ given by
$$f(x) = \begin{cases} x^2 - 1 & \text{if } x \leq 2, \\ 3x + 2 & \text{if } x > 2. \end{cases}$$
Is f $(\mathcal{U} - \mathcal{U})$-continuous? $(\mathcal{L} - \mathcal{U})$-continuous? $(\mathcal{FC} - \mathcal{U})$-continuous? $(\mathcal{L} - \mathcal{FC})$-continuous? $\mathcal{R} - \mathcal{U}$ continuous? $\mathcal{R} - \mathcal{R}$ continuous? $\mathcal{U} - \mathcal{R}$ continuous? [We recall that the topology $\mathcal{R}$ on $\mathbb{R}$ is the right-hand topology, with basis $\{(a, b] : a < b\}$.]

3. Consider the function $f : \mathbb{R} \to \mathbb{R}$ given by
$$f(x) = \begin{cases} x^3 + 1 & \text{if } x < 2, \\ 2x + 12 & \text{if } x \geq 2. \end{cases}$$
Is f $(\mathcal{U} - \mathcal{U})$-continuous? $(\mathcal{L} - \mathcal{U})$-continuous? $(\mathcal{FC} - \mathcal{U})$-continuous? $(\mathcal{L} - \mathcal{FC})$-continuous? $(\mathcal{R} - \mathcal{U})$-continuous? $(\mathcal{R} - \mathcal{R})$-continuous? $(\mathcal{U} - \mathcal{R})$-continuous? [We recall that the topology $\mathcal{R}$ on $\mathbb{R}$ is the right-hand topology, with basis $\{(a, b] : a < b\}$.]

4. Let $A \subset X_\tau$ and let $f : X_\tau \to Y_\nu$ be continuous. If x is a limit point of A, must $f(x)$ be a limit point of $f(A) \subset Y$?

CHAPTER 5

NEW SPACES FROM OLD

5.1 INTRODUCTION

In this chapter, we show how to construct new topological spaces from existing spaces. We've already seen one example of this kind of construction: Given a space X_τ and a subset $A \subset X$, we put the *subspace topology* $\tau_A = \{U \cap A : U \in \tau\}$ on A. This topology works well with functions out of X_τ, as Theorem 3.5.3 shows: if $f : X_\tau \to Y_\nu$ is continuous, then for any subset $A \subset X_\tau$, the restriction map $f|_A : A_{\tau_A} \to Y_\nu$ is continuous. We want this sort of good behavior whenever we construct new spaces from existing spaces. Indeed, this desire for the construction to work properly with respect to continuous functions will motivate our definitions of the topologies on the new spaces.

5.2 PRODUCT SPACES

Given two sets X and Y, we know how to define their Cartesian product: $X \times Y := \{(x, y) : x \in X, y \in Y\}$. We now consider how to deal with this when X and Y are *spaces*. Given topological spaces X_μ and Y_ν, what topology do we put on the *set* $X \times Y$ to ensure that functions into and out of $X \times Y$ work "well"? We have an obvious candidate for a topology: consider the collection $\mu \times \nu \subset \mathcal{P}(X \times Y)$. First, we must ask ourselves: is $\mu \times \nu$ even a topology on $X \times Y$? Well, obviously $X \times Y \in \mu \times \nu$, since $X \in \mu$ and $Y \in \nu$; and $\emptyset = \emptyset \times \emptyset \in \mu \times \nu$ as well, so so far, so good. To check whether the intersection of two (or finitely many) sets from $\mu \times \nu$ is back in the collection, consider the following diagram of sets from $\mathcal{U} \times \mathcal{U}$, our guess at the topology on $\mathbb{R} \times \mathbb{R}$:

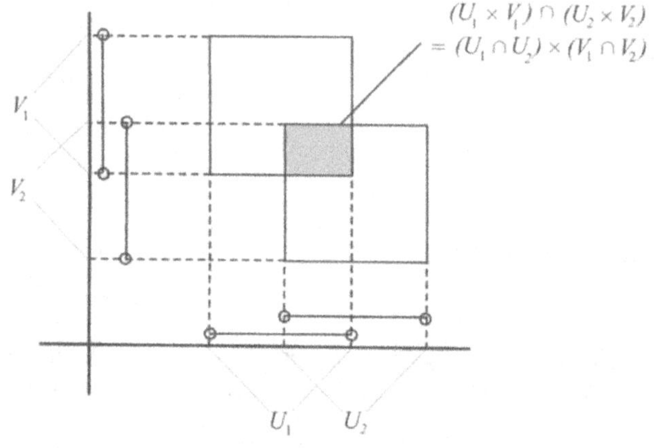

So it seems believable that $\mu \times \nu$ might be a topology, at least so far. However, when we look at unions of sets from $\mu \times \nu$, we run into trouble, as the following diagram of sets from $\mathcal{U} \times \mathcal{U}$ shows:

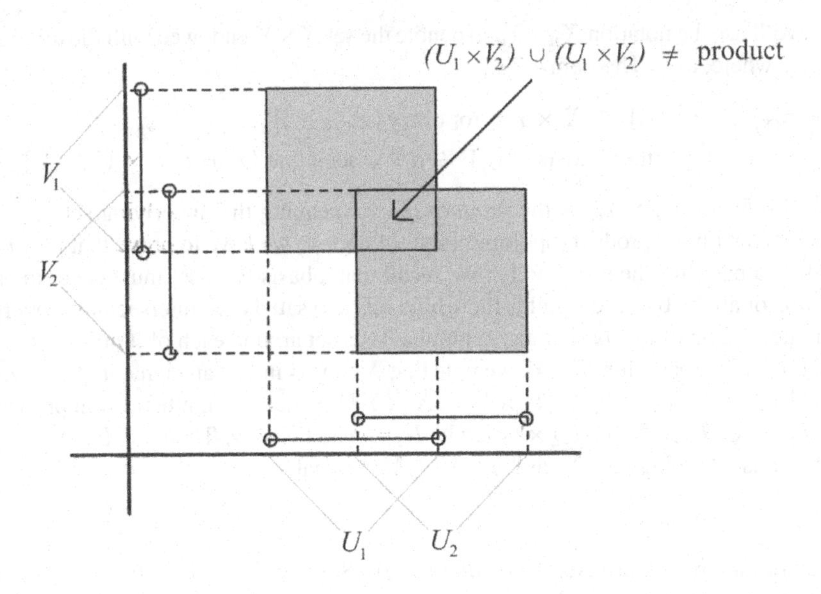

So $\mu \times \nu$ is not, in general, a topology on $X \times Y$. But the diagrams above point us in the right direction: a set of the form $(U_1 \times V_1) \cup (U_2 \times V_2)$, with $U_i \in \mu$ and $V_i \in \nu$, is not back in $\mu \times \nu$, but $(U_1 \times V_1) \cup (U_2 \times V_2)$ *contains* a set from $\mu \times \nu$ around each of its points, as the following diagram of sets from $\mathcal{U} \times \mathcal{U}$ in $\mathbb{R} \times \mathbb{R}$ shows:

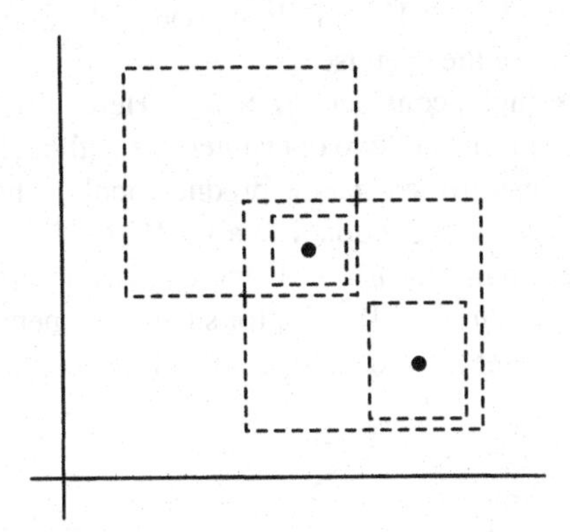

This suggests the following:

Definition 5.2.1. *For X_μ and Y_ν any topological spaces, the product topology on $X \times Y$ is the topology generated by the basis $\mu \times \nu$.*

We'll use the notation $X_\mu \times Y_\nu$ to denote the set $X \times Y$ endowed with the topology $\tau_{\mu\times\nu}$, where, as in Theorem 4.2.1,

$$\begin{aligned}\tau_{\mu\times\nu} : \;=\; & \{W \subset X \times Y : \text{ for every } (x,y) \in W, \\ & \text{there exists } U \times V \in \mu \times \nu \text{ such that } (x,y) \in U \times V \subset W\}.\end{aligned}$$

We'll refer to $X_\mu \times Y_\nu$ as the *product space*, meaning the underlying set $X \times Y$ endowed with the product topology. First, of course, we have to prove that $\mu \times \nu$ is indeed a basis for the set $X \times Y$. We recall that a basis for a set must cover the set (union of all the basis sets yields the whole set) and satisfy the intersection property: an intersection of two basis must contain a basis set around each of it points. To see that $\mu \times \nu$ is a basis for $X \times Y$, we note that $X \times Y$ is in fact an element of $\mu \times \nu$, so that $\bigcup \mu \times \nu = \bigcup_{U \times V \in \mu\times\nu} U \times V = X \times Y$. To check then intersection property, let $(x,y) \in (U_1 \times V_1) \cap (U_2 \times V_2)$, with $U_i \in \mu$ and $V_i \in \nu$. Then $x \in U_1 \cap U_2 \in \mu$, since μ is a topology for X, and $y \in V_1 \cap V_2$, so that

$$(x,y) \in (U_1 \cap U_2) \times (V_1 \cap V_2) \subset (U_1 \times V_1) \cap (U_2 \times V_2)$$

provides the requisite basis set around (x,y). So $\mu \times \nu$ is a basis for $X \times Y$, and we've got our product topology.

The simplest examples deal with the extreme cases of spaces: discrete and indiscrete spaces. Stated tersely: the product of two discrete spaces is discrete and the product of two indiscrete spaces is indiscrete. More verbosely, if $X_\mathcal{D}$ and $Y_\mathcal{D}$ are two sets endowed with the discrete topologies, then the product space $X_\mathcal{D} \times Y_\mathcal{D}$ has as its topology $\mathcal{P}(X \times Y)$, since the basis contains all of the singleton subsets $\{x\} \times \{y\} = \{(x,y)\}$ of $X \times Y$. Similarly, one looks at the basis to analyze the topology on $X_\mathcal{I} \times Y_\mathcal{I}$ (see the exercises).

For a less trivial example, consider $\mathbb{R}_\mathcal{U} \times \mathbb{R}_\mathcal{U}$. Here, a typical basis set is of the form $(a,b) \times (c,d)$, the product of two open intervals (although there are lots of other basis sets, as well). It's easy to see that the product topology here agrees exactly with the "open ball" topology τ_b of Section 4.2; a set $W \subset \mathbb{R}^2$ is open in the open ball topology τ_b if for every point $(x,y) \in W$ there exists an open ϵ-ball $B_\epsilon((x_0,y_0))$ such that $(x,y) \in B_\epsilon((x_0,y_0)) \subset W$. So, for such a τ_b-open set W, for every (x,y) we can get a $\tau_{\mathcal{U}\times\mathcal{U}}$-basis set around (x,y) inside W as follows:

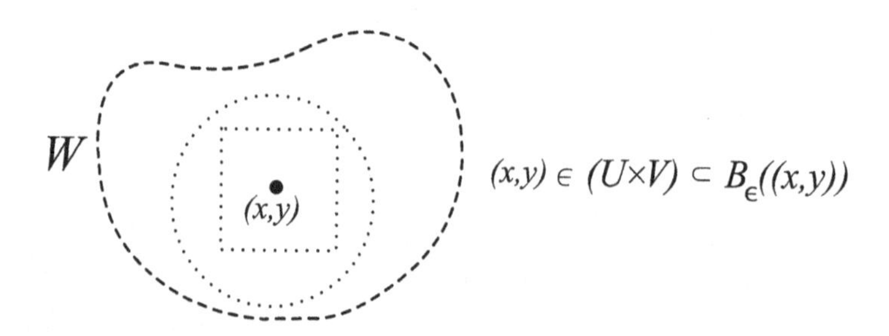

Similarly, every set Z that is open in the product topology $\tau_{\mathcal{U}\times\mathcal{U}}$ is also open in the ball topology:

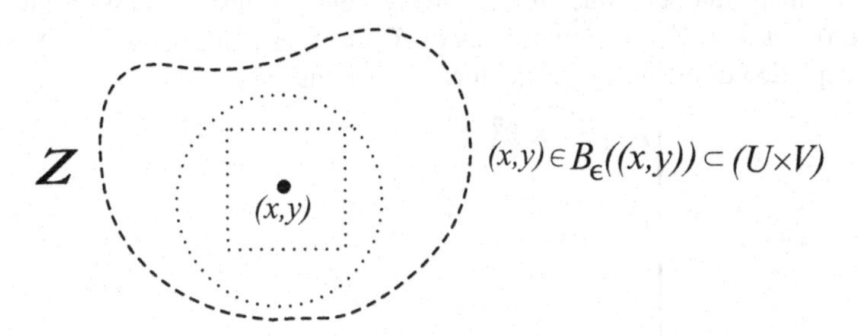

Again, note that lots of sets are open in $\mathbb{R}_{\mathcal{U}} \times \mathbb{R}_{\mathcal{U}} = \mathbb{R}^2_{\mathcal{U}^2}$ aside from the "box" sets:

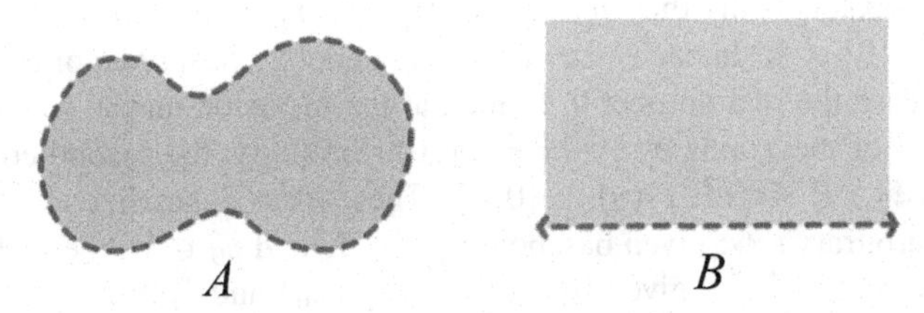

Of course, we can use this construction repeatedly to get a product topology on any *finite* product of spaces: $(X_1)_{\mu_1} \times (X_2)_{\mu_2} \times \cdots \times (X_k)_{\mu_k}$ for any $k \in \mathbb{N}$. For example, when we consider $\mathbb{R}_{\mathcal{U}} \times \mathbb{R}_{\mathcal{U}} \times \cdots \times \mathbb{R}_{\mathcal{U}}$, the n-fold product, the topology should be quite familiar. For every $\epsilon > 0$ and every $(x_1, x_2, \ldots, x_n) \in \mathbb{R}^n$, let

$$\begin{aligned} B_\epsilon((x_1, \ldots, x_n)) \quad := \quad & \{(y_1, y_2, \ldots, y_n) \in \mathbb{R}^n : \\ & \sqrt{(y_1 - x_1)^2 + (y_2 - x_2)^2 + \cdots + (y_n - x_n)^2} < \epsilon\} \\ & (\text{"an open } \epsilon\text{-ball"}). \end{aligned}$$

The topology given by the basis of all such open ϵ-balls in $\mathbb{R}^n$ is equal to the product topology $\tau_{\mathcal{U}^n}$, by the same argument as above.

[*Note:*Be warned that the infinite Cartesian product of topological spaces is *much* more complicated, even when we're lucky enough to be looking at something as simple as the product of a sequence of spaces
$(X_1)_{\mu_1}, (X_2)_{\mu_2}, \ldots, X_k)_{\mu_k}, \ldots$. Here we have a very good idea of what the underlying set $X_1 \times X_2 \times \cdots \times X_k \times \ldots$, looks like, but the "correct" topology is not at all evident. In general, one might be asked to take the Cartesian product of an uncountable collection of spaces, which is much harder to think about even as sets, let alone as spaces. These concepts will be dealt with in Section 5.3.]

The product topology defined above is the "correct" topology because it works well with respect to the inclusions and projection maps into and out of the product. The inclusion maps into a product are intuitively clear, because when we think of a Cartesian product $X \times Y$, we tend to think of X and Y as being subsets of $X \times Y$. Our usual depiction of the plane $\mathbb{R}^2$ reinforces this intuition:

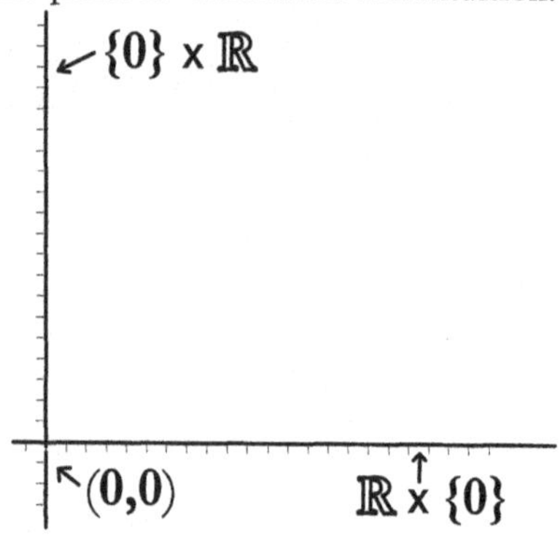

However, even the the case of $\mathbb{R} \times \mathbb{R}$, the real line is *not* really a subset of the plane: the x-axis is really the subset $\mathbb{R} \times \{0\} = \{(x, y) : y = 0\} \subset \mathbb{R}^2$. Similarly, the y-axis is $\{0\} \times \mathbb{R}$. In each case, we had to choose a *basepoint* of the coordinate set, in this case the real number 0, to look at the inclusion maps: $i_1 : \mathbb{R} \hookrightarrow \mathbb{R}^2$ is the inclusion of the x-axis by $i_1(x) = (x, 0)$. Similarly, the second coordinate set includes by $i_2 : \mathbb{R} \hookrightarrow \mathbb{R}^2$, $i_2(y) = (0, y)$. These ideas generalize to any Cartesian product of arbitrary sets; given basepoints $x_0 \in X$ and $y_0 \in Y$, the inclusion maps are $i_X : X \hookrightarrow X \times Y$, given by $i_X(x) = (x, y_0)$ and $i_Y : Y \hookrightarrow X \times Y$, by $i_Y(y) = (x_0, y)$. (When dealing with the Cartesian product of a set with itself, like $\mathbb{R}^2$, we'll use i_1 and i_2 to designate the inclusions, to avoid ambiguity.) We picture X as living inside $X \times Y$ as $Im(i_X) = X \times \{y_0\}$, and Y as $\{x_0\} \times Y = Im(i_Y)$:

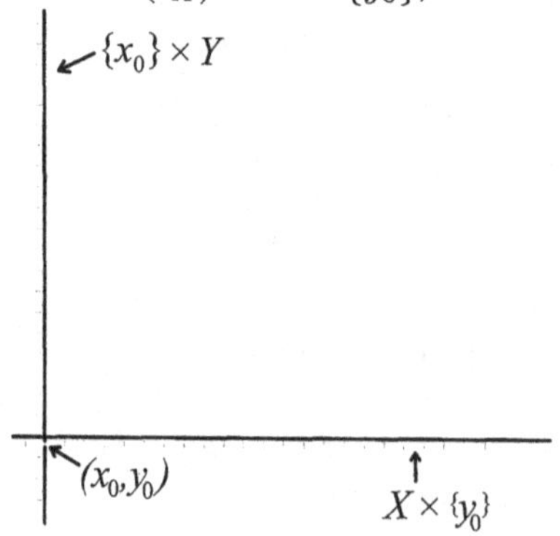

Dual to the inclusion maps are the projection maps: $p_X : X \times Y \to X$ and $p_Y : X \times Y \to Y$ given by $p_X(x, y) = x$ and $p_Y(x, y) = y$. We can picture how these projection maps work on subsets of $X \times Y$ by the following diagram:

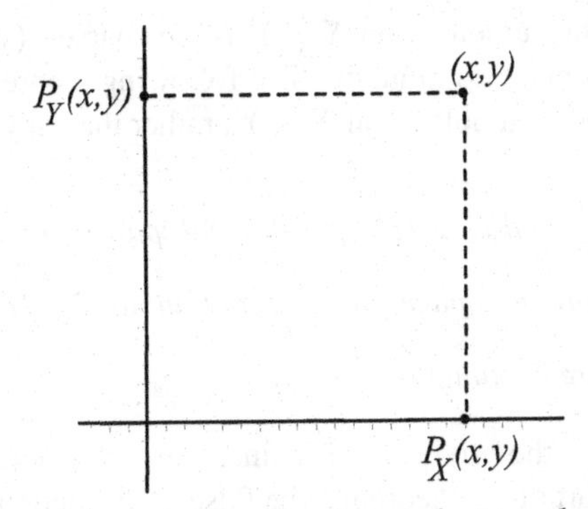

The following theorem is one reason why the product topology is the right one.

Theorem 5.2.1. *If X_μ and Y_ν are any topological spaces, then the projection maps*

$$p_X : X_\mu \times Y_\nu \to X_\mu$$

and

$$p_Y : X\mu \times Y_\nu \to Y_\nu$$

are both continuous, where $X_\mu \times Y_\nu$ denotes $X \times Y$ endowed with the product topology.

Proof: We'll prove p_X is continuous. Let $U \subset X$ be any μ-open set. Then $p_X^{-1}(U) = U \times Y$, as the following plot makes clear:

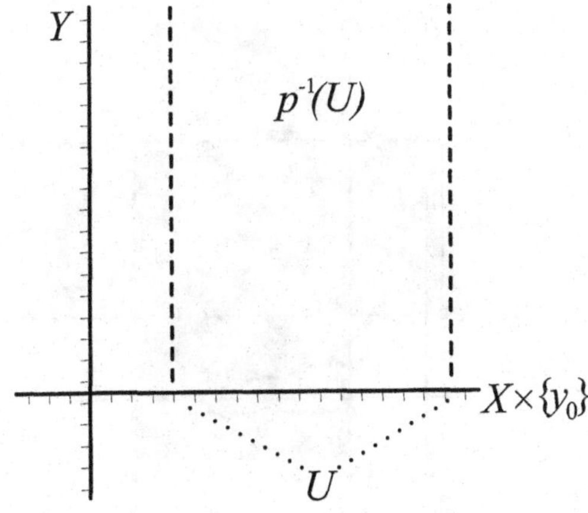

so $p_X^{-1}(U) = U \times Y$ is open in the product space, since it's a basis set. □

Note that the product topology on $X \times Y$ is the coarsest (smallest) topology that makes the projection maps continuous. The following theorem makes even clearer why we want the product topology on $X \times Y$, rather than any other.

Theorem 5.2.2. *If X_μ and Y_ν are any topological spaces, and if $f : Z_\sigma \to X_\mu \times Y_\nu$ is any function into the product space, then f is continuous if and only if the compositions $p_X \circ f$ and $p_Y \circ f$ are continuous.*

We will see during the proof that any finer (larger) topology on $X \times Y$ would cause the "only if" part of the theorem to be false. This theorem nicely characterizes continuous maps into products, so it will go by the nickname "Big Theorem on Maps into Products."

Proof: $\Longrightarrow$) If $f : Z_\sigma \to X_\mu \times Y_\nu$ is continuous, then the compositions are continuous, since Theorem 5.2.1 shows that the projection maps are continuous and the composition of continuous functions is continuous by Theorem 3.5.2.

$\Longleftarrow$) Let $p_X \circ f$ and $p_Y \circ f$ be continuous. To show that f is continuous, we need to start with an arbitrary open set $W \subset X \times Y$ and show that $f^{-1}(W)$ is σ-open in Z. Let $z \in f^{-1}(W)$, with $f(z) = (x, y) \in W$. Since W is open in the product topology, we have a basis set $U \times V$ with $(x, y) \in U \times V \subset W$, with U and V open in X and Y, respectively. In other words, $x = (p_X \circ f)(z) \in U$ and $y = (p_Y \circ f)(z) \in V$. But we know that $p_X \circ f$ and $p_Y \circ f$ are both continuous, so $(p_X \circ f)^{-1}(U)$ and $(p_Y \circ f)^{-1}(V)$ are both open in Z, but neither are likely to be contained in $f^{-1}(W)$, as the following diagram shows:

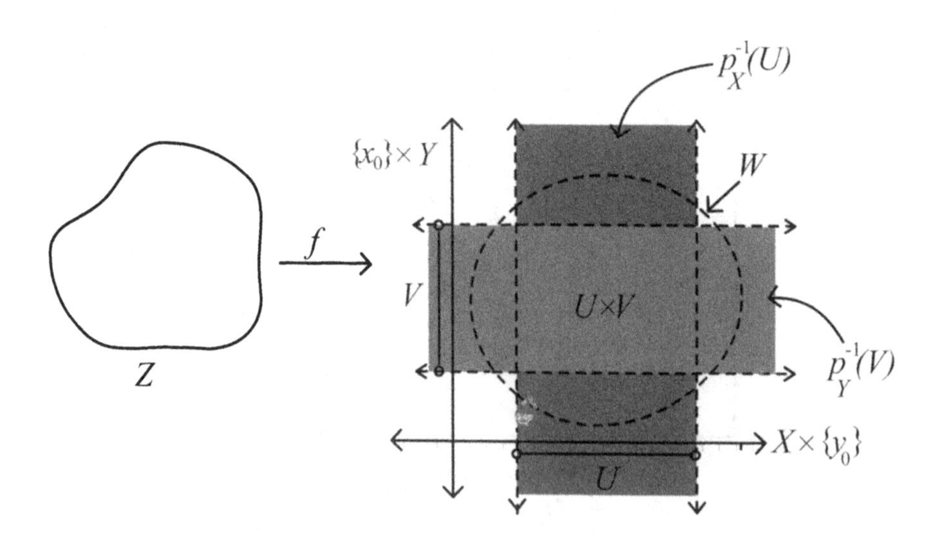

The key point to see is that the basis set $U \times V$ in the product is really the intersection of the inverse images under the projection maps: $U \times V = p_X^{-1}(U) \cap p_Y^{-1}(V)$. Hence, we get the following:

$$\begin{aligned}(p_X \circ f)^{-1}(U) \cap (p_Y \circ f)^{-1}(V) &= f^{-1}(p_X^{-1}(U)) \cap f^{-1}(p_Y^{-1}(V)) \\ &= f^{-1}\left(p_X^{-1}(U) \cap p_Y^{-1}(V)\right) = f^{-1}(U \times V),\end{aligned}$$

which must be σ-open in Z, since it's the intersection of two open sets. We've shown, then, that for each $z \in f^{-1}(W)$, there's a σ-open set $f^{-1}(U \times V)$ with $z \in f^{-1}(U \times V) \subset f^{-1}(W)$. Hence $f^{-1}(W)$ is open, since it contains a neighborhood of each of its points. We conclude that f must be continuous. □

So we have a very simple characterization of when functions into a product space are continuous. As an example of how useful this is, we have the following theorem.

Theorem 5.2.3. *If X_μ and Y_ν are any topological spaces, with basepoints $x_0 \in X$ and $y_0 \in Y$, then the inclusions maps*

$$i_X : X_\mu \hookrightarrow X_\mu \times Y_\nu$$

and

$$i_Y : Y_\nu \hookrightarrow X\mu \times Y_\nu$$

are both continuous, where $X_\mu \times Y_\nu$ denotes $X \times Y$ endowed with the product topology.

The proof is left as an exercise.

We can use induction to show that all of the results of this section hold for k-fold Cartesian products of spaces, for any $k \in \mathbb{N}$. We recall that the k-fold Cartesian product of sets is given by $\prod_{i=1}^k X_i = X_1 \times X_2 \times \cdots \times X_k = \{(x_1, x_2, \ldots, x_k) : x_i \in X_i\}$, the set of ordered k-tuples. If each set X_i has a topology τ_i, then the product space $\prod_{i=1}^k (X_i)_{\tau_i}$ has its topology generated by the basis $\{\prod_{i=1}^k U_i : U_i \in \tau_i\}$. Maps f into such a product space will be continuous if and only $p_i \circ f$ is continuous for each $i = 1, 2, \ldots, k$, where $p_i : \prod_{i=1}^k X_i \to X_i$ is projection onto the ith coordinate space.

Exercises

1. Verify that the product of two indiscrete spaces is indiscrete.
2. Consider the product space $\mathbb{R}_{\mathcal{L}} \times \mathbb{R}_{\mathcal{L}}$.

- Sketch a typical basis set in $\mathbb{R}_{\mathcal{L}} \times \mathbb{R}_{\mathcal{L}}$.
- Sketch several open sets in $\mathbb{R}_{\mathcal{L}} \times \mathbb{R}_{\mathcal{L}}$.
- Sketch several sets which are not open in $\mathbb{R}_{\mathcal{L}} \times \mathbb{R}_{\mathcal{L}}$.

3. Consider the product space $\mathbb{R}_{\tau_1} \times \mathbb{R}_{\tau_1}$, where τ_1 is the distinguished point topology: $\tau_1 = \{U \subset \mathbb{R} : 1 \in U \text{ or } U = \emptyset\}$.

   - Sketch a typical basis set in $\mathbb{R}_{\tau_1} \times \mathbb{R}_{\tau_1}$.
   - Sketch several open sets in $\mathbb{R}_{\tau_1} \times \mathbb{R}_{\tau_1}$.
   - Sketch several sets which are not open in $\mathbb{R}_{\tau_1} \times \mathbb{R}_{\tau_1}$.

4. For the sets $A = \{(x, y) : 2 \leq x < 4, x < y \leq 6\}$ and $B = \{(x, y) : 0 < y \leq x^2, -1 \leq x \leq 1\}$ of $\mathbb{R}^2_{\mathcal{U}^2}$, sketch the sets and their projections onto the axes.

5. Prove Theorem 5.2.3: If X_μ and Y_ν are any topological spaces, with basepoints $x_0 \in X$ and $y_0 \in Y$, then the inclusions maps $i_X : X_\mu \hookrightarrow X_\mu \times Y_\nu$ and $i_Y : Y_\nu \hookrightarrow X\mu \times Y_\nu$ are both continuous. (*Hint*: These are maps into a product.)

5.3 INFINITE PRODUCT SPACES (OPTIONAL)

In this section, we generalize our ideas about the product of two topological spaces (and hence any product of finitely many spaces) to arbitrary Cartesian products of spaces. This leap is not at all as simple as it might appear at first glance. The most obvious kind of Cartesian product of infinitely many sets is pretty easy to understand: $X_1 \times X_2 \times \cdots \times X_n \times \ldots$ should have as its elements all *sequences* of the form $(x_1, x_2, x_3, \ldots)$ where $x_i \in X_i$ for each $i \in \mathbb{N}$. However, this is the simplest possible case. How do we deal with a situation where we wish to look at the product of *uncountably* many sets? What should the elements in such a product look like?

The answers to these questions are hinted at in the observations that the elements of the set $X_1 \times X_2 \times \cdots \times X_n \times \ldots$ should be sequences $(x_1, x_2, x_3, \ldots)$ where $x_i \in X_i$ for each $i \in \mathbb{N}$. Recall that we can regard sequences of real numbers, for example, as *functions* $s : \mathbb{N} \to \mathbb{R}$, where we denote $s(n)$ by s_n. Such a sequence should be regarded, then, as an element in the Cartesian product $\mathbb{R} \times \mathbb{R} \times \ldots$. In fact, we can think of ordered n-tuples $(x_1, x_2, \ldots, x_n) \in X^n$ as functions, too, of the form $x : \{1, 2, \ldots, n\} \to X$. More generally, an element in $X_1 \times X_2 \times \cdots \times X_n$ is a function $x : \{1, 2, \ldots, n\} \to X_1 \cup X_2 \cup \cdots \cup X_n$, where $x(i) \in X_i$. It's easy to see that such a function x determines an n-tuple $(x_1, x_2, \ldots, x_n)$ and vice–versa. We'll use this approach to define what is meant by the Cartesian product of infinitely many sets.

Definition 5.3.1. *Let $\{X_\alpha\}_{\alpha\in\Lambda}$ be an indexed collection of sets. The Cartesian product of the collection, denoted*

$$\prod_{\alpha\in\Lambda} X_\alpha,$$

is defined as the set of all functions

$$f : \Lambda \to \bigcup_{\alpha\in\Lambda} X_\alpha,$$

such that $f(\alpha) \in X_\alpha$. We will refer to an element of the product $f \in \prod_{\alpha\in\Lambda} X_\alpha$ as a Λ-tuple, using the notation $f = (f_\alpha)_{\alpha\in\Lambda}$, where $f_\alpha = f(\alpha)$.

When the indexing set is clear, we'll often use the shorthand notation $\prod X_\alpha$, denoting its elements by (x_α). When the indexing set Λ is finite, then it's easy to see that

$$\prod_{i\in\{1,2,\ldots,n\}} X_i = X_1 \times \cdots \times X_n.$$

However, be warned that our indexing sets may be uncountable. For example, consider the indexed collection of sets $A_r = (-r, r) \subset \mathbb{R}$ for each $r \in \mathbb{R}$. Then the Cartesian product $\prod_{r\in\mathbb{R}} A_r$ has as its elements all possible $\mathbb{R}$-tuples $(f_r)_{r\in\mathbb{R}}$; that is, a typical element is a function $f : \mathbb{R} \to \mathbb{R}$ such that $f(r) \in A_r$. Such products are not as easy to grasp as are finite products. In any product, however, we have obvious projection functions onto each factor

$$p_\beta : \prod_{\alpha\in\Lambda} X_\alpha \to X_\beta$$

is defined by $p_\beta((x_\alpha)_{\alpha\in\Lambda}) = x_\beta = x(\beta)$, the Λ-tuple evaluated on the element $\beta \in \Lambda$.

To put a topology on an arbitrary product of topological spaces is pretty straight forward. However, one has to recall first the most important theorem about the product topology, Theorem 5.2.2: If X_μ and Y_ν are any topological spaces, and if $f : Z_\sigma \to X_\mu \times Y_\nu$ is any function into the product space, then f is continuous if and only if the compositions $p_X \circ f$ and $p_Y \circ f$ are continuous. So, for an arbitrary product of topological spaces, we would want our *product topology* to have this property hold, as well; a function f from a topological space into a product of spaces should be continuous if and only if f composed with the projection map onto each factor is continuous.

The most obvious topology to put on a product is a straightforward generalization of the product topology defined for a finite product of spaces: Given an indexed collection of topological spaces $\{(X_\alpha)_{\tau_\alpha}\}_{\alpha\in\Lambda}$, take as a basis for our topology the product of the topologies on the X_αs; that is, use the basis

$$\mathcal{B} = \left\{ \prod_{\alpha\in\Lambda} U_\alpha : U_\alpha \in \tau_\alpha \right\}.$$

Equivalently, we can write

$$\mathcal{B} = \prod_{\alpha\in\Lambda} \tau_\alpha.$$

One can confirm that $\mathcal{B}$ is, indeed, a basis for $\prod X_\alpha$, much like our proof that the basis for the product topology is a basis (proved after Definition 5.2.1). However, the topology generated by $\mathcal{B}$ is *not* the "correct" topology on $\prod X_\alpha$, as the following example shows.

■ EXAMPLE 5.1

We use the shorthand notation $\mathbb{R}^\infty$ to denote the set

$$\mathbb{R}^\infty = \mathbb{R} \times \mathbb{R} \times \mathbb{R} \times \cdots = \prod_{n\in\mathbb{N}} \mathbb{R},$$

the product of countably many copies of $\mathbb{R}$. Then the "diagonal" function $d : \mathbb{R} \to \mathbb{R}^\infty$ given by $i(x) = (x, x, x, \dots)$ has the composition with the projection maps $(p_n \circ d)(x) = x$ for every $n \in \mathbb{N}$. If we take each copy of the set $\mathbb{R}$ to have the usual topology $\mathcal{U}$, and put the topology generated by the basis $\mathcal{B} = \prod_{n\in\mathbb{N}} \mathcal{U}$ on the product, then $p_n \circ d$ is $(\mathcal{U} - \mathcal{U})$-continuous for every n. However, the function $d : \mathbb{R} \to \mathbb{R}^\infty$ is **not** $(\mathcal{U} - \tau_\mathcal{B})$-continuous; the set

$$U = (-1, 1) \times (-\frac{1}{2}, \frac{1}{2}) \times (-\frac{1}{3}, \frac{1}{3}) \times \cdots$$

is $\tau_\mathcal{B}$-open in $\mathbb{R}^\infty$, but $d^{-1}(U) = \{0\}$, which is not $\mathcal{U}$-open in the domain.

The difficulty that arises here is rather familiar – the intersection of finitely many open sets is open, but the intersection of infinitely many open sets need not be. If one reviews the proof of Theorem 5.2.2, the key point is that for $f : Z \to X \times Y$, we have $(p_1 \circ f)^{-1}(U) \cap (p_1 \circ f)^{-1}(V) = f^{-1}(U \times V)$. For U and V open in X and Y, respectively, then, we see that $f^{-1}(U \times V)$ must be open in Z, as the intersection of two open sets. In our example above, we can see that $d^{-1}(U)$ is the intersection of infinitely many $\mathcal{U}$-open sets, yielding the nonopen set $\{0\}$. So we need another topology on a product of spaces that avoids this difficulty.

Definition 5.3.2. *Given an indexed collection of topological spaces* $\{(X_\alpha)_{\mathcal{T}_\alpha}\}_{\alpha \in \Lambda}$, *we take as a basis for our topology*

$$\mathcal{P} = \{\prod_{\alpha \in \Lambda} U_\alpha\},$$

where each U_α *is an open set in the space* $(X_\alpha)_{\mathcal{T}_\alpha}$ *and* $U_\alpha = X_\alpha$ *for all but finitely many* $\alpha \in \Lambda$. *We refer to the topology generated by the basis* $\mathcal{P}$ *as the product topology on* $\prod X_\alpha$.

We hereafter refer to the topology generated by the basis $\mathcal{B}$ as the *box topology*on $\prod X_\alpha$. Note that the box topology agrees with the product topology for finite products of spaces, but that the box topology is strictly finer than the product topology for infinite products. One can verify that $\mathcal{P}$ is a basis for $\prod X_\alpha$ in a manner similar to that for the basis for the finite product of spaces.
(*Warning:* Whenever we consider the Cartesian product of an indexed collection of topological spaces $\{(X_\alpha)_{\mathcal{T}_\alpha}\}_{\alpha \in \Lambda}$, we will always assume that $\prod X_\alpha$ is given the product topology, unless we explicitly state otherwise.)

The following theorem makes clear why the product topology is the "right" one to put on $\prod X_\alpha$.

Theorem 5.3.1. *Let* $\{(X_\alpha)_{\mathcal{T}_\alpha}\}_{\alpha \in \Lambda}$ *be an indexed collection of topological spaces, and let*

$$f : Y_\nu \to \prod_{\alpha \in \Lambda} X_\alpha$$

be any function, where $\prod X_\alpha$ *has the product topology. Then* f *is continuous if and only if* $p_\alpha \circ f$ *is continuous for every* $\alpha \in \Lambda$, *where* $p_\beta : \prod X_\alpha \to X_\beta$ *is the projection onto the* β*th factor.*

The proof is a straightforward generalization of Theorem 5.2.2.
Proof: $\Longrightarrow$) First, we show that $p_\beta : \prod X_\alpha \to X_\beta$, the projection onto the βth factor, is continuous for each $\beta \in \Lambda$. Let U_β be any subset of X_β which is $\mathcal{T}_\beta$-open. Then

$$p_\beta^{-1}(U_\beta) = \prod_{\alpha \in \Lambda} U_\alpha,$$

where

$$U_\alpha = \begin{cases} X_\alpha & \text{if } \alpha \neq \beta \\ U_\beta & \text{if } \alpha = \beta, \end{cases}$$

which is open in the product topology on $\prod X_\alpha$ since it's a basis set; it's a product of sets, all but one of which is the entire coordinate set X_α, and the other is U_β, which is τ_β-open. Thus each coordinate projection map p_β is continuous, and hence, $(p_\alpha \circ f)$ is continuous for each α, as the composite of two continuous functions.
$\Longleftarrow)$: We're told that for each $\alpha \in \Lambda$, the composite $p_\alpha \circ f$ is continuous. Let $W \subset \prod X_\alpha$ be open (in the product topology). Then W contains a basis set around each of its points, where such a basis set is of the form $\prod U_\alpha$, where $U_\gamma = X_\gamma$ for all but finitely many $\gamma \in \Lambda$ and where U_α is τ_α-open for all αs. We'll assume that $U_\gamma = X_\gamma$ for all $\gamma \in \Lambda$ except for the coordinates $\alpha_1, \alpha_2, \ldots \alpha_n$. Then

$$\begin{aligned} f^{-1}(\prod U_\alpha) &= \bigcap_{\gamma \in \Lambda} (p_\gamma \circ f)^{-1}(U_\gamma) \\ &= \bigcap_{\gamma \in \Lambda} f^{-1}(p_\gamma^{-1}(U_\gamma)) \\ &= (p_{\alpha_1} \circ f)^{-1}(U_{\alpha_1}) \cap (p_{\alpha_2} \circ f)^{-1}(U_{\alpha_2}) \cap \cdots \cap (p_{\alpha_n} \circ f)^{-1}(U_{\alpha_n}), \end{aligned}$$

since $p_\gamma^{-1}(X_\gamma)$ is the whole product $\prod X_\alpha$. Note that each set $(p_{\alpha_i} \circ f)^{-1}(U_{\alpha_i})$ is open in Y, since the composite $p_{\alpha_i} \circ f$ is continuous for each i. Thus we see that f^{-1} of any basis set in $\prod X_\alpha$ is ν-open in Y, since it's the finite intersection of sets that are open. Hence, $f^{-1}(W)$ contains an open set around each of its points and must be open in Y. □

Exercises

1. Prove that the basis for the box topology on $\prod X_\alpha$

$$\mathcal{B} = \prod_{\alpha \in \Lambda} \tau_\alpha,$$

   is indeed a basis.

2. Prove precisely that $i^{-1}(U) = \{0\}$ in Example 5.1.

3. Prove that $\mathcal{P}$ is a basis for $\prod X_\alpha$, where $\mathcal{P}$ is the product basis:

$$\mathcal{P} = \{\prod_{\alpha \in \Lambda} U_\alpha\},$$

   where each U_α is an open set in the space $(X_\alpha)_{\tau_\alpha}$ *and* $U_\alpha = X_\alpha$ for all but finitely many $\alpha \in \Lambda$.

4. Show that for an arbitrary product of spaces, the box topology is finer than the product topology. Show that the inclusion is strict if the collection of spaces is infinite and infinitely many of the spaces are not indiscrete. (*Hint*: Read the second statement of the problem carefully.)

5.4 QUOTIENT SPACES

The quotient topology is a construction that makes the idea of "gluing" in a topological space precise. As we saw in Chapter 1, we can construct certain interesting surfaces, at least intuitively, by gluing together appropriate edges of rectangles. For example, the torus, T^2, is built by identifying edges of a rectangle as follows:

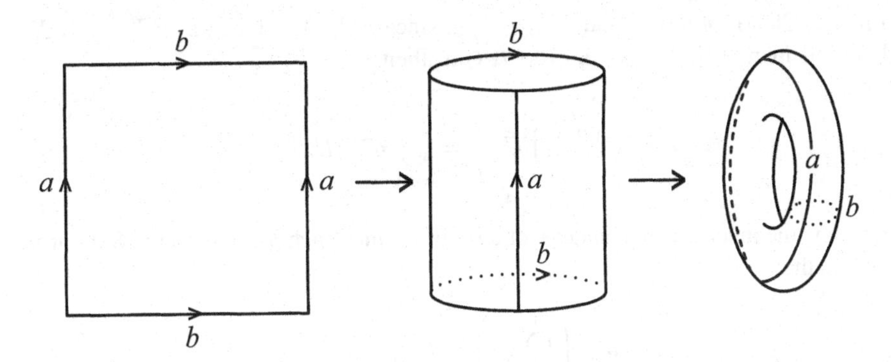

To make this precise, we recall that an *equivalence relation* on a set X is a relation $\sim$ (a subset of $X \times X$) that is *reflexive*, *symmetric* and *transitive* (Definition 2.3.2). We need all three of these properties to guarantee that the relation $\sim$ on X partitions X into distinct *equivalence classes*, denoted by $[x]$ for each $x \in X$. We recall that the set of equivalence classes is denoted by $X_{/\sim}$. Whenever $\sim$ is an equivalence relation on a set X, every point $x \in X$ is equivalent to itself (by reflexivity). With this in mind, we'll often specify an equivalence relation on a set by mentioning only equivalences $x_1 \sim x_2$ for distinct points x_1 and x_2, leaving implicit that every point x is equivalent to itself.

The main idea is this: given a topological space X_τ, we wish to describe how to glue things together in X_τ by defining an appropriate equivalence relation $\sim$ on the set X. If two points x_0 and x_1 are equivalent under $\sim$ ($x_0 \sim x_1$) then the points are to be identified under the "gluing." Then, we'll need to define a topology on the set of equivalence classes $X_{/\sim}$ that preserves the topology on X for those points not identified together, while doing the "gluing" correctly. Here's how it's done. We start with the canonical "projection" function $p : X \to X_{/\sim}$, given by $p(x) = [x]$, the function that sends each point x to its equivalence class (a function of *sets*). This is clearly a surjection, but not an injection if $\sim$ is a nontrivial equivalence relation (Why?) We consider the collection

$$\tau_{/\sim} = \{U \subset X_{/\sim} : p^{-1}(U) \text{ is } \tau\text{-open}\}.$$

Theorem 5.4.1. *For X_τ a topological space with an equivalence relation $\sim$ on the set X, the set*

$$\tau_{/\sim} = \{U \subset X_{/\sim} : p^{-1}(U) \text{ is } \tau\text{-open}\}$$

is a topology on $X_{/\sim}$.

Proof: Clearly $\emptyset \in \tau_{/\sim}$ and $X_{/\sim} \in \tau_{/\sim}$, since $p^{-1}(\emptyset) = \emptyset$ and $p^{-1}(X_{/\sim}) = X$, both sets in τ. If $U_\alpha \in \tau_{/\sim}$ for every $\alpha \in \Lambda$, then

$$p^{-1}\left(\bigcup_{\alpha\in\Lambda} U_\alpha\right) = \bigcup_{\alpha\in\Lambda} p^{-1}\left(U_\alpha\right),$$

which is the inverse image under p of a set in τ, since each $p^{-1}U_\alpha$ is in τ. Similarly, we see that

$$p^{-1}\left(\bigcap_{i=1}^{n} U_i\right) = \bigcap_{i=1}^{n} p^{-1}\left(U_i\right),$$

showing that $\tau_{/\sim}$ is closed under finite intersection.

Definition 5.4.1. *For X_τ a topological space with an equivalence relation $\sim$ on the set X, the set*

$$\tau_{/\sim} = \{U \subset X_{/\sim} : p^{-1}(U) \text{ is } \tau\text{-open}\}$$

is known as the quotient topology on $X_{/\sim}$. The set $X_{/\sim}$ endowed with the topology $\tau_{/\sim}$ is a quotient space.

In fact, the quotient topology on $X_{/\sim}$ is the finest (largest) topology such that the projection map $p : X \to X_{/\sim}$ is continuous, as one can check from the definition.

■ **EXAMPLE 5.2**

Consider the following relation on the real line $\mathbb{R}_{\mathcal{U}}$. Let $x \sim \frac{1}{2}$ for all $x \in (0, 1)$ (with each $x \in \mathbb{R}$ equivalent to itself, of course). Then we have the following depiction of $\mathbb{R}_{/\sim}$:

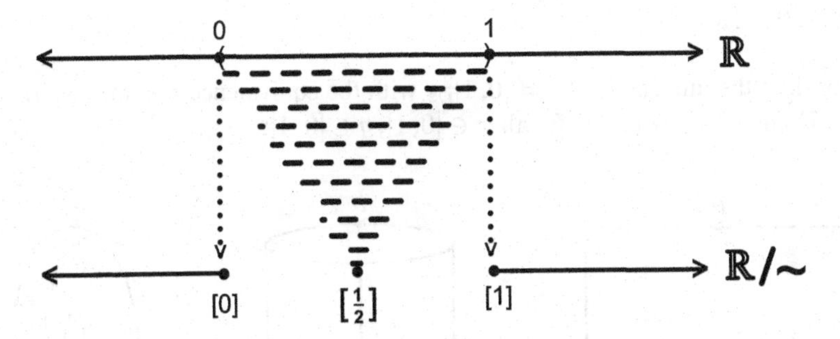

Hence, $(\mathbb{R}_{/\sim})_{\mathcal{U}_{/\sim}}$ is not homeomorphic to $\mathbb{R}_{\mathcal{U}}$, as one can easily see by noting that there is no open neighborhood of [0] in $(\mathbb{R}_{/\sim})_{\mathcal{U}_{/\sim}}$ that does not contain $[\frac{1}{2}]$.

EXAMPLE 5.3

Consider $I_{\mathcal{U}} = [0, 1]_{\mathcal{U}}$, in the usual subspace topology inherited from $\mathbb{R}_{\mathcal{U}}$. (Here and in the sequel, we will often use an abbreviated notation for the subspace topology: $I_{\mathcal{U}}$, rather than $I_{\mathcal{U}_I}$.) We now form a quotient space by defining the following equivalence relation: $0 \sim 1$ (where all other xs are equivalent only to themselves). Then the quotient space $([0, 1]_{/\sim})_{\mathcal{U}_{/\sim}}$ is homeomorphic to $S^1_{\mathcal{U}^2}$, the unit circle in the plane, as one can see from the following diagram:

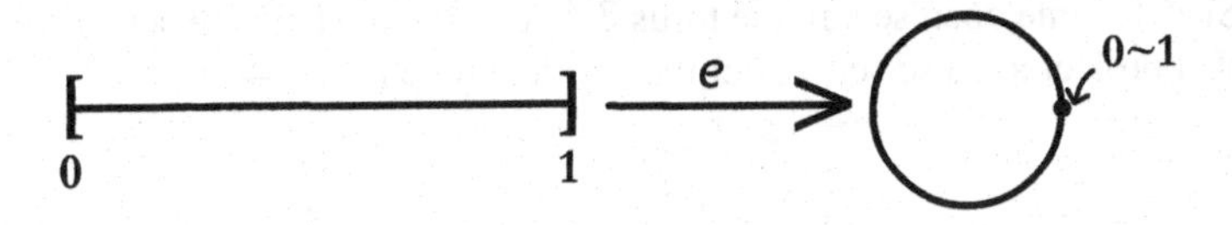

So the homeomorphism is given by the "exponential map" $e : [0, 1] \to \mathbb{R}^2$, where $e(t) = (\cos 2\pi t, \sin 2\pi t)$. Note that e is indeed one-to-one and onto on $[0, 1]_{/\sim}$ (although it's not 1-1 on $[0, 1]$ itself.) One can see that e is continuous and open by using the proofs (from calculus) that the sine and cosine functions are continuous and open. (*Think*: $\sin t$ is an open function because....) The function e is referred to as an "exponential map" because of its expression in complex analysis: $e(t) = (cos 2\pi t, \sin 2\pi t) = e^{2\pi i t}$.

■ EXAMPLE 5.4

Consider the unit square $I^2 = [0,1]^2_{\mathcal{U}^2}$ with the equivalence relation $(x,0) \sim (x,1)$ and $(0,y) \sim (1,y)$, for all $x \in [0,1], y \in [0,1]$:

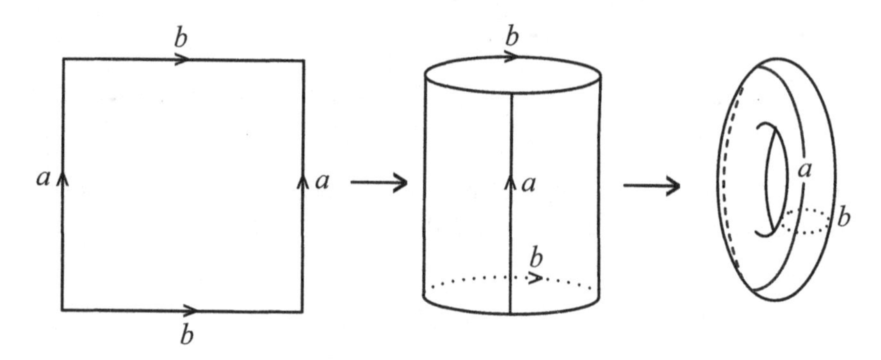

So $([0,1]^2_{/\sim})_{\mathcal{U}^2_{/\sim}} = T^2$, the usual torus.

Exercises

1. Sketch some open sets in the quotient space $\mathbb{R}_{/\sim}$ of Example 5.2. Be sure to show the sets in $\mathbb{R}_{\mathcal{U}}$ that project to these sets under the quotient map $p : \mathbb{R} \to \mathbb{R}_{/\sim}$.

2. Sketch some open sets in the torus T^2. Be sure to show the sets in $I^2 \subset \mathbb{R}^2_{\mathcal{U}^2}$ that project to these sets under the quotient map $p : I^2 \to T^2$.

5.5 UNIONS AND WEDGES

The goal of this section is to show how to "add" two topological spaces, as opposed to the product from Section 5.2. Set theoretically, the operation of union generally works well as an additive operation, with $\emptyset$ as a unit.

Here's what we might hope would be the appropriate way to put two topological spaces together: If Y_ν and Z_σ are topological spaces, we put the following topology on $Y \cup Z$: just take the collection $\nu \cup \sigma$. This won't work, though, when we think about what the union of two sets *of sets* means: we get Y and Z as elements of $\nu \cup \sigma$, not $Y \cup Z$, so that $\nu \cup \sigma$ is not a topology on $Y \cup Z$. We can get around this by letting $\nu \cup \sigma$ be a subbasis[1] for the topology on $Y \cup Z$, so we definitely get a topology, but this approach is still problematic. If Y and Z are disjoint, the construction works well. However, if Y and Z have nontrivial intersection, we run the risk of some incompatibility between the topologies. For example, singleton sets might be ν-open, but not σ-open. Which topology, ν or σ, would "dominate" on the overlap between the two sets?

To avoid this difficulty, we'll look at unions only in the case of two subspaces of the same "ambient" space. Precisely, given A and B, both subsets of a topological space X_τ, we take the topology on the union $A \cup B$ to be $\tau_{(A \cup B)}$, just the subspace topology in the usual sense on the subspace $A \cup B$. If A and B are disjoint, this is the construction above: $\tau_{(A \cup B)}$ is just the topology generated by the basis $\tau_A \cup \tau_B$. If A and B have nontrivial intersection, then this topology is defined consistently, since $\tau_A = \{U \cap A : U \in \tau\}$ and $\tau_B = \{U \cap B : U \in \tau\}$.

As an example of how this concept is useful, we have the following result, which shows how two continuous functions can be "pasted" together to form a continuous function.

Theorem 5.5.1. *(Pasting Lemma) Let $A \cup B = X$ where A and B are τ-closed subsets of X, and let $f : A_{\tau_A} \to Y_\nu$ and $g : B_{\tau_B} \to Y_\nu$ be continuous maps. If $f(x) = g(x)$ for every $x \in A \cap B$, then the function $f \cup g : (A \cup B)_{\tau_{A \cup B}} \to Y_\nu$, defined by*

$$(f \cup g)(x) = \begin{cases} f(x) & \text{if } x \in A, \\ g(x) & \text{if } x \in B. \end{cases}$$

is continuous.

Proof: Let $V \subset Y_\nu$ be any ν-closed set. Then $(f \cup g)^{-1}(V) = f^{-1}(V) \cup g^{-1}(V)$, where both sets in the union are τ-closed since f and g are continuous. By Theorem 4.5.1, we see that $(f \cup g)$ is continuous. □

Note that the function $(f \cup g)$ is defined consistently only if the functions f and g agree on the intersection of their domains. Also note that it's perfectly permissible for $A \cap B$ to be empty. The following functions from $\mathbb{R}_\mathcal{U}$ to itself illustrate how continuous functions can be pasted together to form continuous functions, and how this process can fail if the hypotheses of the Pasting Lemma are not met.

[1] A subbasis for a topology on X is a collection $\mathcal{S}$ of subsets of X whose union is all of X. A subbasis determines a basis: all finite unions of the sets in $\mathcal{S}$ form a basis for the topology.

Let $f : \mathbb{R}_{\mathcal{U}} \to \mathbb{R}_{\mathcal{U}}$ be defined by

$$f(x) = \begin{cases} 2x & \text{if } x \leq 0, \\ x^2 & \text{if } x \geq 0. \end{cases}$$

Then f is continuous because the two component functions $m(x) = 2x$ and $n(x) = x^2$ are both continuous on the subspaces of $\mathbb{R}_{\mathcal{U}}$ that form their domains, and $m = n$ over the intersection of the domains. This is illustrated in the figure below.

Let $g : \mathbb{R}_{\mathcal{U}} \to \mathbb{R}_{\mathcal{U}}$ be defined by

$$g(x) = \begin{cases} 2x - 1 & \text{if } x \leq 0, \\ x^2 + 1 & \text{if } x \geq 0. \end{cases}$$

Then g is not well defined as $(m \cup n)$, for $m(x) = 2x - 1$ and $h(x) = x^2 + 1$, since the functions disagree on the intersection of their domains ($\{0\}$). This is illustrated in the figure below, as well.

Let $h : \mathbb{R}_{\mathcal{U}} \to \mathbb{R}_{\mathcal{U}}$ be defined by

$$h(x) = \begin{cases} 2x - 1 & \text{if } x < 0, \\ x^2 + 1 & \text{if } x \geq 0. \end{cases}$$

Then h is well defined as $(m \cup n)$, for $m(x) = 2x - 1$ and $h(x) = x^2 + 1$, since the functions agree on the intersection of their domains ($\emptyset$). However, $h = (m \cup n)$ is not continuous, since $g^{-1}((\frac{1}{2}, \frac{3}{2})) = [0, \frac{1}{4})$, which is not $\mathcal{U}$-open. The function h is not a counterexample to Theorem 5.5.1, however, since these sets $A = (-\infty, 0)$ and $B = [0, +\infty)$, whose union forms $\mathbb{R}$, are not both $\mathcal{U}$-closed. This is illustrated, too, in the figure below.

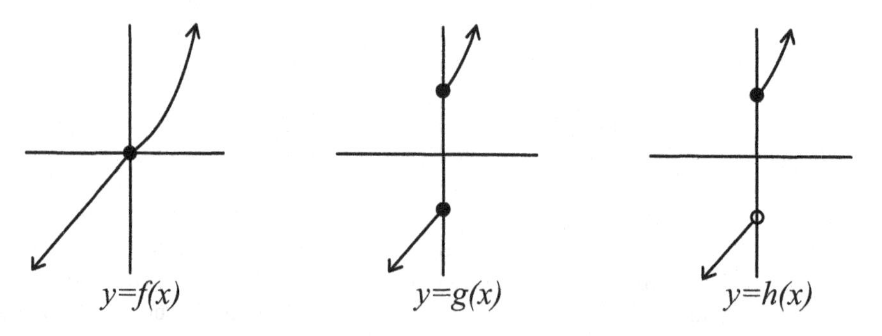

A variant on the union construction considers the "wedge" or "one-point union" of two subspaces of a space X_{τ}. Here, we require $A \cap B = \emptyset$, and we choose basepoints $a_0 \in A$ and $b_0 \in B$.

Definition 5.5.1. *For two disjoint subspaces A and B of X_τ with basepoints $a_0 \in A$ and $b_0 \in B$, we define the wedge $A \vee B$ as the quotient space $(A \coprod B_{/\sim})_{\tau/\sim}$, where the equivalence relation is given by $a_0 \sim b_0$ and $A \coprod B$ means the disjoint union of A with B.*

This is best thought of geometrically. Let A and B be two subspaces of $\mathbb{R}^2_{\mathcal{U}^2}$, both homeomorphic to circles. Then $A \vee B$ is given by the following picture:

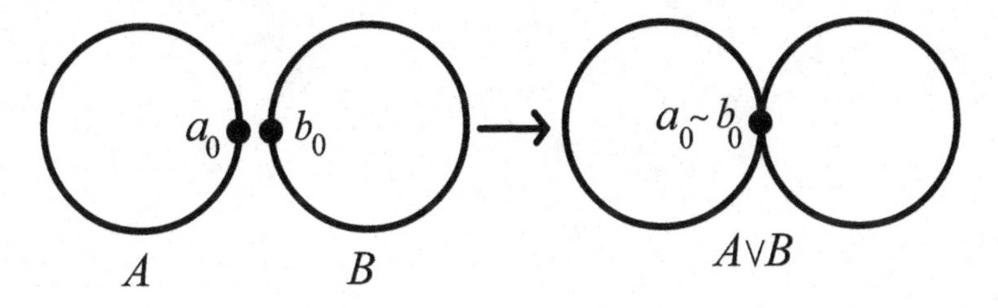

Exercise

1. A *path* in a space X_τ is a continuous function $\alpha : [0,1]_{\mathcal{U}} \to X_\tau$. If α and β are paths in X_τ such that $\alpha(1) = \beta(0)$, then the map $\alpha * \beta : [0,1] \to X$ defined by

$$\alpha * \beta(t) = \begin{cases} \alpha(2t) & \text{if } 0 \le t \le \frac{1}{2} \\ \beta(2t-1) & \text{if } \frac{1}{2} \le t \le 1. \end{cases}$$

is continuous. (*Hint*: draw two such paths, then consider the Pasting lemma.)

CHAPTER 6

CONNECTED SPACES

6.1 INTRODUCTION

Connectedness is our first example of a *topological property*: a property of spaces that is preserved by homeomorphisms. We will define what it means for a space X_τ to be connected is Section 6.2, and prove that if X_τ is connected and $X_\tau \cong Y_\nu$, then Y_ν must be connected as well. Thus, we can use this property to tell spaces "apart": if X_τ is connected and Y_ν is disconnected, then $X_\tau \ncong Y_\nu$.

In fact the property of connectedness is not just a topological property – we'll prove that it's also preserved by continuous functions, meaning that if X_τ is connected and $f : X_\tau \to Y_\nu$ is any continuous function, then $Im(f) \subset Y_\nu$ must be connected as well. Such a property will be called a *strong topological property*. This aspect of connectedness is actually used in calculus in the Intermediate Value Theorem, as we'll show in Section 6.3.

We'll also define a related topological property: *path-connectedness*. Intuitively, a space X_τ is path-connected if any two points in X can be connected by a (continuous) path. Surprisingly, path-connectedness implies connectedness, but not vice-versa. Finally, we'll prove that finite products of path-connected spaces are (path-)connected.

6.2 DEFINITION, EXAMPLES AND PROPERTIES

Definition 6.2.1. *A topological space $X_\mathcal{T}$ is disconnected if there exist disjoint subsets $A, B \subset X$ that are both $\mathcal{T}$-open, both nonempty, and with $X = A \coprod B$. A space is connected if it is not disconnected.*

So a space $X_\mathcal{T}$ is disconnected if X is the disjoint union of two nonempty $\mathcal{T}$-open sets: $X = A \coprod B$. Note that this is equivalent to saying that X has a nonempty proper subset A that is both $\mathcal{T}$-open and $\mathcal{T}$-closed (meaning that $X \setminus A$ is nonempty and $\mathcal{T}$-open as well). We will refer to the subsets A and B as a *disconnection* or *separation* of $X_\mathcal{T}$. The following examples illustrate this concept.

■ **EXAMPLE 6.1**

The set $\mathbb{R}$ in the left-hand topology $\mathcal{L}$ is disconnected, because the set $[0, 1)$ is simultaneously $\mathcal{L}$-open and $\mathcal{L}$-closed. In other words,
$\mathbb{R} = [0, 1) \coprod ((-\infty, 0) \cup ([1, +\infty))$, where both $[0, 1)$ and its complement are $\mathcal{L}$-open. (We'll see later in Section 6.3 that $\mathbb{R}_\mathcal{U}$ is connected.)

■ **EXAMPLE 6.2**

If X is any set with $Card(X) \geq 2$, then $X_\mathcal{D}$ is disconnected, where $\mathcal{D} = \mathcal{P}(X)$ is the discrete topology on X. To see why, if $x_0 \in X$, then both $\{x_0\}$ and $(X \setminus \{x_0\})$ are nonempty and $\mathcal{D}$-open, with X the disjoint union of the two subsets.

At the other extreme, we have the following example.

■ **EXAMPLE 6.3**

If X is any nonempty set, then $X_\mathcal{I}$ is connected, where $\mathcal{I} = \{\emptyset, X\}$ is the indiscrete topology on X. (Why?)

■ EXAMPLE 6.4

Let $X = \{a, b, c\}$, with $\tau = \{\emptyset, X, \{a\}, \{b\}, \{a, b\}, \{b, c\}\}$ and $\sigma = \{\emptyset, X, \{a\}, \{b\}, \{a, b\}\}$. Then X_τ is disconnected, since $X = \{a\} \coprod \{b, c\}$, where both subsets are τ-open, but X_σ is connected (since no proper subset of X is simultaneously σ-open and σ-closed).

■ EXAMPLE 6.5

The set $\mathbb{R}$ in the finite complement topology $\mathcal{FC}$ is connected: If A is any proper, nonempty $\mathcal{FC}$-open subset of $\mathbb{R}_{\mathcal{FC}}$, then $\mathbb{R} \setminus A$ must be finite and nonempty, so $\mathbb{R} \setminus A$ is not $\mathcal{FC}$-open, since it doesn't have a finite complement.

It's easy to see why the following theorem is true:

Theorem 6.2.1. *Let $f : X_\tau \to Y_\nu$ be any homeomorphism. Then X_τ is connected if and only if Y_ν is connected.*

Proof: The proof is simple, using the contrapositive; if A is a nonempty, proper subset of X which is simultaneously τ-open and closed, then $f(A)$ will be a nonempty, proper subset of Y which is simultaneously ν-open and closed, since any homeomorphism f is an open map. Similarly a disconnection of Y_ν yields, after applying the inverse function f^{-1}, a disconnection of X_τ. □

Theorem 6.2.1 can be restated very simply: *Connectedness is a topological property.*

In fact, we have the following, much stronger result, which shows that connectedness is a *strong topological property* :

Theorem 6.2.2. *The continuous image of a connected space is connected; that is, if X_τ is connected and $f : X_\tau \to Y_\nu$ is any continuous and onto function, then Y_ν is also connected.*

Note that this is equivalent to saying that if X_τ is connected and $f : X_\tau \to Y_\nu$ is any continuous function, then $Im(f)$ is connected as a subspace of Y_ν.

Proof: We proceed by contrapositive. Let $f : X_\tau \to Y_\nu$ be any continuous and onto function. Assume that Y_ν is disconnected. Then there exists a disconnection: $Y = C \coprod D$, with both C and D nonempty and ν-open. Since f is onto, both $f^{-1}(C)$ and $f^{-1}(D)$ are nonempty subsets of X. Further, since $f^{-1}(C) \cap f^{-1}(D)$ is just $f^{-1}(C \cap D) = \emptyset$, we see that $f^{-1}(C)$ and $f^{-1}(D)$ are disjoint. Since f is continuous, both $f^{-1}(C)$ and $f^{-1}(D)$ are τ-open subsets of X, disconnecting X_τ. We conclude that if the image of f is disconnected, then the domain is disconnected, which is equivalent to our theorem. □

Given that connectedness is preserved by continuous functions, one might ask whether the property is *inherited* or *hereditary* (preserved by subspaces). The following example shows that connectedness is not inherited.

EXAMPLE 6.6

Consider the subspace $Z = \{1,2,3,4,5\} \subset \mathbb{R}_{\mathcal{FC}}$. Then $A = \{1,2,3\}$ is $\mathcal{FC}_Z$-open (since $A = (\mathbb{R} \setminus \{4,5\}) \cap Z$) and is also $\mathcal{FC}_Z$-closed (since $A = \{1,2,3\} \cap Z$, where $\{1,2,3\}$ is $\mathcal{FC}$-closed as a subset of $\mathbb{R}$). So Z is disconnected in the subspace topology inherited from $\mathbb{R}_{\mathcal{FC}}$.

In the other direction, we can show the following result, which relates connectedness and closure.

Theorem 6.2.3. *Let A be a subset of a space X_τ, with A_{τ_A} connected. Then $Cl(A)$ is also connected in the subspace topology.*

Proof: We proceed by contrapositive. Assume that $Cl(A)$ is disconnected as a subspace of X_τ, so that $Cl(A) = C \coprod D$, where C and D are both nonempty and both $\tau_{Cl(A)}$-open. In other words, $C = U \cap Cl(A)$, where U is τ-open and $D = V \cap Cl(A)$, where V is τ-open. Consider what happens when we intersect the τ-open sets U and V down to A; obviously $U \cap A$ and $V \cap A$ are τ_A-open, by the definition of the subspace topology. Further, $U \cap A$ and $V \cap A$ are disjoint, since $A \subset Cl(A)$ and the two sets $C = U \cap Cl(A)$ and $D = V \cap Cl(A)$ are disjoint. So we have a disconnection of A_{τ_A} *if* we can show that both $U \cap A$ and $V \cap A$ are nonempty. The key is that both $C = U \cap Cl(A)$ and $D = V \cap Cl(A)$ are nonempty, with U and V being τ-open sets. Recall that $Cl(A) = A \cup A'$ (Theorem 4.3.3). So if $U \cap Cl(A) = U \cap (A \cup A') = (U \cap A) \cup (U \cap A') \neq \emptyset$, then U must contain points of the set A itself; if a point $x \in U \cap A'$, then U is a τ-neighborhood of x, a limit point of A, so that $U \cap A$ contains points other than x. Similarly, $V \cap A \neq \emptyset$, so that $U \cap A$ and $V \cap A$ is a disconnection of A. □

The proof of the following generalization of Theorem 6.2.3 is left as an exercise:

Theorem 6.2.4. *Let A_{τ_A} be a connected subspace of a space X_τ, and let B be a subset of X with $A \subset B \subset Cl(A)$. Then B is also connected in the subspace topology.*

Exercises

1. Prove that $(1,2) \cup [3,4)$ is disconnected as a subspace of $\mathbb{R}_{\mathcal{U}}$.

2. Give an example showing that the union of two disconnected spaces can be connected.

3. Give an example showing that the union of two connected spaces can be disconnected.

4. Prove that any infinite set X in the finite complement topology $\mathcal{FC} = \{U \subset X : U = \emptyset \text{ or } X \setminus U \text{ is finite}\}$ is connected.

5. Prove Theorem 6.2.4: Let A_{τ_A} be a connected subspace of a space X_τ, and let B be a subset of X with $A \subset B \subset Cl(A)$. Then B is also connected in the subspace topology.

6. Prove the alternative version of Theorem 6.2.2: If X_τ be connected and $f : X_\tau \to Y_\nu$ is any continuous function, then $Im(f)$ is connected as a subspace of Y_ν.

7. A space X_τ is said to be *totally disconnected* if every subspace of X with more than one element is disconnected (in the subspace topology).

   (a) Show that every discrete space is totally disconnected.

   (b) Is the converse true? (*Hint*: think of a topology on $\mathbb{R}$ where the basis sets are themselves disconnected.)

8. Prove that if X_τ is connected and $\tau' \subset \tau$, then $X_{\tau'}$ is also connected.

6.3 CONNECTEDNESS IN THE REAL LINE

In this section, we prove that intervals in the usual real line, $\mathbb{R}_\mathcal{U}$, are connected, and we derive several useful consequences. The connectedness of the real line follows fairly directly from the *completeness* of $\mathbb{R}_\mathcal{U}$, which we explain below.

We recall that a set $A \subset \mathbb{R}$ is *bounded below* if there exists $b \in \mathbb{R}$ such that $b \leq a$ for every $a \in A$. Similarly, A is *bounded above* if there exists $c \in \mathbb{R}$ such that $c \geq a$ for every $a \in A$. We'll use the term *bounded* to describe A if A is bounded both above and below. A real number m is the *greatest lower bound of* A [denoted $m = glb(A)$] if m is a lower bound for A and if $m \geq n$ for every lower bound n of A. We define the *least upper bound of* A, $n = lub(A)$, in the evident analogous fashion. It's easy to see that for any set A, the numbers $glb(A)$ and $lub(A)$ are unique, if they exist. The following axiom of the real line, $\mathbb{R}_\mathcal{U}$, will be key to understanding the topology of $\mathbb{R}_\mathcal{U}$:

Property 6.3.1. *(Completeness property) If $A \subset \mathbb{R}$ is nonempty and bounded below, then A has a greatest lower bound. Equivalently, if A is nonempty and bounded above, then A has a least upper bound.*

The importance of the completeness property is illustrated in the proof of the following theorem.

Theorem 6.3.2. *The space $\mathbb{R}_{\mathcal{U}}$ is connected.*

Proof: Assume $\mathbb{R}_{\mathcal{U}}$ is disconnected, so that $\mathbb{R} = A \coprod B$, with both subsets $\mathcal{U}$-open and nonempty. Let $a \in A$ and $b \in B$. Without loss of generality, we assume that $a < b$. Let $C = \{x \in A : x < b\} \subset A$. Then C is nonempty (since $a \in C$) and $\mathcal{U}$-open, since $C = (-\infty, b) \cap A$, the intersection of two open sets. Since C is bounded above (by b, among others), the completeness property says that C has a least upper bound, say, $y = lub(C)$. Now we ask ourselves whether $y \in C$. If so, since C is $\mathcal{U}$-open, there exists some $\epsilon > 0$ such that the entire open neighborhood $(y - \epsilon, y + \epsilon) \subset C$. This would imply that $y + \frac{\epsilon}{2} \in C$, which contradicts the fact that y is an upper bound for C, so we see that $y \notin C$. This implies that $y \notin A$, also, since $C = (-\infty, b) \cap A$, unless $y \geq b$, which we can rule out as follows. We already know $b \notin A$, since $b \in B$ and $A \cap B = \emptyset$, so if $y = b$ then $y \notin A$. If $y > b$, we get an easy contradiction: b is an upper bound for the set C, and y is $lub(C)$. So we conclude $y = lub(C) \notin A$, else we get a contradiction.

The only option, then, is that $y \in B$, since $A \cup B = \mathbb{R}$, and we'll show that this leads to a contradiction as well: B is $\mathcal{U}$-open, so if $y \in B$, then there exists a real number $\delta > 0$ such that the entire open neighborhood $(y - \delta, y + \delta) \subset B$. But y is the *least* upper bound for C, so the neighborhood $(y - \delta, y + \delta)$ must intersect C (otherwise numbers like $y - \frac{\delta}{2}$ would be smaller upper bounds for C). So we get a contradiction of $y \in B$, too.

Our assumption that $\mathbb{R}_{\mathcal{U}}$ is disconnected must be false, then, since it leads directly to a contradiction. □

We have the following consequence: $\mathbb{R}_{\mathcal{U}} \not\cong \mathbb{R}_{\mathcal{L}}$, since the usual real line is connected, while the set $\mathbb{R}$ in the left-hand topology is disconnected.

We also can see the following corollary:

Corollary 6.3.3. *Any interval in $\mathbb{R}_{\mathcal{U}}$ is connected in the subspace topology.*

Sketch of Proof: First, we note that any open interval (a, b) is homeomorphic to $(0, 1)$, by using the linear function $f : (0,1) \to (a, b)$ given by $f(x) = (b - a)x + a$:

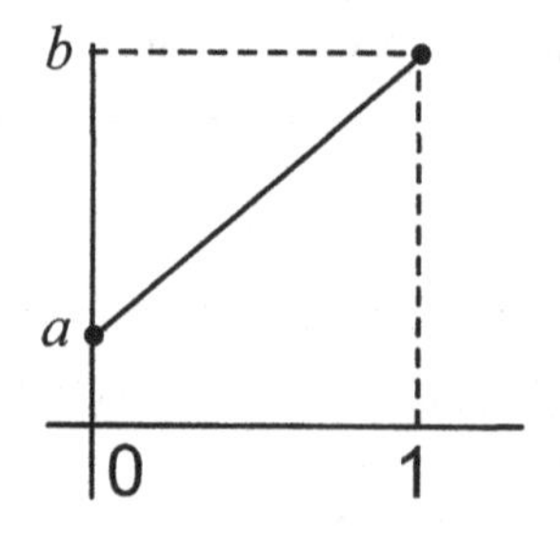

We then show, for example, that $(-\frac{\pi}{2}, \frac{\pi}{2}) \cong \mathbb{R}_{\mathcal{U}}$ by using the arctangent function, and we're done. To show that a closed interval $[a, b]$ is connected as a subspace of $\mathbb{R}_{\mathcal{U}}$, we could recapitulate the proof of Theorem 6.3.2 in this context. It's far easier, though, to construct a function $g : \mathbb{R} \to [a, b]$ as the following diagram shows:

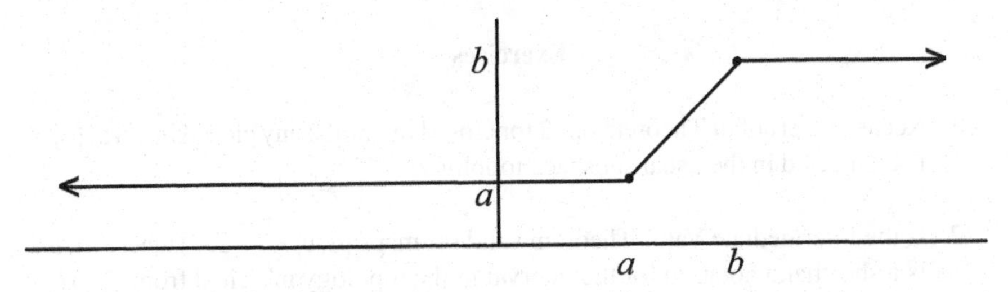

We then see that g is $\mathcal{U} - \mathcal{U}_{[a,b]}$-continuous and use the fact that the continuous image of a connected space is connected (Theorem 6.2.2). Slick, eh? Of course, an even slicker proof is to say that $[a, b] = Cl((a, b))$, so that the connectedness of the open interval implies the connectedness of the closed interval by Theorem 6.2.3. □

These results have some very nice implications, which are particularly useful in calculus settings – finding zeroes of polynomials, figuring out when a function is increasing or decreasing, and so on. Perhaps the most useful way to state these results is the following:

Theorem 6.3.4. *(Intermediate Value Theorem) If $f : [a, b]_{\mathcal{U}_{[a,b]}} \to \mathbb{R}_{\mathcal{U}}$ is continuous and if c is any real number between $f(a)$ and $f(b)$, then there exists a number $z \in (a, b)$ such that $f(z) = c$.*

Note that we can't just write $f(a) < c < f(b)$, since we don't know that $f(a) < f(b)$. The function need not be monotone for the result to hold, as the following plot illustrates nicely:

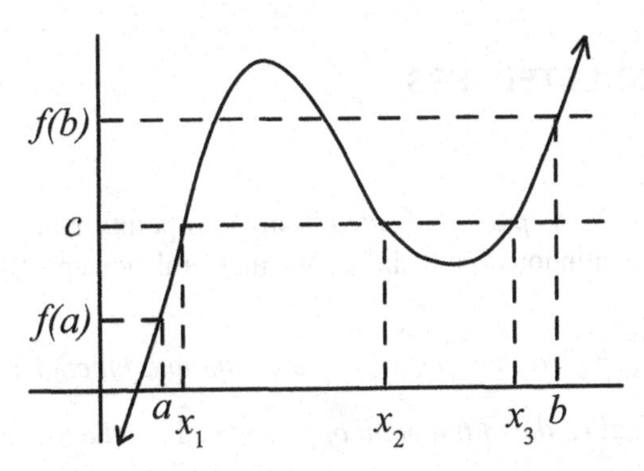

The proof of the Intermediate Value Theorem is left as an exercise.

We often use this result to help find zeroes of polynomials (which are continuous, by $\epsilon - \delta$ considerations). If $f(a)$ is positive and $f(b)$ is negative, the Intermediate Value Theorem asserts that there must be an $x \in (a, b)$ with $f(x) = 0$. One can then "home in" on the root by bisecting the interval iteratively.

Exercises

1. Recast the proof of Theorem 6.3.2 for closed intervals; any closed interval $[a, b]$ is connected in the usual subspace topology.

2. Is the Intermediate Value Theorem valid for maps $[a, b]_{\mathcal{L}} \to \mathbb{R}_{\mathcal{U}}$ (where $[a, b]_{\mathcal{L}}$ is a shorthand notation for the interval in the topology inherited from $\mathbb{R}_{\mathcal{L}}$)?

3. In high school algebra, you learned that a continuous function $f : \mathbb{R}_{\mathcal{U}} \to \mathbb{R}_{\mathcal{U}}$ with $f(a)$ negative and $f(b)$ positive must have a zero between a and b. Why?

4. Prove the Brouwer Fixed-point Theorem for intervals: any continuous function $f : [0, 1]_{\mathcal{U}} \to [0, 1]_{\mathcal{U}}$ has a *fixed point*; that is, there must exist $x_0 \in [0, 1]$ such that $f(x_0) = x_0$. (*Hint*: draw a graph of a typical continuous f and see whether it must cross the diagonal line $y = x$. Then think about an algebraic variant of f that looks at the difference between f and the diagonal line. Can this new function avoid being 0?)

5. Is the Brouwer Fixed-point Theorem true for maps of the open interval $(0, 1)_{\mathcal{U}}$?

6. Prove the Intermediate Value Theorem.

6.4 PATH-CONNECTEDNESS

Intuitively, a space $X_{\mathcal{T}}$ is *path-connected* if any two points x_0 and x_1 in X can be "connected" by a continuous "path" in X. We make this idea precise as follows.

Definition 6.4.1. *A path α in a space $X_{\mathcal{T}}$ is a continuous function $\alpha : [0, 1]_{\mathcal{U}} \to X_{\mathcal{T}}$. The point $\alpha(0)$ is called the initial point of α, and $\alpha(1)$ is the terminal point of α.*

The following figure clarifies this idea.

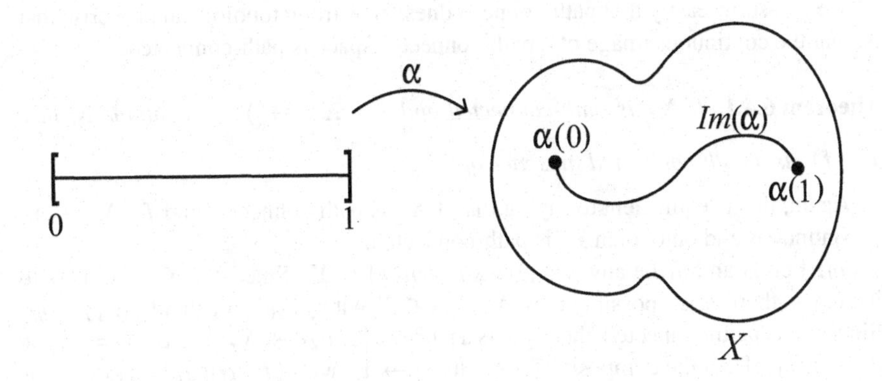

Definition 6.4.2. *A space X_τ is path-connected if for every pair of points $x_0, x_1 \in X$, there exists a path $\alpha : [0,1]_\mathcal{U} \to X$ with $\alpha(0) = x_0$ and $\alpha(1) = x_1$.*

As a trivial example, any indiscrete space X_τ is path-connected, since *any* function into an indiscrete space is continuous. Given any two points x_0 and x_1 in X, define $\alpha(t) = x_0$ for all $t \in [0,1)$, with $\alpha(1) = x_1$. Then this function α is $(\mathcal{U} - \mathcal{I})$-continuous.

As a less trivial example, $\mathbb{R}^2_{\mathcal{U}^2}$ is path-connected. Let (x_0, y_0) and (x_1, y_1) be any two points in the plane, with $x_0 \neq x_1$. We define $\alpha : [0,1] \to \mathbb{R}^2$ as the straight-line function: let $m = \frac{y_1 - y_0}{x_1 - x_0}$, and take $\alpha(t) = (x_0 + mt, y_0 + mt)$ for each $t \in [0,1]$. Then α is continuous by Theorem 5.2.2, since α followed by projection onto each variable is linear. This diagram shows us the image of α in $\mathbb{R}^2$:

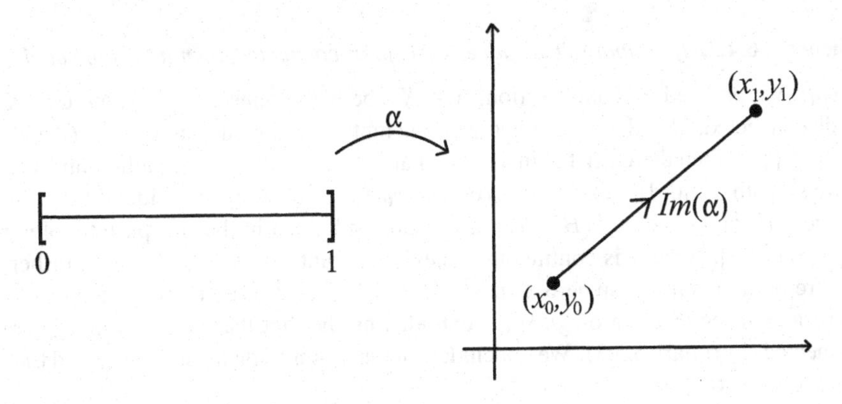

If $x_0 = x_1$, one can define a path α by $\alpha(t) = (x_0, ty_1 + (1-t)y_0)$.

We can show easily that path-connectedness is a strong topological property; that is, that the continuous image of a path-connected space is path-connected.

Theorem 6.4.1. *If X_τ is path-connected and $f : X_\tau \to Y_\nu$ is continuous, then $Im(f)_\nu$ is a path-connected subspace of Y.*

Again, this is equivalent to saying that if X_τ is path-connected and $f : X_\tau \to Y_\nu$ is continuous and onto, then Y_ν is path-connected.

Proof: Let y_0 and y_1 be any two points in $Im(f) \subset Y$. Since $f : X \to Y$ is onto its image, there exist (possibly many) $x_0, x_1 \in X$ with $f(x_0) = y_0$ and $f(x_1) = y_1$. Since X_τ is path-connected, there exists a path $\alpha : [0,1]_\mathcal{U} \to X_\tau$ with $\alpha(0) = x_0$ and $\alpha(1) = x_1$. Then the composite $f \circ \alpha : [0,1] \to Y_\nu$ with $(f \circ \alpha)(0) = f(x_0) = y_0$ and $(f \circ \alpha)(1) = f(x_1) = y_1$, as we wished, with $f \circ \alpha$ continuous as the composition of two continuous maps. □

The following picture should make the proof very clear:

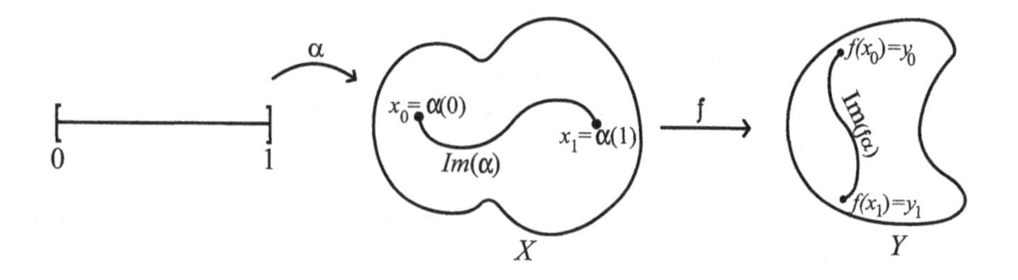

The relationship of connectedness to path-connectedness is given by the following result.

Theorem 6.4.2. *If a topological space X_τ is path-connected, then it is connected.*

Proof: We proceed by contradiction. Let X_τ be path-connected. Assume that X is disconnected. So there are disjoint, nonempty, τ-open subsets A and B with $X = A \coprod B$, then there exist points $x_0 \in A$ and $x_1 \in B$. Since X is path-connected, there's a path $\alpha : [0,1]_\mathcal{U} \to X_\tau$ with $\alpha(0) = x_0$ and $\alpha(1) = x_1$. Consider the inverse images: $\alpha^{-1}(A)$ and $\alpha^{-1}(B)$. These sets are both open in the subspace topology $\mathcal{U}_{[0,1]}$ on $[0,1]$, since α is continuous. They're disjoint, since $A \cap B = \emptyset$. Further, they're both nonempty, since $0 \in \alpha^{-1}(A)$ and $1 \in \alpha^{-1}(B)$. Thus these two sets serve as a disconnection of $[0,1]_\mathcal{U}$, contradicting the fact that intervals in $\mathbb{R}_\mathcal{U}$ are connected (Corollary 6.3.3). We conclude that our assumption must be false and that X_τ is connected. □

Surprisingly, perhaps, the converse of Theorem 6.4.2 is false, as the following examples demonstrate.

■ **EXAMPLE 6.7**

The *comb space* is the following subspace of $\mathbb{R}^2_{\mathcal{U}^2}$:

$$C := ([0,1] \times \{0\}) \cup \bigcup_{n \in \mathbb{N}} \left(\{\frac{1}{n}\} \times [0,1]\right) \cup (\{0\} \times [0,1]).$$

This picture should make the example clear:

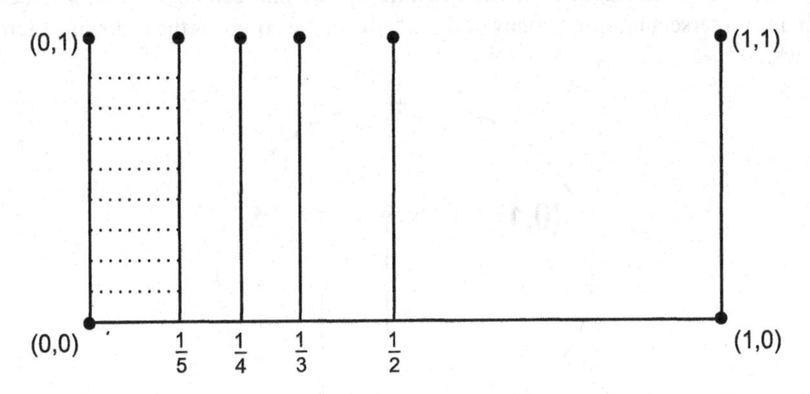

■ **EXAMPLE 6.8**

The *deleted comb space* is the following subspace of $\mathbb{R}^2_{\mathcal{U}^2}$:

$$D := ([0,1] \times \{0\}) \cup \bigcup_{n \in \mathbb{N}} \left(\{\frac{1}{n}\} \times [0,1]\right) \cup \{(0,1)\}.$$

So we delete the interval $\{0\} \times (0,1)$ from the comb space to form the deleted comb space. The following diagram should make the example clear:

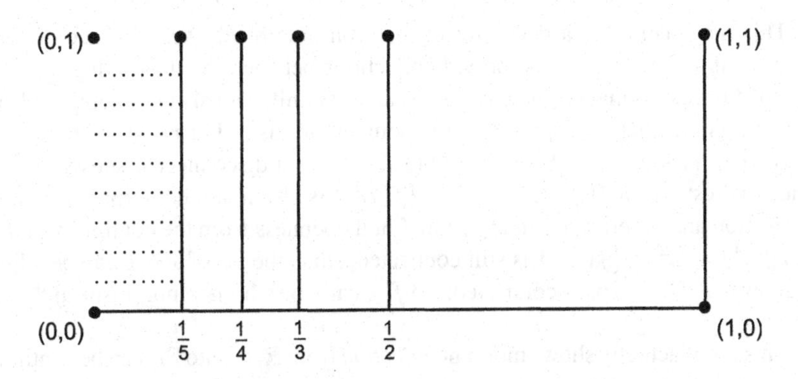

Clearly, the full comb space is path-connected (and hence connected), as one can see by connecting any two points (x_0, y_0) and (x_1, y_1) in C by a path [that goes from (x_0, y_0) down to the x-axis, if necessary, then along the x-axis to $(x_1, 0)$, then up to (x_1, y_1)]. Almost as clearly, the deleted comb space D is not path-connected, since one cannot find a continuous path from $(0, 1)$ to any other point in D. However, D *is* connected, despite our intuition to the contrary. One might think that it's possible to find an open set around $(0, 1)$ that is disjoint from an open set containing the rest of D. However, remember that D is given the subspace topology from $\mathbb{R}^2_{\mathcal{U}^2}$, so that any open set containing $(0, 1)$ contains an entire open ϵ-ball centered at $(0, 1)$. Such a ball must intersect infinitely many of the "teeth" of the comb, as the following picture shows:

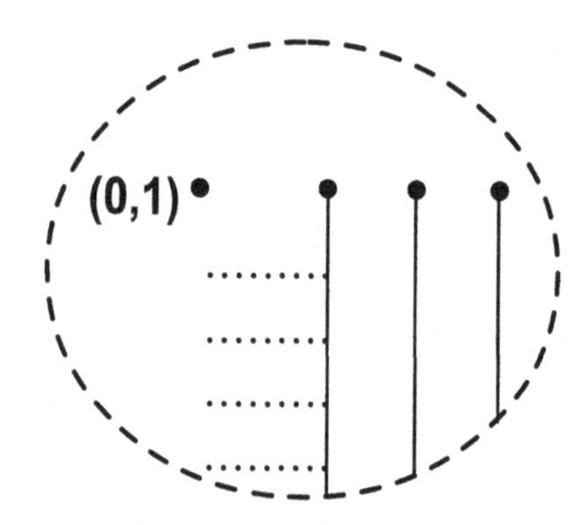

So there can be no disconnection of the deleted comb space D that doesn't also disconnect infinitely many of the teeth, and we know that the teeth are homeomorphic to closed intervals, so that can't occur.

The strong topological properties of connectedness and path-connectedness can supply very simple proofs that certain spaces are not homeomorphic, such as the following.

Proposition 6.4.3. *The real line $\mathbb{R}_{\mathcal{U}}$ and the real plane $\mathbb{R}^2_{\mathcal{U}^2}$ are not homeomorphic.*

This may seem like a trivial result, but you may recall the existence of certain "space-filling" curves, which provide bijections between the underlying sets. This proposition shows that no such bijection can be simultaneously continuous and open.

Proof: Assume that $f : \mathbb{R}_{\mathcal{U}} \to \mathbb{R}^2_{\mathcal{U}^2}$ is a homeomorphism. Let x_0 be any point in $\mathbb{R}_{\mathcal{U}}$. Then $A = \mathbb{R} \setminus \{x_0\} = (-\infty, x_0) \cup (x_0, +\infty)$ is a disconnected subspace of $\mathbb{R}_{\mathcal{U}}$. Then the restriction $f|_A : A \to (\mathbb{R}^2 \setminus \{f(x_0)\}$ is also a homeomorphism (since it's a bijection and inherits continuity from f and openness from the continuity of f^{-1}). But $f(A) = \mathbb{R}^2 \setminus \{f(x_0)\}$ is still connected, since the proof above can be adjusted to show that $f(A)$ is path-connected. So $f|_A$ can't be a homeomorphism, and neither can f. □

This proof actually shows more: no *bijection* from $\mathbb{R}^2_{\mathcal{U}^2}$ onto $\mathbb{R}_{\mathcal{U}}$ can be continuous.

Exercises

1. Prove that $\mathbb{R}_{\mathcal{U}}$ is path-connected.

2. Is $\mathbb{R}_{\mathcal{L}}$ path-connected?

3. Is $\mathbb{R}_{\mathcal{FC}}$ path-connected?

4. Is there a continuous function from $\mathbb{R}_{\mathcal{U}}$ onto $\mathbb{R}_{\mathcal{L}}$? From $\mathbb{R}_{\mathcal{FC}}$ onto $\mathbb{R}_{\mathcal{L}}$?

5. Another example of a space that is connected but not path-connected is the "topologist's sine curve":

$$S := \{(x, y) : y = \sin\left(\frac{1}{x}\right) \text{ with } 0 < x \leq 1\} \cup \{(0, 1)\}$$

as a subspace of $\mathbb{R}^2_{\mathcal{U}^2}$. Sketch S. Show why S is connected but not path-connected.

6. Prove: if X_τ is path-connected and $\tau' \subset \tau$, then $X_{\tau'}$ is also path-connected.

6.5 CONNECTEDNESS OF UNIONS AND FINITE PRODUCTS

In this section, we'll prove the following two results:

- The union of (path-) connected spaces *with a point in common* is (path-)connected.
- The finite product of (path-) connected spaces is (path-) connected.

It turns out to be simple to prove these results for path-connectedness, so we'll do this first. We'll have to work a bit harder to get the analogous results for connected spaces.

Theorem 6.5.1. *For each $\gamma \in \Lambda$, let A_γ be a path-connected subspace of X_τ. If $\bigcap_{\gamma\in\Lambda} A_\gamma \neq \emptyset$, then $\bigcup_{\gamma\in\Lambda} A_\gamma$ is path-connected as a subspace of X.*

Proof: Let $x_0, x_1 \in \bigcup_{\gamma\in\Lambda} A_\gamma$. Since $\bigcap_{\gamma\in\Lambda} A_\gamma \neq \emptyset$, there exists some element a common to all the A_γs. Now $x_0 \in A_\beta$ for some $\beta \in \Lambda$, and a is in that particular A_β as well, so there exists a path $\alpha : [0,1] \to A_\beta$ from x_0 to a. Similarly, $x_1 \in A_\delta$ for some $\delta \in \Lambda$, and a is in that particular A_δ as well, so there exists a path $\epsilon : [0,1] \to A_\delta$ from a to x_1. Then we define our desired path $\xi : [0,1] \to \bigcup_{\gamma\in\Lambda} A_\gamma$ as follows:

$$\xi(t) = \begin{cases} \alpha(2t) & \text{if } 0 \leq t \leq \frac{1}{2}, \\ \epsilon(2t-1) & \text{if } \frac{1}{2} \leq t \leq 1. \end{cases}$$

The function ξ "traverses" the path α from x_0 to a at twice the usual speed, then continues, doing the path ϵ from a to x_1 at twice speed as well. Note that $\xi(\frac{1}{2})$ is defined in two ways, but yields a for both descriptions. We see that the function ξ is continuous, using the Pasting Lemma (Theorem 5.5.1). We've divided the interval $[0,1]_{\mathcal{U}}$ into two $\mathcal{U}$-closed sets: $[0, \frac{1}{2}]$ and $[\frac{1}{2}, 1]$. The functions $\alpha(2t)$ and $\epsilon(2t-1)$ are continuous functions on each piece, and they agree on the intersection $\{\frac{1}{2}\}$. So ξ is continuous, and it's the desired path from x_0 to x_1. The diagram below illustrates the idea of the proof. □

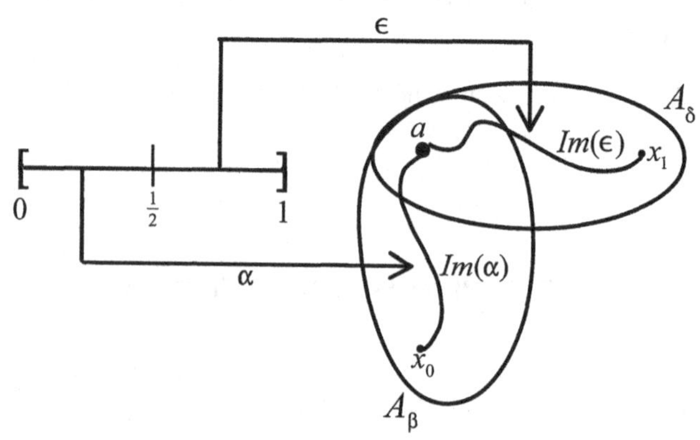

This process of "pasting" paths together will be used to great advantage in Chapter 11, where we'll use it to construct a group for each space X_τ (called the *fundamental group of* X_τ), which can often be used to prove or disprove the existence of continuous maps between spaces. Theorem 6.5.1 provides an easy proof that the comb space C of Example 6.7 is path-connected. We now address the question for products.

Theorem 6.5.2. *The finite product of path-connected spaces is path-connected.*

Proof: Of course, we'll prove this for the product of *two* path-connected spaces, then use induction to conclude the result. Let (x_0, y_0) and (x_1, y_1) be any two points of $X_\tau \times Y_\nu$, where the two coordinate spaces are path-connected. We'll build a path from (x_0, y_0) to (x_1, y_1), using the Pasting Lemma. First, $\{x_0\} \times Y_\nu$ is a subspace of $X_\tau \times Y_\nu$ which is obviously homeomorphic to Y_ν. Since (x_0, y_0) and (x_0, y_1) are points in the path-connected space $\{x_0\} \times Y_\nu$, there's a path $\alpha : [0,1] \to \{x_0\} \times Y_\nu$ from (x_0, y_0) to (x_0, y_1). Similarly, $X \times \{y_1\}$ is a subspace of $X_\tau \times Y_\nu$ that is obviously homeomorphic to X_τ. Since (x_0, y_1) and (x_1, y_1) are points in the path-connected space $X \times \{y_1\}$, there's a path $\beta : [0,1] \to X \times \{y_1\}$ from (x_0, y_1) to (x_1, y_1). We paste the two paths together to get $\gamma : [0,1] \to X \times Y$, defined by

$$\gamma(t) = \begin{cases} \alpha(2t) & \text{if } t \in [0, \frac{1}{2}], \\ \beta(2t-1) & \text{if } t \in [\frac{1}{2}, 1]. \end{cases}$$

It's easy to check whether γ is continuous by the Pasting lLemma, and that $\gamma(0) = (x_0, y_0), \gamma(\frac{1}{2}) = (x_0, y_1)$ and $\gamma(1) = (x_1, y_1)$. The crude picture below illustrates the proof. □

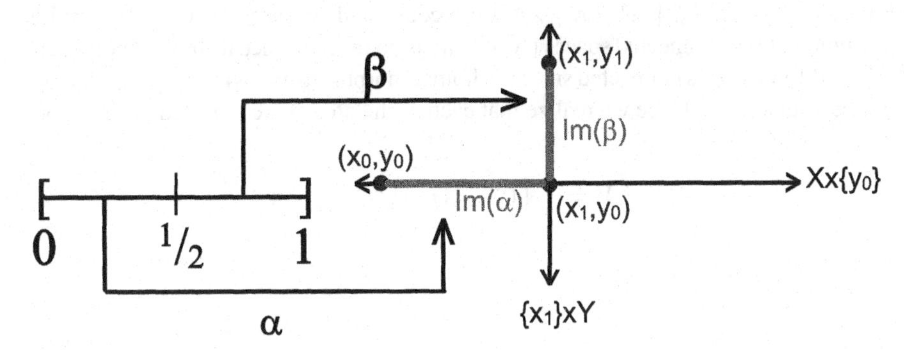

An alternative (and slicker) proof of the theorem is outlined in the exercises. The proof above is worth thinking about, though, because it further illustrates how to "paste" paths together, so we'll be more comfortable with the process in Chapter 11.

As an obvious corollary, we have: $\mathbb{R}^n_{\mathcal{U}^n}$ is path-connected (and hence connected) for all $n \in \mathbb{N}$. (This requires the exercise that $\mathbb{R}_{\mathcal{U}}$ is path-connected, but you've been shamed into doing this by now, right?)

Now we're prepared to deal with the standard connectedness versions of these results.

Theorem 6.5.3. *For each $\gamma \in \Lambda$, let A_γ be a connected subspace of X_τ. If $\bigcap_{\gamma\in\Lambda} A_\gamma \neq \emptyset$, then $\bigcup_{\gamma\in\Lambda} A_\gamma$ is connected as a subspace of X.*

Proof (of a special case): We'll show that the union of two connected subspaces of X_τ, with at least one point in common, is connected. We proceed by contradiction. Let A and B denote the two connected subspaces, with $A \cap B \neq \emptyset$. Assume that $A \cup B$ is disconnected, so that $A \cup B = U \coprod V$ for some pair of disjoint, nonempty and $\tau_{A\cup B}$-open sets U and V. Consider $U \cap A$ and $V \cap A$; if both are nonempty, then they disconnect A_{τ_A}. So one (say, $U \cap A$) must be empty, meaning that $U \subset B$. Now consider $U \cap B$ and $V \cap B$; if both are nonempty, they disconnect B. So one must be empty, meaning that $V \subset A$. But $A \cap B$ is nonempty, so there is at least one point $x \in A \cap B$ that's in neither U nor V, But $U \cup V$ is assumed to be all of $A \cup B$. Since our assumption contradicts itself, it must be false, and $A \cup B$ is connected as a subspace of X_τ. □

The proof of the general case proceeds similarly and is left as an exercise.

We use this in the proof of the following:

Theorem 6.5.4. *The finite product of connected spaces is connected.*

Proof: As before, we'll prove this for the product of two connected spaces and then let induction finish it. Let X_τ and Y_ν be connected spaces. We'll show that $X \times Y$ is the union of connected spaces having a point in common. Let $x_0 \in X$ and $y_0 \in Y$. Then the "cross" $(X \times \{y_0\}) \cup (\{x_0\} \times Y)$ is a connected subspace of $X \times Y$, since it's the union of two connected spaces ($X \times \{y_0\}$ is connected since its homeomorphic to X_τ and $\{x_0\} \times Y$ is connected since it's homeomorphic to Y_ν) with the point (x_0, y_0) in the intersection. Once we realize that each such "cross" is connected, we're about finished, since

$$X \times Y = \bigcup_{y \in Y} (X \times \{y\}) \cup (\{x_0\} \times Y).$$

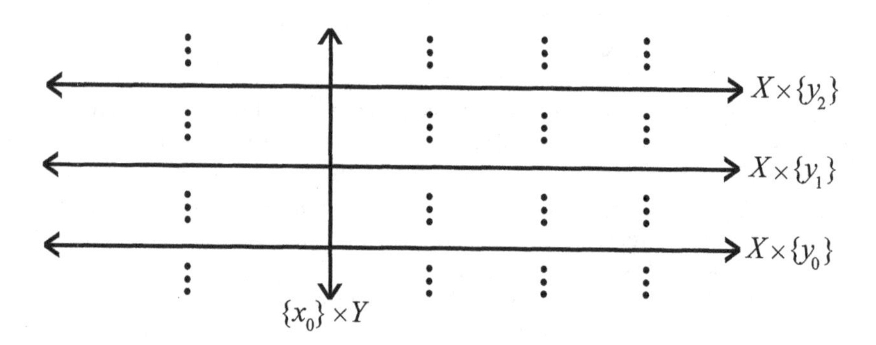

Now apply Theorem 6.5.3 to this union, noting that (x_0, y_0) is common to all the crosses. □

Note that the converses of Theorems 6.5.2 and 6.5.4 are valid as well: The product $X_\tau \times Y_\nu$ is (path-)connected if and only if X_τ and Y_ν are (path-)connected. The proof of the converses is trivial: the coordinate spaces are the continuous image of $X \times Y$ under the projection maps, and the continuous image of a (path-) connected space us (path-) connected.

Exercises

1. Verify that the following provides another proof that the product of two path-connected spaces is path-connected. Let (x_0, y_0) and (x_1, y_1) be any two points

of $X_\tau \times Y_\nu$, where the two coordinate spaces are path-connected. Since X_τ is path-connected, there exists a path $\alpha : [0,1]_\mathcal{U} \to X_\tau$ with x_0 as the initial point and x_1 as the terminal point. Similarly, since Y_ν is path-connected, there exists a path $\beta : [0,1]_\mathcal{U} \to Y_\nu$ with y_0 as the initial point and y_1 as the terminal point. Let $\Delta : [0,1] \to [0,1] \times [0,1]$ denote the diagonal map: $\Delta(t) = (t,t)$. Show that $(\alpha \times \beta) \circ \Delta$ provides the requisite path from (x_0, y_0) to (x_1, y_1). (*Hint*: Δ is a map into a product space. How do you show that such maps are continuous?)

2. Prove the general form of Theorem 6.5.3: If A_γ is a connected subspace of X_τ for every $\gamma \in \Lambda$ and if $\bigcap_{\gamma\in\Lambda} A_\gamma \neq \emptyset$, then $\bigcup_{\gamma\in\Lambda} A_\gamma$ is connected as a subspace of X.

6.6 CONNECTEDNESS OF INFINITE PRODUCTS (OPTIONAL)

In this section we will prove that arbitrary products of (path-) connected spaces are (path-) connected. Surprisingly, this is easier for path-connectedness, so we'll deal with this property first.

Theorem 6.6.1. *Any product of path-connected spaces is path-connected.*

Note that we assume the product topology here. The proof is shockingly simple.
Proof: For each $\alpha \in \Lambda$, let X_α be a path-connected space. Let $\mathbf{x}$ and $\mathbf{y}$ be two points in $\prod_{\alpha\in\Lambda} X_\alpha$; that is, $\mathbf{x}$ and $\mathbf{y}$ are Λ-tuples, with coordinates given for each $\beta \in \Lambda$ by $x_\beta = p_\beta(\mathbf{x})$ and $y_\beta = p_\beta(\mathbf{y})$, points in the coordinate space X_β, which is path-connected. So for each $\beta \in \Lambda$, there exists a path $f_\beta : [0,1]_\mathcal{U} \to X_\beta$ with $f_\beta(0) - x_\beta$ and $f_\beta(1) = y_\beta$. Now we piece these paths together to form our path:

$$f : [0,1]_\mathcal{U} \to \prod_{\alpha\in\Lambda} X_\alpha$$

is defined by $(f(t))_\alpha = f_\alpha(t)$. Then f is continuous since it's a map into a product, with the composition with the projection onto each factor continuous. (See the "Big Theorem on Maps into Products," Theorem 5.3.1.) Then f begins at $\mathbf{x}$ and ends at $\mathbf{y}$, as we wished. □

Theorem 6.6.2. *Any product of connected spaces is connected.*

This theorem is not as straightforward to prove, but it's not too daunting. It involves showing that a certain "almost finite product" subspace of an arbitrary product space is connected, and then proving that this space is dense.
Proof: Let X_α be a connected space for every $\alpha \in \Lambda$, and let $X = \prod_{\alpha \in \Lambda} X_\alpha$. Choose any basepoint $\mathbf{b} \in X$. Let $F = \{\alpha_1, \ldots \alpha_n\}$ be any *finite* subset of the indexing set Λ and define the subspace X_F as the set of all Λ-tuples $\mathbf{x} \in X$ with $x_\alpha = b_\alpha$ whenever $\alpha \in \Lambda \setminus F$; that is, X_F consists of all the points of X which agree with the basepoint $\mathbf{b}$ in all coordinates outside the finite set F. Then the subspace X_F is homeomorphic to the finite product

$$X_{\alpha_1} \times \cdots \times X_{\alpha_n},$$

and is therefore connected by Theorem 6.5.4. Now let W be the subspace of X defined by

$$W = \bigcup_{F\ finite} X_F.$$

Since the basepoint $\mathbf{b}$ is common to all of the X_Fs, we see that W is connected by Theorem 6.5.3. Finally, we note that $Cl(W) = X$, since every neighborhood of every point in X must be the entire coordinate space except for finitely many coordinates, and hence intersects at least one X_F. By Theorem 6.2.3, we're done. □

CHAPTER 7

COMPACT SPACES

7.1 INTRODUCTION

Compactness is a topological property that encodes a certain kind of "finiteness," and hence compact spaces have many useful properties. In $\mathbb{R}_{\mathcal{U}}$ or $\mathbb{R}^n_{\mathcal{U}^n}$, for example, this sort of finiteness occurs in subsets that are both closed and bounded, by a result known as the *Heine–Borel Theorem*. Such subsets lend themselves nicely to the kinds of infinite processes that we use in calculus, because, for example, every infinite subset of a closed and bounded subset of $\mathbb{R}_{\mathcal{U}}$ or $\mathbb{R}^n_{\mathcal{U}^n}$ has a limit point. Similarly, every continuous function $f : [a, b]_{\mathcal{U}} \to \mathbb{R}_{\mathcal{U}}$ attains an absolute maximum and an absolute minimum. So *compact spaces* are good places to work when one is concerned with topics of the field of *analysis* (basically, the field encompassing the more modern versions of classical calculus).

We'll define compactness in general topological spaces, rather than just in $\mathbb{R}_{\mathcal{U}}$ or $\mathbb{R}^n_{\mathcal{U}^n}$, and look at some basic properties and examples of this concept. Closely tied to the idea of compactness is a way of measuring how "fine" a topology is – in other words, how easy it is to "separate" distinct points by open neighborhoods. One particularly useful separation property for a space $X_\mathcal{T}$ to have is to be a *Hausdorff space*. We'll see how this relates to compactness in Section 7.3. Then we'll specialize to $\mathbb{R}_\mathcal{U}$, giving a simple characterization of compact subsets of $\mathbb{R}_\mathcal{U}$ (the Heine–Borel Theorem). Finally, we'll examine how compactness relates to finite products of spaces and some consequences.

7.2 DEFINITION, EXAMPLES AND PROPERTIES

Our concept of compactness will be defined in terms of *covers* of a space $X_\mathcal{T}$.

Definition 7.2.1. *A cover of a set X is a collection $\mathcal{C}$ of subsets of X whose union equals X; that is, $\mathcal{C} \subset \mathcal{P}(X)$ with $\bigcup \mathcal{C} = \bigcup_{C \in \mathcal{C}} C = X$.*

We'll often use the phrase "$\mathcal{C}$ covers X" to indicate that the collection $\mathcal{C}$ is a cover for X. We already know some examples – any basis $\mathcal{B}$ for a set X is a cover for X (see Definition 4.2.1) and any topology $\mathcal{T}$ on a space X covers X.

If $X_\mathcal{T}$ is a topological space, we say that $\mathcal{C}$ is an *open cover* for X if $\mathcal{C}$ covers X and if every set $C \in \mathcal{C}$ is $\mathcal{T}$-open.

Definition 7.2.2. *A space $X_\mathcal{T}$ is compact if every open cover of X has a finite subcover. That is, if $\mathcal{C}$ is a cover for X consisting entirely of $\mathcal{T}$-open sets, then there exists a subcollection $\{C_1, C_2, \ldots, C_k\} \subset \mathcal{C}$ such that*

$C_1 \cup C_2 \cup \cdots \cup C_k = X.$

At first glance, it appears that it would be *very* difficult to check that a space $X_\mathcal{T}$ is compact, because one would have to check *every* possible open cover of X. Just knowing that one open cover of $X_\mathcal{T}$ has a finite subcollection covering X tells us nothing about other open covers.

■ EXAMPLE 7.1

The space $\mathbb{R}_\mathcal{U}$ is not compact. Note that *some* open covers of $\mathbb{R}_\mathcal{U}$ do have finite subcovers, like $\mathcal{C}_1 = \{\mathbb{R}, (0,1)\}$ and $\mathcal{C}_2 = \{(n-1, n+1) : n \in \mathbb{Z}\} \cup \{\mathbb{R}\}$. Both of these are open covers with finite subcovers (namely, $\{\mathbb{R}\}$ in both cases). However, $\mathcal{C}_3 = \{(n-1, n+1) : n \in \mathbb{Z}\}$ is an open cover (since every $x \in \mathbb{R}$ falls into at least one such interval) with *no* finite subcover: every integer n lies in exactly one such interval, so we can't leave any of the sets out of $\mathcal{C}_3$ without failing to cover $\mathbb{R}$.

We have the following more trivial examples.

EXAMPLE 7.2

Any indiscrete space $X_{\mathcal{I}}$ is compact.

So $\mathbb{R}_{\mathcal{I}}$ is compact, in contrast to $\mathbb{R}_{\mathcal{U}}$.

EXAMPLE 7.3

If X is a finite set and τ is any topology on X, then X_τ is compact. More tersely: any finite space is compact.

EXAMPLE 7.4

If X is any infinite set, then the discrete space $X_{\mathcal{D}}$ is not compact, as the open cover $\mathcal{C} = \{\{x\} : x \in X\}$ shows.

Here's an example that might be mildly surprising, at first glance.

EXAMPLE 7.5

The interval $(0,1)_{\mathcal{U}}$ is not compact as a subspace of $\mathbb{R}_{\mathcal{U}}$. To see this, consider the cover $\mathcal{C} = \{(\frac{1}{n}, 1) : n = 2, 3, \ldots\}$, which is a "nested collection" of intervals: $(\frac{1}{2}, 1) \subset (\frac{1}{3}, 1) \subset \cdots$. This is indeed an open cover, since every $x \in (0,1)$ has $x > \frac{1}{n}$ for some n. But there is no finite subcover; if $x > 0$, then $x \notin (\frac{1}{n}, 1)$ for all $n < \frac{1}{x}$. Thus no finite subcollection of $\mathcal{C}$ can contain all $x \in (0,1)$.

Of course, this example shouldn't really surprise us, because $(0,1)_{\mathcal{U}} \cong \mathbb{R}_{\mathcal{U}}$, and we would hope that compactness is a topological property. The following theorem establishes this, and more.

Theorem 7.2.1. *The continuous image of a compact space is compact; that is, if X_τ is compact and $f : X_\tau \to Y_\nu$ is continuous and onto, then Y_ν is compact.*

Proof: Let $\mathcal{C}$ be any open cover of Y_ν. Then $\mathcal{C}' = \{f^{-1}(C) : C \in \mathcal{C}\}$ is a collection of τ-open subsets of X, since f is continuous. Further, since $\mathcal{C}$ covers Y, $\mathcal{C}'$ must cover $X = Domain(f)$. Hence $\mathcal{C}'$ has a finite subcover:
$f^{-1}(C_1) \cup f^{-1}(C_2) \cup \cdots \cup f^{-1}(C_k) = X$. Applying f, we see that
$C_1 \cup C_2 \cup \cdots \cup C_k = Y$, since $f(X) = Y$ (because f is onto). □

So compactness is a strong topological property, like connectedness and path-connectedness.

Corollary 7.2.2. *If X_τ is compact and τ' is any topology on X which is coarser than τ, then $X_{\tau'}$ is also compact. If Y_ν is not compact and ν' is any topology on Y which is finer than ν, then $Y_{\nu'}$ is not compact.*

The proof is left for the exercises. Note that $\mathbb{R}_{\mathcal{L}}$ is now known to be noncompact. (Why?)

One might ask if compactness is a hereditary property: Is a subspace of a compact space also compact? This turns out to be false, but the easiest counterexample requires us to use a result from Section 7.4, where we prove (in Theorem 7.4.1) that any closed interval in $\mathbb{R}$ is compact in the usual subspace topology. Then the example of $(0,1)_{\mathcal{U}}$ as a subspace of $[-1,34]_{\mathcal{U}}$ shows that subspaces of compact spaces need not be compact.

However, it turns out that we get some degree of heredity for compactness:

Theorem 7.2.3. *If X_τ is compact and A is any closed subset of X, then A_{τ_A} is compact.*

So compactness is preserved in *closed* subspaces. The proof is quite nice.

Proof: Let X_τ be compact and let A be any τ-closed subspace. We need to show that every τ_A-open cover of A has a finite subcover. Let $\mathcal{C}$ be any τ_A-open cover of A. Then for each $U_\alpha \in \mathcal{C}$, there exists a corresponding $V_\alpha \in \tau$ with $U_\alpha = V_\alpha \cap A$. (Of course, there may be many choices for each V_α, but we'll choose *one* τ-open set V_α for each τ_A-open set $U_\alpha \in \mathcal{C}$.) Then let $\mathcal{C}' = \{V_\alpha : U_\alpha \in \mathcal{C}\} \cup \{X \smallsetminus A\}$. Since $X \smallsetminus A$ is τ-open (A is τ-closed) and every V_α is τ-open, the collection $\mathcal{C}'$ consists entirely of τ-open sets. Further, $\mathcal{C}'$ covers X, since $A = \bigcup_{U_\alpha \in \mathcal{C}} U_\alpha \subset \bigcup_{U_\alpha \in \mathcal{C}} V_\alpha$, so adding in the set $X \smallsetminus A$ guarantees we cover all of X. So $\mathcal{C}'$ has a finite subcollection that covers X: $V_{\alpha_1} \cup \ldots V_{\alpha_k} \cup (X \smallsetminus A) = X$, where the last set in the union may be redundant. So the corresponding union of τ_A-open sets $U_{\alpha_1} \cup \cdots \cup U_{\alpha_k} = A$, showing that A_{τ_A} is compact. □

One might ask if the converse to this theorem is true: is a compact subspace of a space always closed? In other words, if A_{τ_A} is compact as a subspace of X_τ, need A be τ-closed? This is false, as the following example shows.

■ **EXAMPLE 7.6**

Consider the subspace $[0,1]_{\mathcal{I}} \subset \mathbb{R}_{\mathcal{I}}$. Then $[0,1]_{\mathcal{I}}$ is compact, since the only $\mathcal{I}_{[0,1]}$-open sets are $\emptyset$ and $[0,1]$. However, $[0,1]$ is *not* $\mathcal{I}$-closed in $\mathbb{R}_{\mathcal{I}}$.

So the converse of Theorem 7.2.3 is false, in general. However, it turns out that this converse is true, if we add an additional property to the space X_τ; we need X_τ to be *Hausdorff*, which we'll define precisely in Section 7.3.

Exercises

1. Verify that the spaces in Examples 7.2 and 7.3 are compact, using the definition.
2. Prove Corollary 7.2.2: If X_τ is compact and τ' is any topology on X that is coarser that τ, then $X_{\tau'}$ is also compact. If Y_ν is not compact and ν' is any topology on Y which is finer than ν, then $Y_{\nu'}$ is not compact. (*Hint*: use the identity maps on X and on Y.)
3. Is $\mathbb{R}_{\mathcal{RR}}$ compact?
4. Is $\mathbb{R}_{\mathcal{FC}}$ compact?
5. Is $\mathbb{R}_{\tau_1}$ compact, where τ_1 is the distinguished point topology?
6. Show, from the definition, that $\mathbb{R}^2_{\mathcal{U}^2}$ is not compact.
7. Consider the following topology on $\mathbb{R}$:

$$\mathcal{CC} = \{V \subset \mathbb{R} : \mathbb{R} \setminus V \text{ is countable, or } V = \emptyset.\}$$

This is the *countable complement* topology. Is $\mathbb{R}_{\mathcal{CC}}$ compact?

7.3 HAUSDORFF SPACES AND COMPACTNESS

In this section, we define what it means for a space to be *Hausdorff* and show how this property is intimately intertwined with compactness.

Definition 7.3.1. *A space X_τ is Hausdorff if for every pair of points $x_1, x_2 \in X$ with $x_1 \neq x_2$, there exist disjoint τ-open sets U and V containing x_1 and x_2, respectively.*[1]

In other words, a space X_τ is Hausdorff if one can find disjoint neighborhoods of any two distinct points. We'll often describe this by using the phrase "the neighborhoods U and V separate x_1 from x_2." A Hausdorff space is often called a T_2 space, for reasons that will be clear in Chapter 8. The following diagram illustrates the idea:

[1]Felix Hausdorff, who in 1914 first gave the precise axiomatic definition of a topological space, much like the definition we use, required that all spaces have this property. Later, it became clear that certain objects that lacked this property were quite interesting, and the requirement was dropped.

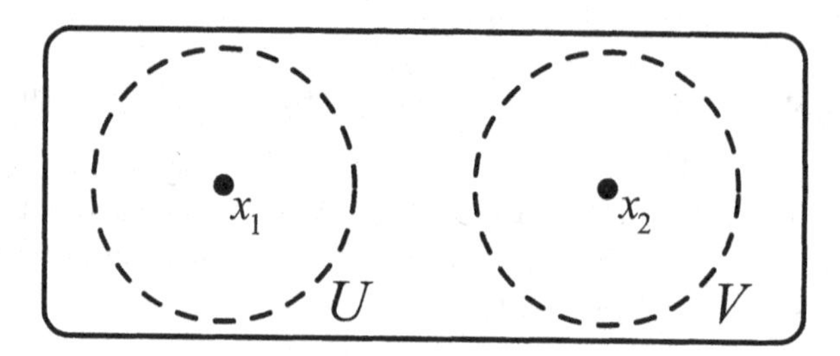

Some examples are in order.

■ **EXAMPLE 7.7**

Let X be any set. Then the discrete space $X_{\mathcal{D}}$ is Hausdorff, as one can see by noting that $\{x_1\}$ and $\{x_2\}$ are $\mathcal{D}$-open neighborhoods that separate x_1 from x_2 whenever $x_1 \neq x_2$. [If $Card(X) = 1$, $X_{\mathcal{D}}$ is still Hausdorff. Be sure that you understand why.]

■ **EXAMPLE 7.8**

Let X be any set with $Card(X) \geq 2$. Then the indiscrete space $X_{\mathcal{I}}$ is not Hausdorff, since there are points $x_1, x_2 \in X$, with $x_1 \neq x_2$, that can't be separated (since the only $\mathcal{I}$-open neighborhood of any point is all of X).

■ **EXAMPLE 7.9**

$\mathbb{R}_{\mathcal{U}}$ is Hausdorff, as one can see by noting that whenever $x_1 \neq x_2$, the neighborhoods $(x_1 - \frac{\epsilon}{2}, x_1 + \frac{\epsilon}{2})$ and $(x_2 - \frac{\epsilon}{2}, x_2 + \frac{\epsilon}{2})$ separate x_1 from x_2, where $\epsilon = |x_1 - x_2|$.

■ **EXAMPLE 7.10**

The set $\mathbb{R}$ in the finite-complement topology $\mathcal{FC}$ is not Hausdorff, since whenever $x_1 \neq x_2$, any $\mathcal{FC}$-neighborhood of x_1 contains all but finitely many real numbers, so it must intersect any $\mathcal{FC}$-neighborhood of x_2.

We'll prove many properties of Hausdorff spaces in Chapter 8. For now, we're interested in how the properties of compactness and Hausdorff interact. The following is the promised partial converse to Theorem 7.2.3.

Theorem 7.3.1. *A compact subspace of a Hausdorff space is closed. More verbosely, if X_τ is Hausdorff and if $A \subset X$ with A_{τ_A} compact, then A is τ-closed.*

So this theorem provides the *partial* converse to Theorem 7.2.3. The proof of this theorem is pretty cool in its own right, and the ideas here will resurface in several other contexts, so the reader should pay particularly close attention to this argument. The picture below will summarize the proof.

Proof: Let A_{τ_A} be a compact subspace of a Hausdorff space X. We'll show that $X \setminus A$ is τ-open by showing that it contains a neighborhood of each of its points. If $X \setminus A$ is empty, we're done. If not, let $x \in X \setminus A$. For every $a \in A$, then, $a \neq x$, so we can find disjoint τ-open neighborhoods that separate a and x. We'll let U_a denote the neighborhood of a and V_a denote the corresponding neighborhood of x (since the point x will be the same in all that follows, but a will change). For a different $a' \in A$, we get τ-neighborhoods $U_{a'}$ of a and $V_{a'}$ of x with $U_{a'} \cap V_{a'} = \emptyset$, although it's possible that this neighborhood $U_{a'}$ of a' might intersect some other neighborhood of x, like the original V_a. Since we can do this for every point $a \in A$, we get a collection $\mathcal{C} = \{U_a : a \in A\}$ of τ-open sets whose union $\bigcup \mathcal{C} = \bigcup_{a \in A} U_a$ contains A. Since A_{τ_A} is compact, and since taking each U_a intersected with A would yield a τ_A-open cover of A, there must exists a finite subcollection $\{U_{a_1}, U_{a_2}, \ldots, U_{a_k}\}$ of $\mathcal{C}$ with $A \subset U = U_{a_1} \cup U_{a_2} \cup \cdots \cup U_{a_k}$. Each of these sets U_{a_i} has a corresponding V_{a_i}, a τ-open neighborhood of x, with $U_{a_i} \cap V_{a_i} = \emptyset$. Now we're in a position to find our τ-neighborhood of x: let $V = V_{a_1} \cap V_{a_2} \cap \cdots \cap V_{a_k}$. Then V is τ-open, since it's a *finite* intersection of τ-open sets (which explains where the compactness of A comes into play). Further, $x \in V$, since $x \in V_{a_i}$ for each $i = 1, 2, \ldots, k$. Finally, $V \cap A = \emptyset$, since for each i, $U_{a_i} \cap V_{a_i} = \emptyset$, so that

$$V \cap A \subset V \cap U = (V_{a_1} \cap V_{a_2} \cap \cdots \cap V_{a_k}) \cap (U_{a_1} \cup U_{a_2} \cup \cdots \cup U_{a_k}) = \emptyset.$$

Thus we have our τ-neighborhood V of x, with $V \subset X \setminus A$. We conclude that $X \setminus A$ is τ-open and hence, A is τ-closed. □

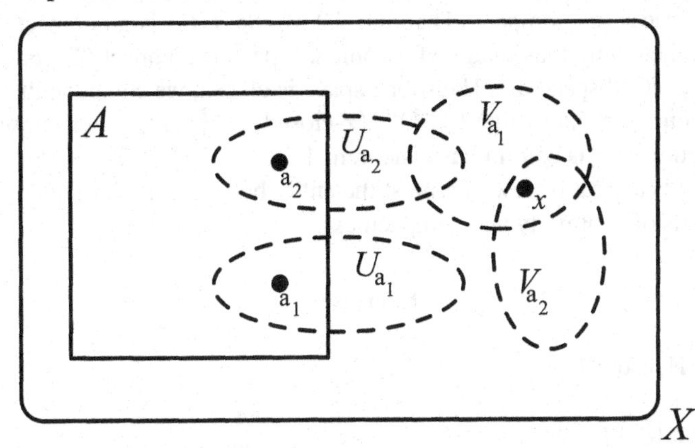

Because of this theorem, compactness and Hausdorff are very closely tied together. Note that a space X_τ is compact if it has "few enough" open sets so that every open cover has a finite subcover. On the other hand, X_τ is Hausdorff if it has "enough"

open sets to enable separation of distinct points. These two properties, then, work in opposition to each other, so that to say that X_τ is both compact and Hausdorff is a very powerful statement. The following theorem illustrates this nicely.

Theorem 7.3.2. *Let X_τ be compact and Hausdorff. If τ' is any topology on X with τ' strictly finer than τ, then $X_{\tau'}$ is not compact. If τ'' is any topology on X with τ'' strictly coarser than τ, then $X_{\tau''}$ is not Hausdorff.*

So removing even a single open set from a topology that makes a space compact and Hausdorff will render the space non-Hausdorff (although most likely one would need to remove many more sets to make the resulting collection a topology). Adding even a single open set to a topology that makes a space compact and Hausdorff will make the resulting space noncompact (with the same reservation that one most likely would need to add many more sets to make the collection a topology). The proof of the theorem is left as an exercise (with hints).

Theorem 7.3.2 is a special case of the following, which seems rather technical at first glance, but is actually a very neat summary of what we know about the relationship between compactness and Hausdorff.

Theorem 7.3.3. *Let $f : X_\tau \to Y_\nu$ be a continuous bijection. If X_τ is compact and Y_ν is Hausdorff, then f is a homeomorphism.*

Proof: We'll show that f is an open map (i.e., show that its inverse f^{-1} is continuous). Let C be any τ-closed subset of X. Then $C_{\tau C}$ is compact, since closed subspaces of compact spaces are compact (Theorem 7.2.3). So $f(C)$ is a compact subspace of Y_ν, since the continuous image of a compact space is compact (Theorem 7.2.1). Since a compact subspace of a Hausdorff space is closed, we conclude that $f(C)$ is ν-closed. Hence, we know that $(f^{-1})^{-1}(\tau\text{-closed set})$ is always a ν-closed set, so that the function f^{-1} is continuous (Theorem 4.5.1). □

This proof is a great review of almost the entire book so far – it includes many of the high points of continuity and compactness.

Exercises

1. Is $\mathbb{R}_{\mathcal{L}}$ Hausdorff?

2. Is $\mathbb{R}_{\mathcal{RR}}$ Hausdorff?

3. Is $\mathbb{R}_{\tau_1}$ Hausdorff?

4. Prove that a subspace of a Hausdorff space is Hausdorff.

5. Is the continuous image of a Hausdorff space Hausdorff? (*Hint*: for the identity function $i_X : X_\tau \to X_{\tau'}$ to be continuous, which topology needs more open sets?)

6. Prove Theorem 7.3.2: Let X_τ be compact and Hausdorff. If τ' is any topology on X with τ' finer than τ, then $X_{\tau'}$ is not compact. If τ'' is any topology on X with τ'' coarser than τ, then $X_{\tau''}$ is not Hausdorff. (*Hint*: Consider the hint for the above problem, then use everything you know about closed subsets of compact spaces, continuous images of compact spaces and compact subspaces of Hausdorff spaces.)

7. Prove the closed graph theorem: If $f : X_\tau \to Y_\nu$ is continuous, with Y both compact and Hausdorff, then the graph

$$G_f = \{(x, y) \in X \times Y : y = f(x)\}$$

is closed in $X_\tau \times Y_\nu$.

7.4 COMPACTNESS IN THE REAL LINE

We've already seen that the real line, $\mathbb{R}$ in the usual topology $\mathcal{U}$, is not compact. In this section, we'll investigate what is needed for subsets of $\mathbb{R}_\mathcal{U}$ to be compact, and see some consequences.

Our first main result is the following theorem.

Theorem 7.4.1. *Any closed interval is compact as a subspace of* $\mathbb{R}_\mathcal{U}$.

Proof: Let $[a, b]$ be any closed interval and let $\mathcal{C}$ be any covering of $[a, b]$ consisting entirely of $\mathcal{U}_{[a,b]}$-open sets. We need to show that $\mathcal{C}$ has a finite subcover. Let

$$A = \{x \in (a, b] : [a, x] \text{ is covered by finitely many sets from } \mathcal{C}\}.$$

Then A is nonempty: $\mathcal{C}$ is a cover for $[a, b]$, so there exists a set $U \in \mathcal{C}$ with $a \in U$. Since U is $\mathcal{U}_{[a,b]}$-open, we know that there exists an $\epsilon > 0$ such that $a \in [a, a+\epsilon) \subset U$. Thus, $a + \frac{\epsilon}{2} \in A$.

Let c denote the least upper bound $lub(A)$, which exists since A is nonempty and bounded above (by b among others). We claim that $c \in A$: since $c \in [a, b]$, which is covered by $\mathcal{C}$, there exists some set $V \in \mathcal{C}$ containing c. Since V is $\mathcal{U}_{[a,b]}$-open, there exists a $\delta > 0$ such that $c \in (c - \delta, c] \subset V$ (since we don't know if $c < b$ or not at this point). So V contains points to the left of c, such as $c - \frac{\delta}{k}$ for some $k \in \{2, 3, \ldots\}$.

But $c - \frac{\delta}{k} \in A$, since $c = lub(A)$, so that $[a, c - \frac{\delta}{k}]$ is covered by finitely many sets $C_1, C_2, \ldots, C_k$ from $\mathcal{C}$. Hence $c \in A$, since the single additional $\mathcal{C}$-set U contains everything from $c - \frac{\delta}{k}$ to c.

Finally, we show $c = b$, which will prove the theorem. Assume $c < b$, then consider any $\mathcal{C}$-set U containing c. Since U is $\mathcal{U}_{[a,b]}$-open and $c < b$, there exists $\gamma > 0$ such that $(c - \gamma, c + \gamma) \subset U$. Since $[a, c]$ is covered by finitely many sets from $\mathcal{C}$ already, adding the $\mathcal{C}$-set U shows that $[a, c + \frac{\gamma}{2}]$ is covered by finitely many sets from $\mathcal{C}$, so that $c + \frac{\gamma}{2} \in A$. This contradicts our definition of $c = lub(A)$, so our assumption (that $c < b$) must be false. □

We've now accumulated everything we need to prove the following theorem, which provides a convenient characterization of compactness for subsets of the real line.

Theorem 7.4.2. *(Heine–Borel[2] Theorem) A subset $A \subset \mathbb{R}_\mathcal{U}$ is compact if and only if A is $\mathcal{U}$-closed and bounded.*

Proof: $\Longrightarrow$) Let A be a compact subset of the real line. To see that A must be bounded, consider the cover $\mathcal{C} = \{(-n, n) : n \in \mathbb{N}\}$. If A is unbounded, then we'd have no finite subcover of $\mathcal{C}$ whose union contains A. Further, A is closed, since A is a compact subspace of the Hausdorff space $\mathbb{R}_\mathcal{U}$.

$\Longleftarrow$) Assume A is closed and bounded. Since A is bounded, $A \subset [-n, n]$ for some $n \in \mathbb{N}$. Then A is closed as a subspace of $[-n, n]_\mathcal{U}$, a compact space, since every closed interval is compact in the usual topology. Thus, A is compact, since closed subspaces of compact spaces are compact. □

So one can easily determine whether $A \subset \mathbb{R}_\mathcal{U}$ is compact: if A is unbounded or if A fails to contain any of its limit points, then $A_\mathcal{U}$ is not compact. We use this to prove the following basic result about the topology of the real line.

Theorem 7.4.3. *(Bolzano–Weierstrass Theorem) Every bounded infinite subset of $\mathbb{R}_\mathcal{U}$ has at least one limit point.*

In fact, this result is true in much more generality:

Theorem 7.4.4. *If $X_\mathcal{T}$ is compact and A is any infinite subset of X, then $A' \neq \emptyset$.*

The proof of Theorem 7.4.4 is left as an exercise.

Proof of the Bolzano–Weierstrass Theorem: We proceed by contrapositive. Let A be bounded in $\mathbb{R}_\mathcal{U}$, where A has no limit points. Since A is bounded, $A \subset [-n, n]$ for

[2]This theorem is "sort of" due to Eduard Heine (1821–1881) and Emile Borel (1871–1956). The theorem was called this by A. Schoenflies, who observed that it followed from separate work by Heine and Borel.

some $n \in \mathbb{N}$, where $[-n, n]_{\mathcal{U}}$ is obviously compact. Since $A' = \emptyset$, then for every $x \in [-n, n]$, there exists a $\mathcal{U}$-neighborhood U_x of x with $U_x \cap A$ either empty or $\{x\}$. But the set $\{U_x : x \in [-n, n]\}$ is an open cover of $[-n, n]$, so there's a finite subcover $\{U_{x_1}, U_{x_2}, \ldots, U_{x_k}\}$. Hence $A \subset U_{x_1} \cup U_{x_2} \cup \cdots \cup U_{x_k}$. However, for each x_i, $U_{x_i} \cap A \subset \{x_i\}$, so that we've shown $A \subset A \cap (U_{x_1} \cup U_{x_2} \cup \cdots \cup U_{x_k})$, so that A must be finite.

Exercises

1. Give an explicit example of a cover of $\mathbb{R}_{\mathcal{L}}$ that has no finite subcover.

2. Is $[a, b]_{\mathcal{L}}$ compact? If not, why does the proof of the Heine–Borel Theorem fail for subsets of $\mathbb{R}_{\mathcal{L}}$? (The second question suggests that there's an elegant way to justify your answer to the first question. However, you'll still need to see how the proof works out in $\mathbb{R}_{\mathcal{L}}$.)

3. Can you speculate on what a version of the Heine–Borel Theorem should be for subsets of $\mathbb{R}^2_{\mathcal{U}^2}$? For subsets of $\mathbb{R}^n_{\mathcal{U}^n}$?

4. Prove Theorem 7.4.4: If $X_\mathcal{T}$ is compact and A is any infinite subset of X, then $A' \neq \emptyset$. Hint: look at the proof of the Bolzano-Weierstrass Theorem.

5. Prove the Extreme Value Theorem, which was so useful in calculus: If $f : [a, b]_{\mathcal{U}} \to \mathbb{R}_{\mathcal{U}}$ is continuous, then f attains an absolute maximum and an absolute minimum.

6. Formulate and prove as general a version of the Extreme Value Theorem as you can.

7.5 COMPACTNESS OF PRODUCTS

The goal of this section is to prove the following theorem.

Theorem 7.5.1. *If $(X_1)_{\mathcal{T}_1}, (X_2)_{\mathcal{T}_2}, \ldots, (X_k)_{\mathcal{T}_k}$ are all compact spaces, then the product space $\prod_{i=1}^{k} (X_i)_{\mathcal{T}_i}$ is also compact.*

Of course, we'll prove it for the product of two compact spaces, $X_\mathcal{T} \times Y_\mathcal{V}$, then use induction. The corresponding result for an infinite product of compact spaces is also true, perhaps surprisingly. This theorem, known as Tychonoff's Theorem for the mathematician who proved it, is beyond the scope of this text. A fairly accessible proof can be found in Munkres' book (See Ref. [13].).

We'll separate out the first step in the proof, just to keep the main part of the proof as uncluttered as possible.

Lemma 7.5.2. *(Tube Lemma) Let Y_ν be compact and let W be any set open in the product $X_\tau \times Y_\nu$, such that W contains the "slice" $\{x_0\} \times Y$. Then there exists a τ-neighborhood U of x_0 such that $U \times Y \subset W$. (We refer to $U \times Y$ as the "tube" around $\{x_0\} \times Y$).*

Proof: The diagram that follows below will illustrate the construction. Let W be an open set in $X_\tau \times Y_\nu$, with W containing $\{x_0\} \times Y$. We recall that the product topology on $X_\tau \times Y_\nu$ has a basis consisting of $\tau \times \nu$. Since W is open in the product topology, and since the slice $\{x_0\} \times Y$ is a subset of W, then W must contain a basis set around each point of the slice. Hence, the slice $\{x_0\} \times Y$ can be covered by sets of the form $U \times V$, with $U \in \tau$ and $V \in \nu$, with $U \times V \subset W$. Since $\{x_0\} \times Y$ is homeomorphic to the compact space Y, there must be some finite collection of these basis sets whose union contains the slice:

$$\{x_0\} \times Y \subset (U_1 \times V_1) \cup (U_2 \times V_2) \cup \cdots \cup (U_k \times V_k),$$

for some $k \in \mathbb{N}$. This finite subcollection can be made "nonredundant," in that we discard any set $U_i \times V_i$ that does not intersect $\{x_0\} \times Y$. So $x_0 \in U = U_1 \cap U_2 \cap \cdots \cap U_k$, a τ-open set (since it's the intersection of finitely many τ-open sets). Then

$$(\{x_0\} \times Y) \subset (U \times Y) \subset W,$$

since $U \times V$ was contained in W in the first place. □

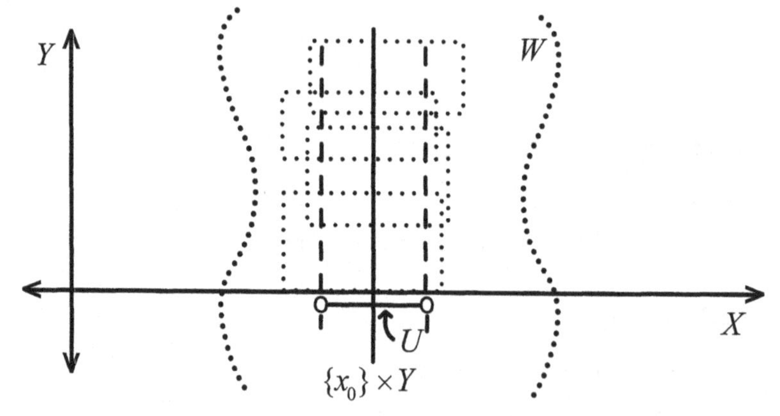

Proof of Theorem 7.5.1: Let X_τ and Y_ν be compact spaces. Let $\mathcal{C}$ be any open cover of $X_\tau \times Y_\nu$. For each $x_0 \in X$, the slice $\{x_0\} \times Y$ is homeomorphic to the compact space Y, so that some finite subcollection of $\mathcal{C}$ will cover $\{x_0\} \times Y$:

$$(\{x_0\} \times Y) \subset C_1 \cup C_2 \cup \cdots \cup C_k,$$

for some $k \in \mathbb{N}$. But $C = C_1 \cup C_2 \cup \cdots \cup C_k$ is an open set containing the slice $\{x_0\} \times Y$, so the Tube Lemma implies that there exists a τ-open neighborhood U_{x_0} of x_0 with $(\{x_0\} \times Y) \subset (U_{x_0} \times Y)$. Thus the entire tube $U_{x_0} \times Y$ is covered by finitely many sets from $\mathcal{C}$:

$$(U_{x_0} \times Y) \subset C_1 \cup C_2 \cup \cdots \cup C_k.$$

So for each $x \in X$, there exists a τ-neighborhood U_x of x with the tube $U_x \times Y$ covered by finitely many sets from $\mathcal{C}$. Since we can do this for every $x \in X$, we have an open cover of X_τ:

$$\mathcal{F} = \{U_x : x \in X, U_x \times Y \text{ is covered by finitely many sets from } \mathcal{C}\}.$$

Since X_τ is compact, $\mathcal{F}$ must contain a finite subcover: $U_1, U_2, \ldots U_m$, with $X = U_1 \cup \cdots \cup U_m$, with each resulting tube $U_i \times Y$ covered by finitely many sets from $\mathcal{C}$. Then

$$X \times Y = (U_1 \times Y) \cup (U_2 \times Y) \cup \cdots \cup (U_m \times Y)$$

must be covered by finitely many sets from $\mathcal{C}$. □

In our familiar setting of $\mathbb{R}^n_{\mathcal{U}^n}$, we have the following obvious corollary of Theorem 7.4.1:

Corollary 7.5.3. *The product of n closed intervals is compact as a subspace of $\mathbb{R}^n_{\mathcal{U}^n}$.*

This is the key point in the proof of the following theorem. We recall that a subspace $A \subset \mathbb{R}^n_{\mathcal{U}^n}$ is *bounded* is there exists m such that $A \subset [-m, m]^n$.

Theorem 7.5.4. *(Higher-dimesional Heine–Borel Theorem) A subspace $A \subset \mathbb{R}^n_{\mathcal{U}^n}$ is compact if and only if A is $\mathcal{U}^n$-closed and bounded.*

The proof is left as an exercise.

1. Prove the higher-dimensional Heine–Borel theorem: A subspace $A \subset \mathbb{R}^n_{\mathcal{U}^n}$ is compact if and only if A is $\mathcal{U}^n$-closed and bounded.

2. Show that a finite union of compact spaces is compact. (Of course, you need to assume that these spaces are all subspaces of some "ambient space." See Section 5.5 for reasons why.)

3. Is $\mathbb{R}_{\mathcal{L}} \times \mathbb{R}_{\mathcal{U}}$ compact?

4. Is the graph $G = \{(x, y) : y = x^2\}$ a compact subspace of $\mathbb{R}^2_{\mathcal{U}^2}$?

5. Prove the converse to the closed graph theorem: for $f : X_\tau \to Y_\nu$, with Y both compact and Hausdorff, if the graph G_f is closed, then f is continuous. [*Hint*: If the graph G_f is closed, then for any ν-neighborhood $V_{f(x_0)}$ of $f(x_0)$, the slice $\{x_0\} \times (Y \setminus V)$ has a tube that doesn't intersect G_f.]

7.6 FINITE INTERSECTION PROPERTY (OPTIONAL)

In the introduction to this chapter, we saw that compactness is a sort of "finiteness" property that certain spaces possess. In subsets of $\mathbb{R}^n_{\mathcal{U}^n}$, compactness is characterized by being closed and bounded, by the Heine–Borel theorem. In more general spaces, the definition seems a bit harder to deal with: a space is finite if every open cover has a finite subcover.

The Finite Intersection Property is an alternative approach to compactness that makes the idea of finiteness clearer. This concept seems particularly useful in analysis, where it's always nice to be able to check only finitely many things to see if an infinite process will work. The following example will help explain this idea.

EXAMPLE 7.11

Consider the following nested sequence of intervals in $\mathbb{R}$: $\mathcal{O} = \{U_n = (0, \frac{1}{n}) : n = 2, 3, \ldots\}$. Then the intersection of any finite subcollection of $\mathcal{O}$ is nonempty, but $\bigcap \mathcal{O} = \bigcap_{n=2}^{\infty} U_n = \emptyset$.

Contrast Example 7.11 with the following example.

EXAMPLE 7.12

Consider the following nested sequence of intervals in $\mathbb{R}$: $\mathcal{C} = \{C_n = [0, \frac{1}{n}] : n = 2, 3, \ldots\}$. Then the intersection of any finite subcollection of $\mathcal{C}$ is nonempty, and $\bigcap \mathcal{C} = \bigcap_{n=2}^{\infty} C_n = \{0\} \neq \emptyset$.

These examples should help explain where the following definition comes from.

Definition 7.6.1. *A collection $\mathcal{C}$ of subsets of a set X has the Finite Intersection Property if for every finite subcollection $\{C_1, C_2, \ldots, C_k\}$ of $\mathcal{C}$, the intersection $C_1 \cap C_2 \cap \cdots \cap C_k \neq \emptyset$.*

Note that both $\mathcal{O}$ and $\mathcal{C}$ above have the Finite Intersection Property. In fact, any nested sequence of nonempty sets $A_1 \supset A_2 \supset \cdots \supset A_k \supset \cdots$ automatically satisfies the Finite Intersection Property.

Theorem 7.6.1. *A topological space X_τ is compact if and only if for every collection of τ-closed sets $\mathcal{C} \subset \mathcal{P}(X)$ satisfying the Finite Intersection Property, the intersection $\bigcap \mathcal{C} = \bigcap_{C \in \mathcal{C}} C \neq \emptyset$.*

Proof: $\Longrightarrow$) Let X_τ be compact, and let $\mathcal{C}$ be any collection of τ-closed sets satisfying the Finite Intersection Property. Assume that $\bigcap \mathcal{C} = \bigcap_{C \in \mathcal{C}} C = \emptyset$. Consider $\mathcal{O} = \{X \setminus C : C \in \mathcal{C}$. Then each set $X \setminus C \in \mathcal{O}$ is τ-open and, further, $\mathcal{O}$ is an open cover for X: by DeMorgan's law, we see that

$$\bigcup \mathcal{O} = \bigcup_{C \in \mathcal{C}} (X \setminus C) = X \setminus \bigcap_{C \in \mathcal{C}} C = X \setminus \emptyset = X.$$

Since X is compact, the open cover $\mathcal{O}$ must have a finite subcover: $\{X \setminus C_1, X \setminus C_2, \ldots, X \setminus C_k\}$ such that $(X \setminus C_1) \cup (X \setminus C_2) \cup \cdots \cup (X \setminus C_k) = X$, which implies, by DeMorgan, that $C_1 \cap C_2 \cap \cdots \cap C_k = \emptyset$, contradicting the hypothesis that $\mathcal{C}$ satisfies the Finite Intersection Property.

$\Longleftarrow$) The other implication is similar and is left as an exercise. □

Note that compactness is essential to the conclusion of the theorem, as Example 7.13 demonstrates.

■ **EXAMPLE 7.13**

For $n \in \mathbb{N}$, let $G_n = [n, +\infty) \subset \mathbb{R}_\mathcal{U}$. Then $\{G_n\}_{n \in \mathbb{N}}$ satisfies the Finite Intersection Property (since it's nested), yet $\bigcap_{n \in \mathbb{N}} [n, +\infty) = \emptyset$.

Don't panic – $\mathbb{R}_\mathcal{U}$ is not compact.

Theorem 7.6.1 has the following corollary for collections of not-necessarily-closed sets, which is sometimes quite useful.

Corollary 7.6.2. *A space X_τ is compact if and only if for every collection of subsets $\mathcal{A}$ of X satisfying the Finite Intersection Property, the intersection $\bigcap_{A \in \mathcal{A}} Cl(A) \neq \emptyset$.*

he proof is an exercise.

The next corollary of Theorem 7.6.1 is due to Cantor.

Corollary 7.6.3. *(Cantor's Theorem of Deduction) If $\mathcal{C}$ is any nested sequence of bounded and closed subsets of $\mathbb{R}_\mathcal{U}$, then $\bigcap_{C \in \mathcal{C}} C \neq \emptyset$.*

This proof is left as an exercise. Please don't be insulted if it seems too easy for your attention.

Exercises

1. Prove that any nested sequence of sets $A_1 \supset A_2 \supset \cdots \supset A_k \supset \cdots$ satisfies the Finite Intersection Property.

2. Prove the other implication of Theorem 7.6.1: If X_τ is a space where every collection of τ-closed sets $\mathcal{C} \subset \mathcal{P}(X)$ satisfying the Finite Intersection Property has the intersection $\bigcap \mathcal{C} = \bigcap_{C \in \mathcal{C}} C \neq \emptyset$, then X_τ is compact.

3. Which of the following collections of subsets of the set $\mathbb{R}$ satisfy the Finite Intersection Property?

   - $A_n = (-n, n)$, $n \in \mathbb{N}$
   - $B_n = (n, n+2)$, $n \in \mathbb{N}$
   - $C_n = (0, n)$, $n \in \mathbb{N}$
   - $D_n = (n, n^2)$, $n \in \mathbb{N}$
   - $E_n = (-\infty, n]$, $n \in \mathbb{N}$

4. Prove Cantor's Theorem of Deduction (Corollary 7.6.3).

CHAPTER 8

SEPARATION AXIOMS

8.1 INTRODUCTION

The topological properties we have studied so far, connectedness and compactness, can be regarded as arising from the study of analysis. On the other hand, the one separation axiom we have studied so far, the Hausdorff axiom, deals with ideas of closeness of points that are far removed from those of calculus. In this chapter, we'll look at a number of separation axioms that fit nicely into a hierarchy, with the Hausdorff axiom roughly in the middle of this organizational principle.

8.2 DEFINITION AND EXAMPLES

Definition 8.2.1. *A topological space $X_\mathcal{T}$ is said to be*

- T_0 *if for every pair of points* $x, y \in X$ *with* $x \neq y$*, there exists a* τ*-neighborhood* U_x *of* x *such that* $y \notin U_x$*, or there exists a* τ*-neighborhood* V_y *of* y *with* $x \notin V_y$.

- T_1 *if for every pair of points* $x, y \in X$ *with* $x \neq y$*, there exist* τ*-neighborhoods* U_x *of* x *and* V_y *of* y *with* $y \notin U_x$ *and* $x \notin V_y$.

- T_2 *(or Hausdorff) if for every pair of points* $x, y \in X$ *with* $x \neq y$*, there exist* τ*-neighborhoods* U_x *of* x *and* V_y *of* y *with* $U_x \cap V_y = \emptyset$.[1]

The following diagrams may help make this definition clearer.

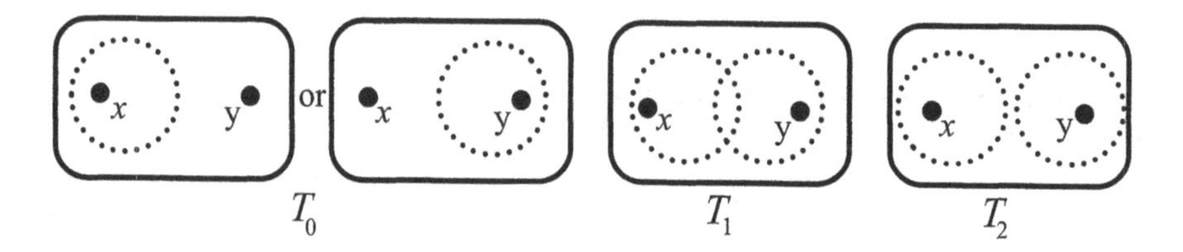

The following theorem is easy to prove.

Theorem 8.2.1. *Any* T_1 *space is also* T_0 *and any* T_2 *space is also* T_1.

So we see that we have the following inclusions of sets:

$$(\text{ topological spaces }) \supset (T_0\text{-spaces}) \supset (T_1\text{-spaces}) \supset (T_2\text{-spaces})$$

We can see that these inclusions are strict by the following examples.

■ **EXAMPLE 8.1**

Any set X with $Card(X) \geq 2$, given the indiscrete topology $\mathcal{I}$, is not T_0, since there are *no* nonempty open sets that miss *any* points of X.

[1] Felix Hausdorff's original definition of topological spaces (in 1914) required the property now known as T_2 or by his name. The property referred to as T_0 above was introduced by A. Kolmolgorov in 1907, the same year Frechet first discussed the property we now call T_1. These properties were given the T_i notation (for *Trennungsaxiomen*) in 1923 by H. Tietze.

EXAMPLE 8.2

The set $\mathbb{R}$ in the right ray topology $\mathcal{RR}$ is T_0 but not T_1: given any $x \neq y$ in $\mathbb{R}$, with $x < y$, we see that for $0 < y - x < \epsilon$, the set $(y - \epsilon, +\infty)$ is an $\mathcal{RR}$-open neighborhood of y not containing x. There is no $\mathcal{RR}$-neighborhood of x that doesn't also contain y.

EXAMPLE 8.3

The set $\mathbb{R}$ in the finite complement topology $\mathcal{FC}$ is T_1 but not T_2; given any $x \neq y$ in $\mathbb{R}$, we see that the set $\mathbb{R} \setminus \{x\}$ is a $\mathcal{FC}$-open neighborhood of y not containing x. Similarly, the set $\mathbb{R} \setminus \{y\}$ is a $\mathcal{FC}$-open neighborhood of x not containing y. However, $\mathbb{R}_{\mathcal{FC}}$ is not T_2, since there are no disjoint nonempty $\mathcal{FC}$-open sets.

One can find interesting examples showing the strict inclusions of the sets of T_i-spaces even among finite spaces, as long as $i \leq 2$, as we'll see in Section 8.4. We recall that $\mathbb{R}_{\mathcal{U}}$ is T_2 (and hence T_1 and T_0) and it's easy to see that any finer topology on $\mathbb{R}$ would yield a space with the same properties.

Theorem 8.2.2. *The T_i-property is a topological property for $i = 0, 1$ and 2.*

Proof: We'll prove this for T_1, leaving the others as an exercise. Let X_τ be T_1 and let $f : X_\tau \to Y_\nu$ be a homeomorphism. If $Card(Y) = Card(X) < 2$, we're done, since any single-point space is vacuously T_1. If not, let $y_1 \neq y_2$ in Y. Since f is a bijection, there exist distinct points $x_1, x_2 \in X$ with $f(x_1) = y_1$ and $f(x_2) = y_2$. Since X is T_1, there exist τ-neighborhoods U_1 of x_1 and U_2 of x_2 with $x_1 \notin U_2$ and $x_2 \notin U_1$. Then $f(U_1)$ and $f(U_2)$ are ν-open, since f is an open map, and $y_1 = f(x_1) \notin f(U_2)$ and $y_2 = f(x_2) \notin f(U_1)$, as we wished. □

One might reasonably ask, then, whether any T_i property a strong topological property; that is, is the continuous image of a T_i-space always T_i? When we think about this, it's clear that a space X_τ needs lots of open sets to be T_2, say, and fewer to be T_0, so this seems unlikely. In fact, any discrete space $X_{\mathcal{D}}$ is T_i for all t (even for $i > 2$, as we'll see in the next section) but for $Card(X) \geq 2$, the indiscrete space $X_{\mathcal{I}}$ fails even to be T_0, so the identity map $i_X : X_{\mathcal{D}} \to X_{\mathcal{I}}$ furnishes the expected counterexample.

One might hope that a quotient space of a T_i-space would be T_i. The following example sheds some light on this.

EXAMPLE 8.4

Recall the relation on the real line $\mathbb{R}_{\mathcal{U}}$: let $x \sim \frac{1}{2}$ for all $x \in (0, 1)$ (with each $x \in \mathbb{R}$ equivalent to itself, of course). Then we have the following depiction of $\mathbb{R}_{/\sim}$:

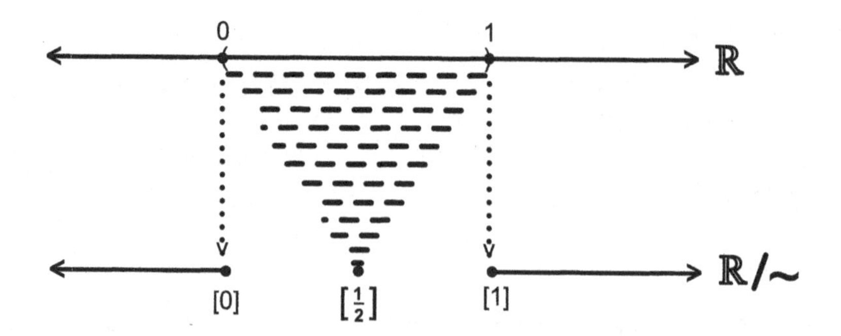

Then $(\mathbb{R}_{/\sim})_{\mathcal{U}_{/\sim}}$ is not T_1, as one can easily see by noting that there is no open neighborhood of [0] in $(\mathbb{R}_{/\sim})_{\mathcal{U}_{/\sim}}$ that does not contain $[\frac{1}{2}]$. So a quotient of a Hausdorff space need not be Hausdorff (or even T_1).

The following theorem gives a very useful characterization of T_1-spaces.

Theorem 8.2.3. *A space X_τ is T_1 if and only if every singleton subset of X is τ-closed.*

Note that this gives a very easy proof that $\mathbb{R}_{\mathcal{RR}}$ is not T_1.
Proof: $\Longrightarrow$) Let X_τ be a T_1-space and let $x \in X$. If $X \setminus \{x\}$ is nonempty, let $y \in X \setminus \{x\}$. Since X is T_1, there exists a τ-neighborhood U_y of y such that $x \notin U_y$. Hence $y \in U_y \subset X \setminus \{x\}$. Thus $X \setminus \{x\}$ contains an open neighborhood of each of its points, so that $X \setminus \{x\}$ is τ-open.
$\Longleftarrow$) Assume that all singleton subsets of X_τ are closed. Let x and y be distinct points in X. [If $Card(X) = 1$, X_τ is T_1 by default.] Since $\{x\}$ is closed, $X \setminus \{x\}$ is a τ-open set, which happens to contain y and miss x, as we wish. Similarly, $X \setminus \{y\}$ is a τ-neighborhood of x that doesn't contain y. □

These separation properties are preserved by subspaces and finite products (although this will not be the case for some more complicated separation axioms, as well see in Section 8.3).

Theorem 8.2.4. *For $i = 0, 1$ and 2,*

- *A subspace of a T_i-space is T_i.*
- *A finite product of T_i-spaces is T_i.*

We remark that the theorem is also true for infinite products of T_i spaces. Further, it's easy to show that $X_\tau \times Y_\nu$ is T_i if and only if X_τ and Y_i are both T_i, for $i = 0, 1$ and 2.

Proof: We'll prove both results for T_2 spaces, leaving the results for T_0- and T_1-spaces as exercises. Let $A \subset X$, where X_τ is Hausdorff. If $a_1, a_2 \in A$, with $a_1 \neq a_2$, then a_1 and a_2 are distinct points of X, as well. Since X_τ is Hausdorff, there exist τ-neighborhoods U_{a_1} and U_{a_2} of a_1 and a_2, respectively, with $U_{a_1} \cap U_{a_2} = \emptyset$. Then $(A \cap U_{a_1})$ and $(A \cap U_{a_2})$ are disjoint τ_A-open neighborhoods of a_1 and a_2, respectively, as we wished.

Let X_τ and Y_ν be T_2-spaces. If (x_1, y_1) and (x_2, y_2) are distinct points in $X \times Y$, then $x_1 \neq x_2$ or $y_1 \neq y_2$ (but not necessarily both). If $x_1 \neq x_2$, since X_τ is Hausdorff, we can separate the points in X by disjoint τ-neighborhoods U_{x_1} and U_{x_2}, with $x_1 \in U_{x_1}$ and $x_2 \in U_{x_2}$. Then $(U_{x_1} \times Y)$ and $(U_{x_2} \times Y)$ are disjoint sets which are open in the product space $X_\tau \times Y_\nu$, with $(x_1, y_1) \in (U_{x_1} \times Y)$ and $(x_2, y_2) \in (U_{x_2} \times Y)$, as the following diagram shows.

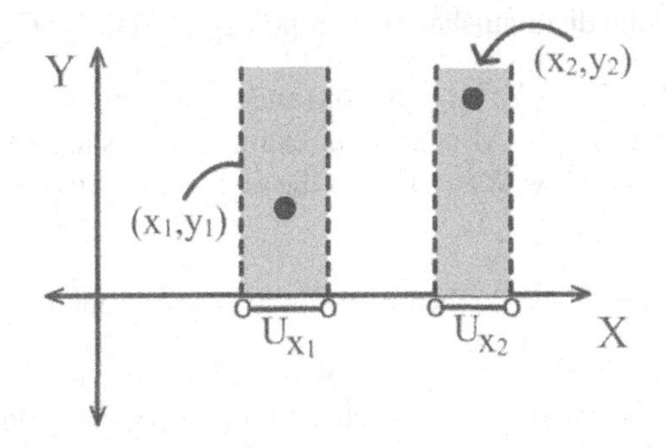

If $x_1 = x_2$, then $y_1 \neq y_2$, so we can separate these points by disjoint neighborhoods in Y_ν. Take the Cartesian product of X with these neighborhoods to separate (x_1, y_1) and (x_2, y_2) in $X \times Y$. (This yields "horizontal stripes", rather than "vertical stripes," as in the diagram above.) □

We close this section with a theorem and corollary closely related to the Closed Graph Theorem (see the exercises of Section 7.3).

Theorem 8.2.5. *Let $f, g : X_\tau \to Y_\nu$ be two continuous functions. If Y_ν is Hausdorff, then the set $A = \{x \in X : f(x) = g(x)\}$ is τ-closed.*

Proof: Let $x \in X \setminus A$, so that $f(x) \neq g(x)$ in the T_2-space Y_ν. So we can separate $f(x)$ from $g(x)$ by disjoint ν-open neighborhoods $U_{f(x)}$ and $U_{g(x)}$. Since f and g are both continuous, the sets $f^{-1}(U_{f(x)})$ and $g^{-1}(U_{g(x)})$ are both τ-open, and both contain x, so that $f^{-1}(U_{f(x)}) \cap g^{-1}(U_{g(x)})$ is a τ-neighborhood of x. We claim that $f^{-1}(U_{f(x)}) \cap g^{-1}(U_{g(x)}) \subset X \setminus A$: if $z \in f^{-1}(U_{f(x)}) \cap g^{-1}(U_{g(x)})$, then $f(z) \in U_{f(x)}$ and $g(z) \in U_{g(x)}$, which are disjoint in Y, so that $f(z) \neq g(z)$, implying $z \notin A$. So $X \setminus A$ contains a τ-neighborhood of each of its points, so that A is closed. □

The following corollary is an obvious consequence of the theorem.

Corollary 8.2.6. *Let $f, g : X_\tau \to Y_\nu$ be two continuous functions. If Y_ν is Hausdorff, and if $f(x) = g(x)$ on a dense subset of X_τ, then $f = g$.*

Exercises

1. Prove that any T_2 space is also T_1 and any T_1 space is T_0.
2. Let $X = \{a, b, c, d\}$. Find a topology τ on X so that X_τ is not T_0. So that X_τ is T_0 but not T_1. So that X_τ is T_1 but not T_2.
3. Prove that the T_i-property is a topological property for $i = 0$ and 2.
4. Consider $\mathbb{R}$ in the distinguished point topology τ_1. Is $\mathbb{R}_{\tau_1}$ T_0? T_1? T_2?
5. Prove that if $f : X_\tau \to Y_\nu$ is continuous and one-to-one, and if Y_ν is Hausdorff, then X_τ is Hausdorff. Also, give an example that shows that the injective hypothesis is necessary. Why isn't this the same as showing that the continuous image of a T_2-space is T_2?
6. If X is finite and X_τ is T_1, then $\tau = \mathcal{D}$, the discrete topology.
7. Prove that if X_τ is T_1 and $A \subset X$, and if $x \in A'$ (i.e., x is a limit point of A) then any neighborhood of x intersects A in infinitely many points.
8. Prove that if X_τ is T_1 and $A \subset X$, then A' is τ-closed.
9. Prove Theorem 8.2.4 for T_0 and T_1-spaces.
10. Prove that if $X_\tau \times Y_\nu$ is T_2 , then both X_τ and Y_ν must be T_2 (the converse of Theorem 8.2.4).

8.3 REGULAR AND NORMAL SPACES

We extend the separation axioms further, determining whether a space has sufficient open sets to separate closed sets from points or from other closed sets.

Definition 8.3.1. *A space X_τ is said to be regular if for every τ-closed set $A \subset X$ and for every $x \notin A$, there exist τ-open sets U_x and V_A with $x \in U_x$, $A \subset V_A$ and $U_x \cap V_A = \emptyset$. A space that is both regular and T_1 is called a T_3-space.*[2]

So a space is regular if one can separate any closed set from any point not in it by open sets. We can see why we require the addition of the T_1 property to be called T_3: we recall that T_1-spaces are exactly those with all singleton subsets closed (Theorem 8.2.3). So if a space X is regular and T_1, then we can separate any singleton subset from any point not in it – i.e. we can separate any two distinct points. So we see that T_3 implies T_2, as we would hope.

The following examples will help clarify the definition.

EXAMPLE 8.5

Let $X = \{a, b, c, d\}$ with topology $\tau = \{\emptyset, X, \{a, b\}, \{c, d\}\}$. Then X_τ is regular, since the only τ-closed sets are $\{a, b\}$ and $\{c, d\}$. For example, since $a \notin K = \{c, d\}$, we must be able to separate the point a from the closed set K: $a \in \{a, b\} = U_a$ and $\{c, d\} = V_K$ are both open, providing the desired separation. Note that X_τ is *not* T_1, since no singleton subsets are closed. So X_τ is regular but not T_3.

EXAMPLE 8.6

Consider the set $\mathbb{R}$ with the distinguished point topology $\tau_1 = \{U \subset \mathbb{R} : 1 \in U \text{ or } U = \emptyset\}$. Then $\mathbb{R}_{\tau_1}$ is not regular, since the set $\{0\}$ is τ_1-closed and cannot be separated from the point 1; there is no open set containing $\{0\}$ that doesn't contain 1.

EXAMPLE 8.7

Any discrete space $X_\mathcal{D}$ is regular and T_1, and hence T_3.

EXAMPLE 8.8

Any indiscrete space $X_\mathcal{I}$ is regular, by default, since there are no proper nonempty closed sets.

[2]The property of regularity was first considered by Leopold Vietoris in 1921 (when he was 30 years old). Vietoris lived to the astounding age of 110 years old, still active as a mathematician and as an athlete, having won Alpine ski titles into his late 90s. His last mathematical paper was published in 1994. (See Ref. [15] for more details.)

■ **EXAMPLE 8.9**

The real line $\mathbb{R}_{\mathcal{U}}$ is regular (and hence T_3): let C be any $\mathcal{U}$-closed subset of $\mathbb{R}$, and let $x \notin C$. Since $\mathbb{R} \setminus C$ is $\mathcal{U}$-open, there exists an $\epsilon > 0$ with $(x - \epsilon, x + \epsilon) \subset \mathbb{R} \setminus C$. So $\mathbb{R} \setminus [x - \frac{\epsilon}{2}, x + \frac{\epsilon}{2}]$ is a $\mathcal{U}$-open set containing C, disjoint from $(x - \frac{\epsilon}{2}, x + \frac{\epsilon}{2})$, a $\mathcal{U}$-open set containing x, as we wished.

We have one more level of separation to think about.

Definition 8.3.2. *A space X_τ is said to be normal if for every pair of disjoint τ-closed sets $C_1, C_2 \subset X$, there exist τ-open sets U_1 and U_2 with $C_1 \subset U_1$, $C_2 \subset U_2$ and $U_1 \cap U_2 = \emptyset$. A space that is both normal and T_1 is called a T_4-space.*[3]

So, for a space X_τ to be normal, we need to be able to separate any disjoint closed sets. It's not at all clear that normal spaces need be regular, since singleton sets need not be closed in general. However, when we add in the T_1 axiom, it's easy to see that T_4-spaces are always T_3. The following examples should make this clear.

■ **EXAMPLE 8.10**

The set $\mathbb{R}$ in the right ray topology $\mathcal{RR}$ is normal, also by default, since there are no disjoint closed sets. Recall that $\mathbb{R}_{\mathcal{RR}}$ is not regular, so we have a counterexample to any claim that normality implies regularity.

■ **EXAMPLE 8.11**

Any discrete space $X_{\mathcal{D}}$ is normal and T_1, and hence T_4, since any $\mathcal{D}$-closed sets are also $\mathcal{D}$-open (and hence separate themselves).

■ **EXAMPLE 8.12**

Any indiscrete space $X_{\mathcal{I}}$ is normal, by default, since there are no proper nonempty closed sets. Of course, $X_{\mathcal{I}}$ is not T_4.

■ **EXAMPLE 8.13**

The real line $\mathbb{R}_{\mathcal{U}}$ is normal. Let C_1 and C_2 be any disjoint nonempty $\mathcal{U}$-closed sets. (If either are empty, we can separate easily.) For each $x \in C_1$, $x \notin C_2$, so we can separate them by disjoint open sets, say, U_x and V_x, since $\mathbb{R}_{\mathcal{U}}$ is

[3] The concept of normality was also introduced by Vietoris in 1921.

regular by Example 8.9, with $x \in U_x$ and $C_2 \subset V_x$. So there exists an $\epsilon_x > 0$ such that $(x - \epsilon_x, x + \epsilon_x) \cap C_2 = \emptyset$. Similarly, for each $y \in C_2$, there exists $\epsilon_y > 0$ such that $(y - \epsilon_y, y + \epsilon_y) \cap C_1 = \emptyset$. We'll use these to get the open sets around all of C_1 and C_2 as follows. Let $U_1 = \bigcup_{x \in C_1}(x - \frac{\epsilon_x}{2}, x + \frac{\epsilon_x}{2})$ and let $U_2 = \bigcup_{y \in C_2}(y - \frac{\epsilon_x y}{2}, y + \frac{\epsilon_y}{2})$. Then $C_1 \subset U_1$ and $C_2 \subset U_2$. Further, $U_1 \cap U_2 = \emptyset$, since any point z common to U_1 and U_2 would be simultaneously within $\frac{\epsilon_x}{2}$ of a point $x \in C_1$ and within $\frac{\epsilon_y}{2}$ of some $y \in C_2$, so that

$$|x - y| \leq |x - z| + |y - z| < \frac{\epsilon_x}{2} + \frac{\epsilon_y}{2} < \epsilon = min\{\epsilon_x, \epsilon_y\},$$

violating the definitions of ϵ_x and ϵ_y above. Note that $\mathbb{R}_\mathcal{U}$ is hence T_4.

We have the following inclusions:

(topological spaces) $\supset$ (T_0-spaces) $\supset$ (T_1-sp) $\supset$ (T_2-sp) $\supset$ (T_3-sp) $\supset$ (T_4-sp)

We've already seen that the first three inclusions are proper, by Examples 8.1–8.3. To see that the other two inclusions are proper, consider the following examples.

EXAMPLE 8.14

Let $K = \{\frac{1}{n} : n \in \mathbb{N}\} \subset \mathbb{R}$. We define a topology σ on $\mathbb{R}$ by taking as a basis all sets of these two forms :

- Every open interval (a, b)
- Every set $(a, b) \setminus K$

This defines a topology σ that is finer that $\mathcal{U}$, since the basis contains all open intervals, so we see that $\mathbb{R}_\sigma$ is T_2. However, $\mathbb{R}_\sigma$ is not regular; the set K is σ-closed, and $0 \notin K$, but 0 cannot be separated from K. If one chooses a basis set around 0 that misses K, it has to be of the form $(a, b) \setminus K$. However, any open set containing K automatically contains 0, since it has to contain a basis set (of the usual open interval form) around $\frac{1}{n}$ for all $n \in \mathbb{N}$. So $\mathbb{R}_\sigma$ is T_2 but not T_3.

One might be tempted to construct a topology on a finite set like $X = \{a, b, c, d\}$, so that the resulting space is T_2 but not T_3 or T_3 but not T_4, but such an enterprise can't succeed, as Theorem 8.4.1 will show. An example of a space that is T_3 but not T_4 will follow shortly, once we've proved a few necessary theorems. First, we note that regularity is hereditary:

Theorem 8.3.1. *Any subspace of a regular space is regular.*

The proof of this result is straightforward and left as an exercise.

Surprisingly, the analogous result is not true for normal spaces. However, if we require the subset to be closed, we get normality for subspaces:

Theorem 8.3.2. *If X_τ is normal and A is a τ-closed subset of X, then A_{τ_A} is normal.*

Proof: Let C_1 and C_2 be disjoint τ_A-closed subsets of A. Then there exist τ-closed sets K_1 and K_2 with $C_1 = A \cap K_1$ and $C_2 = A \cap K_2$. At this point, one is tempted to separate K_1 from K_2 in X and intersect the open sets down to A. However, there's no way to know that K_1 and K_2 are disjoint in X. All is not lost, however, since $C_1 = A \cap K_1$ and $C_2 = A \cap K_2$ are both the intersection of τ-closed sets (remember that A is closed here!), so that C_1 and C_2 are τ-closed, not just τ_A-closed. So we can separate C_1 and C_2 in X by disjoint τ-open sets U_1 and U_2. Intersecting these with A yields the separation of C_1 and C_2 by τ_A-open sets. □

Most counterexamples to the extension of Theorem 8.3.2 to non-closed subsets are quite subtle. (See Ref. [13] for an accessible explanation of such an example.)

Theorem 8.3.3. *A finite product of T_3-spaces is T_3.*

In fact, even an infinite product of T_3-spaces is T_3, and the proof used here carries over nicely to that case. To prove this, we use the following characterization of the T_3 property.

Lemma 8.3.4. *A space X_τ is T_3 if and only if X is T_1 and for every $x \in X$ and every τ-neighborhood U_x of x, there exists another τ-neighborhood V_x of x with $Cl(V_x) \subset U_x$.*

Proof of Lemma: $\Longrightarrow$) Suppose X_τ is a T_3-space, with a given point x and neighborhood U_x. Then $C = X \setminus U_x$ is a τ-closed set not containing x, so we can separate them: $C \subset V_C$ and $x \in V_x$ for some open sets V_C and V_x, with $V_C \cap V_x = \emptyset$. Then $Cl(V_x)$ must be disjoint from C, since any point $c \in C$ has V_C as a neighborhood, disjoint from V_x, so no such c can be a limit point of V_x. Thus, $Cl(V_x) \subset U$.
$\Longleftarrow$) Suppose that for every $x \in X$ and every τ-neighborhood U_x of x, there exists another τ-neighborhood V_x of x with $Cl(V_x) \subset U_x$. Let C be any closed set, with $x \notin C$. Then $U = X \setminus C$ is a τ-open neighborhood of x, so there exists V_x, a neighborhood of x, with $x \in Cl(V_x) \subset U$. Then the open sets V_x and $X \setminus Cl(V_x)$ form the desired separation of x from C. □

Proof of Theorem 8.3.3: We'll prove this for a product of two T_3 spaces $X_\tau \times Y_\nu$, then appeal to induction for all finite products. We already know that $X \times Y$ is Hausdorff, by Theorem 8.2.4, so that singleton subsets of $X \times Y$ are closed in the product topology. Let $(x_0, y_0) \in X \times Y$, with N any neighborhood of (x_0, y_0). Then N contains a basis set around (x_0, y_0), say, $U \times W$. Since both X and Y are T_3, by Lemma 8.3.4, there exist V_{x_0} and V_{y_0}, with V_{x_0} a τ-neighborhood of x_0 such that $Cl(V_{x_0}) \subset U$ and with V_{y_0} a ν-neighborhood of y_0 such that $Cl(V_{y_0}) \subset W$. Then $V_{x_0} \times V_{y_0}$ is a neighborhood of (x_0, y_0), open in the product topology, with $Cl(V_{x_0} \times V_{y_0}) = Cl(V_{x_0}) \times Cl(V_{y_0}) \subset U \times W \subset N$. Lemma 8.3.4 tells us, then, that $X_\tau \times Y_\nu$ is T_3. □

Perhaps surprisingly, the product of T_4 spaces need not be T_4, as the following example shows.

■ EXAMPLE 8.15

$\mathbb{R}_\mathcal{L} \times \mathbb{R}_\mathcal{L}$ is not T_4, despite the fact that $\mathbb{R}_\mathcal{L}$ is T_4. First, we sketch a proof that $\mathbb{R}_\mathcal{L}$ is T_4: it's clearly T_1, since singletons are closed in $\mathbb{R}_\mathcal{U}$ and $\mathcal{L}$ is finer than $\mathcal{U}$. Let A and B be disjoint closed subsets of $\mathbb{R}_\mathcal{L}$. For each $a \in A$, chose a basis set $[a, a + \epsilon_a)$ that doesn't intersect B, and do the same for each point $b \in B$. Then $A \subset \bigcup_{a \in A}[a, a + \epsilon_a)$ and $B \subset \bigcup_{b \in B}[b, b + \epsilon_b)$ demonstrate the separation of A and B by disjoint $\mathcal{L}$-open sets. To see why the product is not T_4, consider the negative diagonal

$$-\Delta = \{(x, -x) : x \in \mathbb{R}\} \subset \mathbb{R}_\mathcal{L} \times \mathbb{R}_\mathcal{L}.$$

Then $-\Delta$ is closed in $\mathbb{R}_\mathcal{L} \times \mathbb{R}_\mathcal{L}$,and, further, $-\Delta$ inherits the discrete topology as a subspace of $\mathbb{R}_\mathcal{L} \times \mathbb{R}_\mathcal{L}$ as one can see from the following illustration:

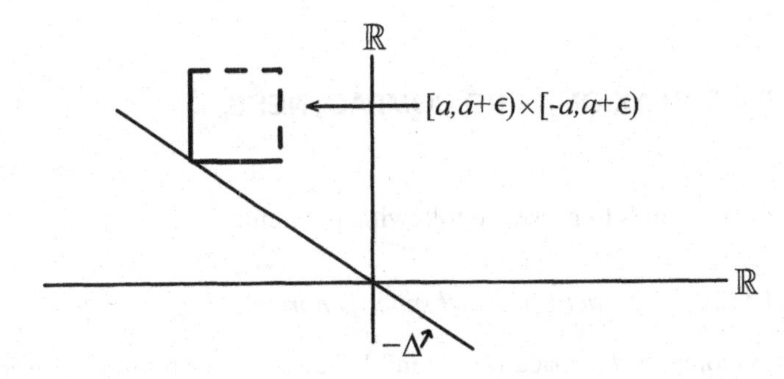

Since all singleton subsets of $-\Delta$ are open in the subspace topology, any subset of $-\Delta$ is open (and closed) as a subspace of $-\Delta$. Let

$$C_1 = \{(x, -x) : x \in \mathbb{Q}\} \subset -\Delta$$

and

$$C_2 = \{(x, -x) : x \notin \mathbb{Q}\} \subset -\Delta.$$

Both of these are closed in $-\Delta$ (and hence in $\mathbb{R}_\mathcal{L} \times \mathbb{R}_\mathcal{L}$, since $-\Delta$ is itself closed in $\mathbb{R}_\mathcal{L} \times \mathbb{R}_\mathcal{L}$). But one cannot separate C_1 and C_2 by means of sets open in $\mathbb{R}_\mathcal{L} \times \mathbb{R}_\mathcal{L}$: it's intuitively clear that $\mathcal{L} \times \mathcal{L}$-basis sets around the rational points in $-\Delta$ must intersect basis sets around the irrational points in $-\Delta$. To make this precise requires a delicate cardinality argument, but the idea is pretty clear. (See Ref. [13] for full details.)

Exercises

1. Is $\mathbb{R}_{\mathcal{RR}}$ T_1? Regular? Normal?
2. Is $\mathbb{R}$ with the distinguished point topology $\tau_1 = \{U \subset \mathbb{R} : 1 \in U \text{ or } U = \emptyset\}$ normal?
3. Show that the collection of sets in Example 8.14 is, indeed, a basis for $\mathbb{R}$.
4. Prove Theorem 8.3.1 Any subspace of a regular space is regular.
5. Prove that $\mathbb{R}_{\mathcal{R}}$ is T_4, where $\mathcal{R}$ is the "right-hand" topology on $\mathbb{R}$.
6. Prove the following useful characterization of the separation axiom T_3: if a space X_τ is T_3, then every pair of distinct points in X can be separated by neighborhoods whose closures are disjoint.

8.4 SEPARATION AXIOMS AND COMPACTNESS

Our goal in this section is to prove the following theorem:

Theorem 8.4.1. *Any compact Hausdorff space is normal.*

Thus, every compact T_2 space is T_4 (and hence T_3). So those of you looking for examples of finite spaces that are T_2 but not T_3 or T_3 but not T_4 are doomed to failure, since every finite space is compact. To prove this theorem, we'll first prove the following result:

Theorem 8.4.2. *Any compact Hausdorff space is regular (and hence T_3).*

Proof: The idea of the proof is hauntingly familiar (see the proof of Theorem 7.3.1). Let X_τ be compact and Hausdorff. Let C be any τ-closed subset of X, with $x \notin C$. For each $c \in C$, $c \neq x$, so there exist disjoint τ-neighborhoods U_c of c and V_c of x. Then $\mathcal{C} = \{U_c : c \in C\}$ is a cover of C by τ-open sets. Since C is a closed subset of a compact space, $C_{\tau C}$ is compact, so there exists a finite subcollection $\{U_{c_1}, U_{c_2}, \ldots, U_{c_k}\} \subset \mathcal{C}$ such that $C \subset U_{c_1} \cup U_{c_2} \cup \cdots \cup U_{c_k}$. For each c_i, there exists a neighborhood V_{c_i} around x, such that V_{c_i} is disjoint from U_{c_i}. Then $U = U_{c_1} \cup U_{c_2} \cup \cdots \cup U_{c_k}$ and $V = V_{c_1} \cap V_{c_2} \cap \cdots \cap V_{c_k}$ are the desired τ-open sets around C and x, respectively, with $U \cap V = \emptyset$. Note that we needed the *finite* subcollection of τ-open sets of $\mathcal{C}$ precisely because their intersection is still τ-open (as opposed to an infinite intersection). □

Now we can prove Theorem 8.4.1. The proof is quite similar that for Theorem 8.4.2.

Proof of Theorem 8.4.1: Let C_1 and C_2 be disjoint τ-closed subsets of a compact Hausdorff space X_τ. If either C_i is empty, the separation is obvious (choose the open set to be $\emptyset$). If both are nonempty, let $x \in C_1$, so that $x \notin C_2$. Since X_τ is regular by Theorem 8.4.2, we can separate x from C_2 by disjoint open sets, with U_x containing x and V_x containing C_2. Since we can do this for every $x \in C_1$, the collection $\mathcal{C} = \{U_x : x \in C_1\}$ is an open cover of C_1. Because C_1 is compact (as a closed subset of a compact space), there exists a finite subcollection $\{U_{c_1}, U_{c_2}, \ldots, U_{c_k}\} \subset \mathcal{C}$ such that $C_1 \subset U_{c_1} \cup U_{c_2} \cup \cdots \cup U_{c_k}$. For each c_i, there exists a τ-open set V_{c_i} around C_2, such that V_{c_i} is disjoint from U_{c_i}. Then $U = U_{c_1} \cup U_{c_2} \cup \cdots \cup U_{c_k}$ and $V = V_{c_1} \cap V_{c_2} \cap \cdots \cap V_{c_k}$ are the desired τ-open sets around C_1 and C_2, respectively, with $U \cap V = \emptyset$. □

Exercises

1. You know that compact and Hausdorff implies normal. Is the converse true?

2. Let X_τ be compact and Hausdorff. Show that if τ' is any strictly coarser topology on X, then $X_{\tau'}$ is not Hausdorff. Show that if τ'' is any topology on X that is strictly finer than τ, then $X_{\tau''}$ is not compact. [*Hint*: Consider the identity function on X. What topologies do you need on the domain and the range to make i_X continuous? What do you know about how compactness and Hausdorffness interact? (Yes, this was a theorem earlier, proved in an earlier exercise. If you haven't done it yet, do it now, please.)]

CHAPTER 9

METRIC SPACES

9.1 INTRODUCTION

Metric spaces are in many ways the most intuitive of topological spaces, in that metric spaces possess an actual function that measures distances numerically. For example, in the usual (or Euclidean) real line, the distance from x to y is $|x - y|$. In the usual Euclidean $\mathbb{R}^n$, the distance from $(x_1, x_2, \ldots, x_n)$ to $(y_1, y_2, \ldots, y_n)$ is $\sqrt{(x_1 - y_1)^2 + \cdots + (x_n - y_n)^2}$. These measures of distance provide a convenient approach to topology, by using a basis consisting of the open ϵ "balls," which consist of all points of distance less than ϵ from some center. Such topological spaces, called *metric spaces*, form the appropriate setting for much of modern analysis, as $\mathbb{R}_{\mathcal{U}}$ and $\mathbb{R}^n_{\mathcal{U}^n}$ were the setting for classical analysis or calculus.

We begin by defining what is meant by a *metric* on a set X and examining the topology it defines on X. In Section 9.3, we'll look more closely at the properties that all metric spaces possess, and we'll consider the question of which topological spaces can have metrics compatible with their topologies.

9.2 DEFINITION AND EXAMPLES

Definition 9.2.1. *A metric on a set X is a function $d : X \times X \to \mathbb{R}$ satisfying the following properties:*

1. $d(x,y) \geq 0$ *for all* $x, y \in X$.

2. $d(x,y) = d(y,x)$ *for all* $x, y \in X$ *(i.e., d is a symmetric function.)*

3. $d(x,y) \leq d(x,z) + d(z,y)$ *for all elements x, y and z of X (triangle inequality).*

4. $d(x,x) = 0$ *for all* $x \in X$.

5. *If* $d(x,y) = 0$ *then* $x = y$.

The pair (X, d) is called a metric space.

We leave the property that $d(x,y) = 0$ implies $x = y$ for last, since it's sometimes useful to consider functions d that satisfy only the first four properties. Such functions are usually called *pseudometrics* on X. The following examples illustrate Definition 9.2.1.

■ **EXAMPLE 9.1**

The function $d : \mathbb{R} \times \mathbb{R} \to \mathbb{R}$ given by $d(x,y) = |x - y|$ is a metric on $\mathbb{R}$. This metric is known as the *usual* or *Euclidean metric* metric on $\mathbb{R}$.

■ **EXAMPLE 9.2**

The function $d : \mathbb{R}^n \times \mathbb{R}^n \to \mathbb{R}$ given by

$$d_1\left((x_1, x_2, \ldots, x_n), (y_1, y_2, \ldots, y_n)\right) = \sqrt{(x_1 - y_1)^2 + \cdots + (x_n - y_n)^2}$$

is a metric on $\mathbb{R}^n$, known as the *usual* or *Euclidean metric* on $\mathbb{R}^n$.

The following examples are less familiar.

EXAMPLE 9.3

Let X be any set. The function $d : X \times X \to \mathbb{R}$ defined by

$$d(x,y) = \begin{cases} 0 & \text{if } x = y, \\ 1 & \text{if } x \neq y \end{cases}$$

is a metric on X, as one can see easily. This d is referred to as the *discrete metric* on X.

The reason why this metric is called "discrete" will be clear shortly. We recall some examples of bases for $\mathbb{R}^2$ in the language of metrics.

EXAMPLE 9.4

The following functions $\mathbb{R}^2 \times \mathbb{R}^2 \to \mathbb{R}$ are metrics on $\mathbb{R}^2$:

$$\begin{aligned} d_1\left((x_1,y_1),(x_2,y_2)\right) &:= \sqrt{(x_2-x_1)^2+(y_2-y_1)^2} \text{ ("Euclidean")} \\ d_2\left((x_1,y_1),(x_2,y_2)\right) &:= |x_2-x_1|+|y_2-y_1| \text{ ("Taxicab")} \\ d_3\left((x_1,y_1),(x_2,y_2)\right) &:= \max\{|x_2-x_1|+|y_2-y_1| \text{ ("square").} \end{aligned}$$

Note that metric d_1 is a special case of Example 9.2. The metrics d_2 and d_3 have obvious analogs in $\mathbb{R}^n$, as well. The function d_2 is called the "taxicab" metric because it measures the distance between (x_1, y_1) and (x_2, y_2) as if one can travel only on the perpendicular lines of an urban grid, rather than along diagonal paths, as the Euclidean metric measures.

A metric on a set X generates a topology using a basis built from the following construction.

Definition 9.2.2. *Given a metric d on a set X and a real number $\epsilon > 0$, we define the open ϵ-ball with center x_0 and radius ϵ as*

$$B^d_\epsilon(x_0) = \{x \in X : d(x,x_0) < \epsilon\}.$$

If the metric d on X is unambiguous, we'll use the notation $B_\epsilon(x_0)$ as a shorthand. Here are some pictures of the ϵ "balls" in $\mathbb{R}^2$ given by the metrics of Example 9.4:

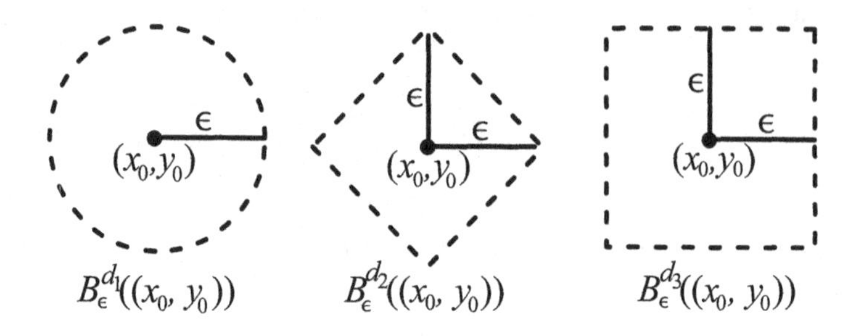

Definition 9.2.3. *Given a metric d on a set X, we define the metric basis on X as the collection*

$$\mathcal{B} = \{B_\epsilon(x_0) : x_0 \in X, \epsilon > 0\}.$$

Theorem 9.2.1. *If d is a metric on X, then the metric basis $\mathcal{B}$ is a basis for X.*

Proof: First, we must show that $\mathcal{B}$ covers X. But any $x \in X$ is guaranteed to be an element of $B_1(x)$, so that $\bigcup \mathcal{B} = \bigcup_{B \in \mathcal{B}} B = X$. Next, we need to show that $\mathcal{B}$ satisfies the intersection condition: If $x \in B_{\epsilon_1}(x_1) \cap B_{\epsilon_2}(x_2)$, then there exists a third basis set $B_{\epsilon_3}(x_3) \in \mathcal{B}$ such that

$$x \in B_{\epsilon_3}(x_3) \subset B_{\epsilon_1}(x_1) \cap B_{\epsilon_2}(x_2).$$

Here, we take

$$\epsilon_3 = min\{\epsilon_1 - d(x, x_1), \epsilon_2 - d(x, x_2)\},$$

so that $\epsilon_3 > 0$ since x is of distance less than ϵ_1 from x_1 and less than ϵ_2 from x_2. Further, we'll center our third ball at x itself, so that $B_{\epsilon_3}(x)$ obviously contains x. To see that $B_{\epsilon_3}(x) \subset B_{\epsilon_1}(x_1) \cap B_{\epsilon_2}(x_2)$, let $z \in B_{\epsilon_3}(x)$, so that $d(x, z) < \epsilon_3 = min\{\epsilon_1 - d(x, x_1), \epsilon_2 - d(x, x_2)\}$. Then the triangle inequality shows that

$$d(x_1, z) < d(x_1, x) + d(x, z) < d(x_1, x) + \epsilon_3 \leq d(x_1, x) + (\epsilon_1 - d(x, x_1)) = \epsilon_1,$$

so that $z \in B_{\epsilon_1}(x_1)$. Similarly, we check that $z \in B_{\epsilon_2}(x_2)$, so that $z \in B_{\epsilon_1}(x_1) \cap B_{\epsilon_2}(x_2)$, as we wished to show. □

The following diagram illustrates the proof.

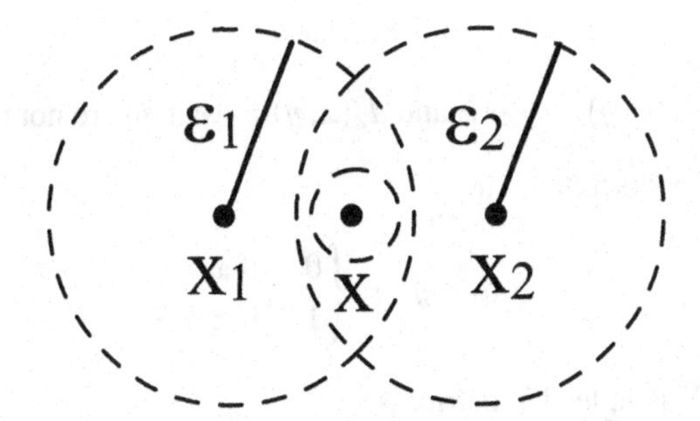

So if a set X has a metric d, the metric basis $\mathcal{B}$ generates a topology:

Definition 9.2.4. *If (X, d) is a metric space, the metric topology on X is given by*

$$\tau_d := \{U \subset X : \text{ if } x \in U, \text{ there exists } \epsilon > 0 \text{ and } x_0 \in X \text{ such that } x \in B_\epsilon(x_0) \subset U\}.$$

If (X, d) is a metric space, we'll often refer to the topological space X_{τ_d} as a *metric space*, as well, although the term X with the metric topology would be more accurate. Note that U is τ_d-open if it contains an open ϵ-ball around each of its points, where we don't specify the center of the ball. The following theorem makes this condition easier to check.

Theorem 9.2.2. *If (X, d) is any metric space, then a set U is τ_d-open if and only if for each $x \in U$, there exists a real number $\epsilon > 0$ such that $B_\epsilon(x) \subset U$.*

So to see if U is τ_d-open. it suffices to check for each $x \in U$ whether some ϵ-ball centered at x is contained in U. The proof is left as an exercise.

The discrete metric d on a set X of Example 9.3 has the following property: if $\epsilon > 1$, then $B_\epsilon(x) = X$, but if $0 < \epsilon < 1$, then $B_\epsilon(x) = \{x\}$. For this reason, singleton subsets of X are τ_d-open, so that $\tau_d = \mathcal{D}$. Hence, the metric is aptly named.

We note that the metric topologies given by the metrics in Examples 9.1 and 9.4 are also familiar: $\tau_d = \mathcal{U}$, where $d(x, y) = |x - y|$ on $\mathbb{R}$, and $\tau_{d_1} = \tau_{d_2} = \tau_{d_3} = \mathcal{U}^2$ for the metrics on $\mathbb{R}^2$, as we saw in Section 4.2. However, the metrics are quite different, even if they generate the same topology. If one wished to work only with metric spaces, rather than with the topology generated by a metric, the appropriate kind of equivalence would be an *isometry*:

Definition 9.2.5. *An isometry of metric spaces (X, d_1) to $Y, d_2)$ is a bijective function $f : X \to Y$ such that $d_2(f(x_1), f(x_2)) = d_1(x_1, x_2)$ for all $x_1, x_2 \in X$.*

It's easy to see that an isometry is always a homeomorphism, but that some homeomorphisms are not isometries.

Exercises

1. Show why $d_1(x, y) = y - x$ and $d_2(x, y) = |x + y|$ are not metrics on $\mathbb{R}$.

2. Prove that the discrete metric

$$d(x,y) = \begin{cases} 0 & \text{if } x = y, \\ 1 & \text{if } x \neq y \end{cases}$$

on any set X is indeed a metric.

3. Prove Theorem 9.2.2: If (X, d) is any metric space, then a set U is τ_d-open if and only if for each $x \in U$, there exists a real number $\epsilon > 0$ such that $B_\epsilon(x) \subset U$. [*Hint*: let $x \in B_\delta(x_0)$. Use a sketch to explain why $B_\epsilon(x) \subset B_\delta(x_0)$ for some $\epsilon > 0$. What's the largest ϵ that will work?]

4. Show that $i_{\mathbb{R}^2} : \mathbb{R}^2_{\tau_{d_1}} \to \mathbb{R}^2_{\tau_{d_2}}$ is a homeomorphism but not an isometry, where d_1 is the Euclidean metric and d_2 is the taxicab metric on $\mathbb{R}^2$, as defined in Example 9.4.

9.3 PROPERTIES OF METRIC SPACES

Theorem 9.3.1. *Every metric space is Hausdorff.*

Proof: Left as an exercise. Please don't be insulted by its simplicity.

One has to work a bit harder to prove the following.

Theorem 9.3.2. *Every metric space is normal.*

Proof: The proof is quite similar to our proofs that $\mathbb{R}_\mathcal{U}$ is regular and normal (Examples 8.9 and 8.13). Let (X, d) be any metric space, and let C_1 and C_2 be disjoint, nonempty τ_d-closed sets in X. (If either were empty, we could separate easily, so we'll ignore this case.) For each $x \in C_1$, $x \notin C_2$, a closed set, so there exists a τ_d-open neighborhood U_x of x such that $U_x \cap C_2 = \emptyset$. Thus there must be a real number $\epsilon_x > 0$ such that $B_{\epsilon_x}(x) \cap C_2 = \emptyset$. Similarly, for each $y \in C_2$, there exists $\epsilon_y > 0$ such that $B_{\epsilon_y}(y) \cap C_1 = \emptyset$. Let $U = \bigcup_{x \in C_1} B_{(\frac{\epsilon_x}{2})}(x)$, a τ_d-open set containing C_1, and let $V = \bigcup_{y \in C_2} B_{(\frac{\epsilon_y}{2})}(x)$, a τ_d-open set containing C_2. Then $U \cap V = \emptyset$, by a triangle inequality argument exactly like that in Example 8.9. □

Clearly, then, every metric space is T_4 and T_3, as well.

The added structure of a metric on a set allows one to do calculus:

Theorem 9.3.3. *Let (X, d_X) and $Y(, d_Y)$ be metric spaces. A function $f : X \to Y$ is $(\tau_{d_X} - \tau_{d_Y})$-continuous if and only if for every $\epsilon > 0$ and for every $x_1 \in X$ there exists $\delta > 0$ such that if $d_X(x, x_1) < \delta$, then $d_Y(f(x), f(x_1)) < \epsilon$.*

The proof is exactly analogous to that of Theorem 4.5.2, and is left as an exercise.

Given that metric spaces seem to act so much like the usual real line $\mathbb{R}_{\mathcal{U}}$, one might expect analogs of the Heine–Borel and Bolzano–Weierstrass Theorems. The property of compactness doesn't work exactly in an arbitrary metric space as it does in the real line, but the following result shows that we get something close.

Theorem 9.3.4. *If A is a compact subspace of a metric space (X, d), then A must be τ_d-closed and bounded [i.e., $A \subset B_N^d(x)$ for some $x \in X$ and some $N > 0$].*

Proof: The proof is straightforward. Since a metric space is Hausdorff, any compact subspace of X_{τ_d} must be closed, by Theorem 7.3.1. Further, A must be bounded, else for any $x \in X$, the collection $\{B_n(x) : n \in \mathbb{N}\}$ consists entirely of open sets and covers A, yet has no finite subcover. □

So we have one direction of a Heine–Borel-type theorem for arbitrary metric spaces. However, the following counterexample shows that this is the best we can do, in general. Recall that for X and set, the discrete metric on X is given by

$$d(x, y) = \begin{cases} 0 & \text{if } x = y, \\ 1 & \text{if } x \neq y. \end{cases}$$

In the case where $X = \mathbb{R}$, then, we see that the subset $A = (0, 1)$ is bounded [since this and any other subset fit in a ball of radius 1.1, because $B_{1.1}(0) = B_{73.4}(2) = \mathbb{R}$]. Further, A is closed, since the topology given by the metric d is $\tau_d = \mathcal{D}$, the discrete topology. But A is not compact, since the $\mathcal{D}$-open cover consisting of all the singleton subsets of A has no finite subcover. This example also shows that the Bolzano–Weierstrass Theorem fails to hold for a general metric space, because A is infinite and bounded but has no limit points.

One might reasonably ask, given that so many familiar topological spaces can be obtained as metric spaces, whether *every* topological space X_τ has τ given by a metric d on X. This turns out to be false, but the property so described is quite interesting.

Definition 9.3.1. *A topological space X_τ is metrizable if there exists a metric $d : X \times X \to \mathbb{R}$ such that the metric topology $\tau_d = \tau$.*

As we've seen, $\mathbb{R}_{\mathcal{U}}$, $\mathbb{R}^n_{\mathcal{U}^n}$ and $X_{\mathcal{D}}$ for any set X are all metrizable. However, there are many nonmetrizable spaces, as the following theorem will allow us to see.

Theorem 9.3.5. *Let X be a finite set and let d be any metric on X. Then $\tau_d = \mathcal{D}$; that is, any metric on a finite set induces the discrete topology.*

Proof: Let X be finite, with, say, $n+1$ distinct elements: $X = \{x_0, x_1, \ldots, x_n\}$. Then the set $\{d(x_0, x_1), d(x_0, x_2), \ldots, d(x_0, x_n)\}$ consists entirely of positive numbers, since $x_0 \neq x_i$ for all $i = 1, 2, \ldots, n$, so that $\epsilon = min\{d(x_0, x_1), d(x_0, x_2), \ldots, d(x_0, x_n)\}$ is also a positive number. Then $B_\epsilon(x_0) = \{x_0\}$, since no other x_i is within distance less than ϵ of x_0. So the singleton set $\{x_0\}$ is τ_d-open for all $x_0 \in X$. □

Thus we see that any finite set X with a non-discrete topology τ is not metrizable. Metrizability is well worth understanding, however. First, we see that it a topological property.

Theorem 9.3.6. *Metrizability is a topological property.*

Sketch of Proof: Let X_τ be metrizable, and let $f : X_\tau \to Y_\nu$ be a homeomorphism. Since X_τ is metrizable, there exists a metric $d_X : X \times X \to \mathbb{R}$ such that a set $U \in \tau$ if and only if U contains an open ϵ ball around each of its points. We will define what we hope will be a metric of Y using d_X. Let $d_Y : Y \times Y \to \mathbb{R}$ be defined by $d_Y(y_1, y_2) = d_X(f^{-1}(y_1), f^{-1}(y_2))$, This is well-defined, since the inverse function f^{-1} is a bijection. We need to show that d_Y is a metric on Y and that $\tau_{d_Y} = \nu$. The proof that d_y is a metric is straightforward and left as an exercise. To see that $\tau_{d_Y} = \nu$, let $V \in \nu$ and $y \in V$. Since f is a homeomorphism, $f^{-1}(V) \in \tau = \tau_{d_X}$. $f^{-1}(y) = x \in f^{-1}(V)$, there exists $\epsilon > 0$ such that $B_\epsilon^{d_X}(f^{-1}(y)) \subset f^{-1}(V)$. Applying f, we see that $f(B_\epsilon^{d_X}(f^{-1}(y)) = B_\epsilon^{d_Y}(y) \subset V$, because of how we defined d_Y. This shows that any ν-open set is τ_{d_Y}-open. The other inclusion is similar. □

Next, we'll see that metrizability is hereditary.

Theorem 9.3.7. *Any subspace of a metrizable space X_τ is metrizable.*

The proof is left as an exercise. It's also true that metrizability is preserved by finite products (but not infinite products), although we don't need this result.

The metrizability property has been extensively studied, since metric spaces are such natural places in which to do calculus-type activities. A fundamental result on metrizability is the following:

Theorem 9.3.8. *(Urysohn's Metrization Theorem) A T_3 space X_τ, where τ has a basis with countably many sets, is metrizable.*

The proof is beyond the scope of this text, but can be found in Ref. [13]. A large chunk of the proof is closely related to our study of separation axioms:

Theorem 9.3.9. *(Urysohn's Lemma[1]) If X_τ is T_4 and if A and B are disjoint τ-closed sets, then there exists a continuous function $f : X_\tau \to [0,1]_\mathcal{U}$ such that $f(A) \subset \{0\}$ and $f(B) \subset \{1\}$.*

This theorem shows up in many guises. It shows that separating closed sets by disjoint open sets (our definition of normality) implies that we can separate closed sets by a continuous function like f above. The converse is easily seen to be valid; if we can separate disjoint closed sets by a function $f : X_\tau \to [0,1]_\mathcal{U}$ such that $f(A) \subset \{0\}$ and $f(B) \subset \{1\}$, then $f^{-1}([0, \frac{1}{10}))$ and $f^{-1}((\frac{9}{10}, 1])$ provide the open sets that separate A from B. The proof uses a characterization of normality analogous to the way Lemma 8.3.4 characterizes regularity. Again, Munkres' book (Ref. [13]) has an accessible proof.

One can give a complete characterization of metrizability in terms of two variants of properties that we've studied so far:

Theorem 9.3.10. *(Smirnov[2] Metrization Theorem) A space X_τ is metrizable if and only if X is paracompact and locally metrizable.*

See Ref. [13] for details.

Exercises

1. Prove Theorem 9.3.1: Every metric space is Hausdorff.
2. Fill in the details of the argument that $U \cap V$ is empty in the proof of Theorem 9.3.2.
3. Prove Theorem 9.3.3.
4. Prove that the function $d_Y : Y \times Y \to \mathbb{R}$ induced from d_X in the proof of Theorem 9.3.6 is indeed a metric.
5. Prove that for the metric d_Y induced from d_X in the proof of Theorem 9.3.6, we have $f(B_\epsilon^{d_X}(f^{-1}(y)) = B_\epsilon^{d_Y}(y)$.
6. Prove Theorem 9.3.7: Any subspace of a metrizable space X_τ is metrizable.

[1] Pavel Urysohn (1898–1924) was best known for his work in *dimension theory* (primarily with K. Menger).

[2] Yuri Smirnov published this result in 1951. A later generalization is usually known as the Nagata-Smirnov Theorem.

9.4 BASICS ON SEQUENCES

Sequences form a large part of a typical introductory calculus course. This tool carries over nicely into metric spaces, and even into arbitrary topological spaces.

Definition 9.4.1. *A sequence in a set X is a function $s : \mathbb{N} \to X$. The point $s(n)$ is the nth term of the sequence, usually denoted by s_n.*

■ EXAMPLE 9.5

The function $a : \mathbb{N} \to \mathbb{R}$ be given by $a_n = \frac{1}{n}$. is a sequence in $\mathbb{R}$.

■ EXAMPLE 9.6

The function $b : \mathbb{N} \to \mathbb{R}$ be given by $a_n = \frac{n+1}{n}$, is a sequence in $\mathbb{R}$.

Of course, your calculus background leads you to think about the numbers that these sequences converge to. The ideas behind convergence of sequences are among the most fundamental in calculus, and all the usual results on derivatives and integrals that involve limits can be defined in terms of sequences (after a bit of work). To think about convergence for sequences in sets other than $\mathbb{R}$, we need the set to have some extra structure. The most familiar setting to make sense of this is when the set X has a metric.

Definition 9.4.2. *Let $s : \mathbb{N} \to X_d$ be a sequence in a metric space. We say that s converges to $x \in X$ if for every $\epsilon > 0$ there exists $N >> 0$ such that $s_k \in B_\epsilon(x)$ for every $n \geq N$. We'll denote this by $s_k \to x$, and say that x is the limit of the sequence s.*

In other words, the sequence $\{s_n\}_{n \in \mathbb{N}}$ converges to the point $x \in X$ if, given any positive distance, all the terms of the sequence past some N are within that distance of x. [We'll often use the term "eventually in" $B_\epsilon(x)$ as a shorthand for this.] This should look quite familiar to anyone who stayed awake during calculus. Note that our examples above converge nicely in the usual metric on $\mathbb{R}$: $a_n \to 0$ and $b_n \to 1$.

We can make this idea much more general:

Definition 9.4.3. *Let $s : \mathbb{N} \to X_\tau$ be a sequence in a topological space. We say that s converges to $x \in X$ if for every τ-open neighborhood U of x, there exists $N >> 0$*

such that $s_k \in U$ *for every* $n \geq N$. *We'll denote this by* $s_k \to x$, *and say that* x *is the limit of the sequence* s.

Note that we've replaced the idea of distance by requiring that we can get all the terms of the sequence past some N into *any* neighborhood of the limit. To give you an idea of how complicated this can be, just take another topology on $\mathbb{R}$. In the indiscrete space $\mathbb{R}_{\mathcal{I}}$, $a_n \to x$ and $b_n \to x$ for *any* $x \in \mathbb{R}$. In the discrete space $\mathbb{R}_{\mathcal{D}}$, neither a nor b converges to any $x \in \mathbb{R}$, since the sequences fail to be eventually in the open set $\{x\}$.

The following result should be expected:

Theorem 9.4.1. *Let* $s : \mathbb{N} \to X_d$ *be any sequence in a metric space. If* s *converges to a limit in* X, *then that limit is unique.*

Rather than prove this, we'll establish the following more general result:

Theorem 9.4.2. *Let* $s : \mathbb{N} \to X_\tau$ *be any sequence in a Hausdorff space. If* s *converges to a limit in* X, *then that limit is unique.*

Proof: Let s be a sequence in a Hausdorff space X_τ. If s converges to two distinct points x and y in X, then separate the points by disjoint neighborhoods, U_x and U_y. Since $s_n \to x$, all the s_k values must be in U_x for all $k \geq N$, and since $s_n \to y$, all the s_k values must be elements of U_y for $k \geq M$, say, which is impossible when $U_x \cap U_y = \emptyset$. □

The following result shows why sequences can be quite useful in topology:

Theorem 9.4.3. *Let* $A \subset X_\tau$, *for any topological space* X. *If there exists a sequence* $a : \mathbb{N} \to A$ *such that* $a_n \to x$, *then* $x \in Cl(A)$.

This nicely links limit points to sequences.

Proof: Let $x \in A'$. Then every neighborhood of x contains (infinitely many) terms of the sequence $\{a_n\}$, and hence contains at least one point of A, so that $x \in Cl(A)$. □

Of more interesting is the partial converse:

Theorem 9.4.4. *Let* $A \subset X_d$, *for any metric space* X. *Then* $x \in Cl(A)$ *if and only if there exists a sequence* $a : \mathbb{N} \to A$ *such that* $a_n \to x$.

This result provides a complete characterization of limit points in terms of sequences, but it only holds in metric spaces.

Proof: $\Longrightarrow$) This direction of the proof is already done, by Theorem 9.4.3.

$\Longleftarrow$) Let $x \in Cl(A)$. For each $n \in \mathbb{N}$, the open ball $B_{\frac{1}{n}}(x)$ intersects A nontrivially: let s_n be any point in $B_{\frac{1}{n}}(x) \cap A$. The resulting sequence is easily seen to converge to x: given $\epsilon > 0$, take $N > \frac{1}{\epsilon}$. Then $s_k \in B_\epsilon(x)$ for every $k \geq N$, by our definition of the sequence $\{s_n\}$. □

Exercises

1. Consider the sequence $c_n = n$ in $\mathbb{R}_{\mathcal{RR}}$, the set of reals with the right ray topology. Show that $c_n \to x$ for any $x \in \mathbb{R}$. Then, find a sequence that fails to converge to any point in $\mathbb{R}_{\mathcal{RR}}$.

2. Prove that no nonconstant sequence in a discrete space converges.

3. Prove that any sequence converges to any point in an indiscrete space.

CHAPTER 10

THE CLASSIFICATION OF SURFACES

10.1 INTRODUCTION

This chapter is devoted to the statement and proof of one of the rarest beasts in the mathematical world – a classification theorem. A classification theorem gives a complete list of all possible forms that a particular mathematical object can take. For example, the most widely studied of all groups are the simple groups, which have no nontrivial normal subgroups. Perhaps the highest achievement of mathematics in the second half of the twentieth century was the Classification Theorem for Finite Simple Groups[1], which lists all possible such groups (up to isomorphism, of course).

Here, we're concerned with a theorem which determines, up to homeomorphism, all the possible compact, connected surfaces, where a surface is a Hausdorff space

[1]The proof of the Classification Theorem for Finite Simple Groups is due to many algebraists, working over several decades, and the published version is scattered over many books and journals. A summary of these efforts can be found in [7]. More recently, a concerted effort to write a streamlined (and well-checked)

where each point has a neighborhood homeomorphic to the open unit disk $B_1((0,0))$ in the usual Euclidean plane. A surface is a particular case of an n-dimensional *manifold*, a Hausdorff space where each point has a neighborhood homeomorphic to the open unit disk $B_1((0,0,\ldots,0))$ in the usual Euclidean $\mathbb{R}^n$. Some familiar examples of surfaces were described in Chapter 1, such as the torus T^2, obtained as a quotient space by identifying the opposite edges of a rectangle:

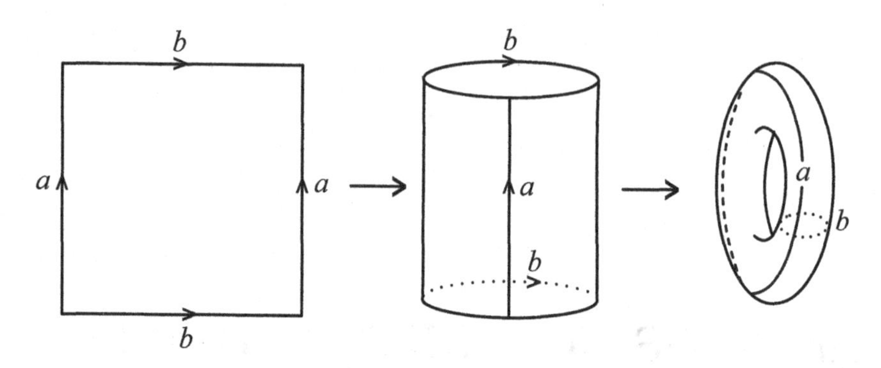

To gain some intuition about what such spaces are like, it's instructive to work at a fairly naive level, as Edwin Abbott[2] did in his classic tale *Flatland*. *Flatland* depicts a 2-dimensional world, where the inhabitants navigate North/South and East/West, but have no ideas about a third direction (which we might call up/down). One Flatlander is visited by a 3-dimensional being, a sphere, who takes him "out of the plane" into "spaceland," dazzling the Flatlander's perception. The Flatlander then proposes that the sphere continue further, into 4 or higher dimensions, which the sphere can comprehend no better than the Flatlanders could 3 dimensions.

One can use *Flatland* as an analogy for understanding higher-dimensional phenomena, as Jeff Weeksdoes very nicely in *The Shape of Space*, [19]. There, Weeks outlines a series of exploratory missions by the Flatlanders, attempting to ascertain the global structure of their world. Locally, Flatland looks like our planet: a flat, 2-dimensional world. Globally, however, our planet is spherical (at least topologically). To determine the global structure of Flatland from within that world, one needs to send out exploratory parties to attempt to "circumnavigate" Flatland. If it's possible

version of the proof has been launched. Thus far, this work encompasses 5 volumes, with several more planned [8].

[2]Be warned that Abbott's *Flatland* contains language that might offend some readers, depicting a rigid caste system, with women at the bottom (due to perceived intellectual weakness). Abbott, however, was actually a strong proponent of women's rights, and was one of the few men active in the movement for women's suffrage at the end of the nineteenth century.

to leave a particular point of Flatland, travel in a single consistent direction and return to the same point, then Flatland cannot be a Euclidean plane, as the following picture shows:

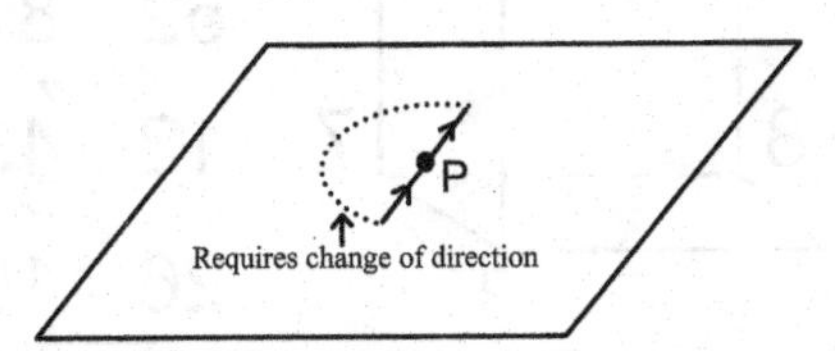

However, this does not show definitively that Flatland is a sphere, since this experimental result is consistent with lots of surfaces:

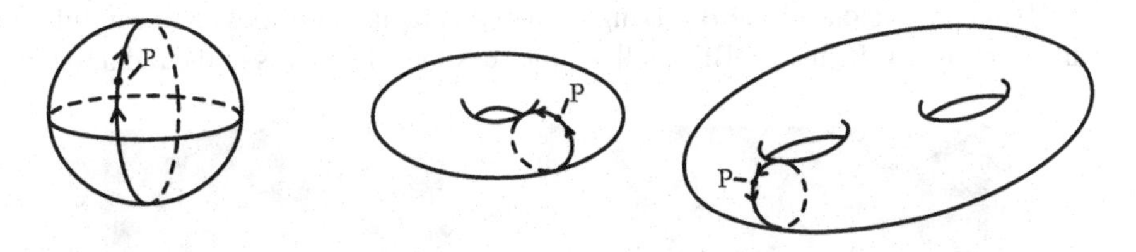

This situation is even more complicated if Flatland is homeomorphic to a non-orientable surface, such as a Klein bottle. Explorers who traverse such a surface would have their sense of up and down, as well as left and right, reversed.

The idea of analyzing the global structure of *our* universe is quite intriguing. Locally, our universe is 3-dimensional, looking like the ideal Euclidean $\mathbb{R}^3$ in a neighborhood of any point. Globally, however, our universe might be "closed": that is, it might curve around on itself, just as our planet does, so that our universe might be homeomorphic to, for example,

$$S^3 = \{(x, y, z, w) : x^2 + y^2 + z^2 + w^2 = r^2\} \subset \mathbb{R}^4.$$

To test this, one might try to "circumnavigate" the universe, but certain physical limitations prevent this. Instead, one might try to send a beam of light in a particular direction, and see if it can be detected from the "opposite" direction (many years later). Even if this could be done, however, it would fail to distinguish between S^3 and the 3-dimensional version of a torus, T^3, which would be obtained by identifying the opposite faces of a three dimensional cube:

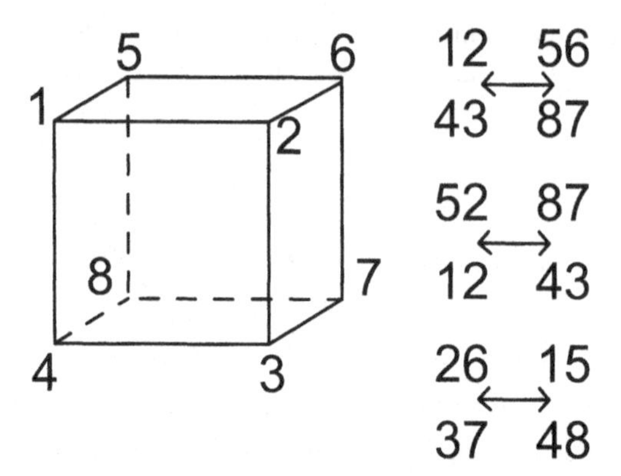

Surprisingly, there is a practical way to make a determination of *which* global structure our universe has, *if* it's closed. Instead of sending out a beam of light, one looks out into the universe for the Cosmic Microwave Background Radiation (CMBR), the residue of the Big Bang. When this radiation is measured carefully in all directions, as by the COBE satellite, one sees lots of patterns in the radiation[3]:

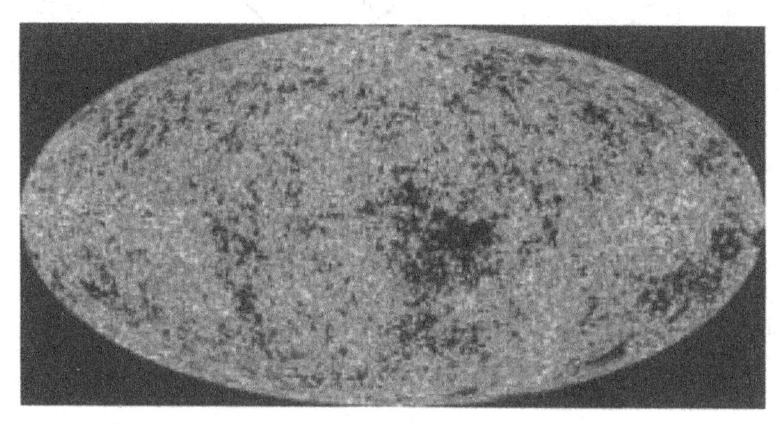

If one sees the exact same pattern in two different directions, then those two directions are "identified" in the global structure of the universe. As a simple example, if one were to look in three orthogonal directions and see the exact matching patterns to the East and West, a different matching pattern both to the North and South, and a third matching pattern both Up and Down (and no other matches), then we would conclude that our universe was homeomorphic to T^3, a (rectangular) three-torus.

This program is actually being carried out as we speak. The successor to the COBE satellite is the Microwave Anisotropy Probe (MAP), (rechristened the Wilkinson Microwave Anisotropy Probe), which was launched in 2001 to get more detailed maps of the CMBR. WMAP resides in the L_3 locus, orbiting the Sun in the shadow

[3]This image from `http://lambda.gsfc.nasa.gov/product/cobe`.

of the Earth, so as to avoid the Sun's radiation. This orbit is unstable, so it requires power to maintain it (as opposed to the L_1 orbit "inside" the Earth).

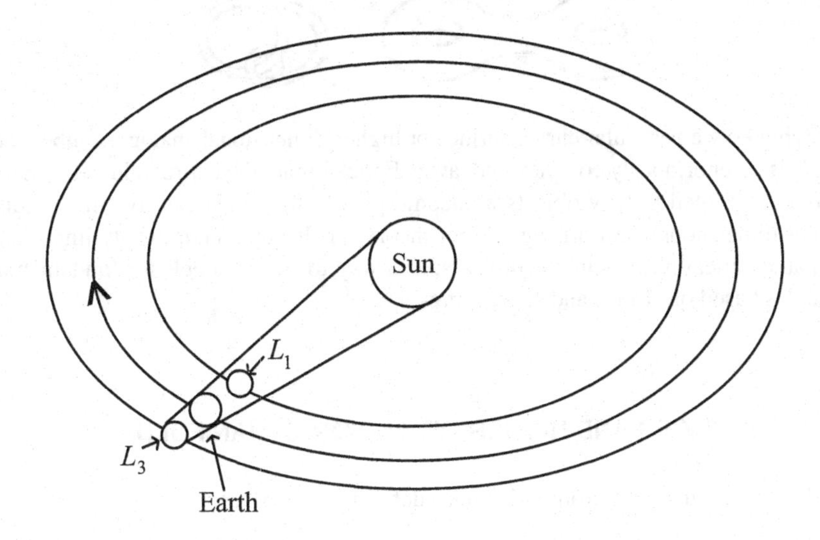

The CMBR pictures from MAP should be accurate enough to distinguish matches clearly. Jeff Weeks has written an interesting software package called SnapPea, which allows one to input the matches from satellite data and have it compute certain information about the resulting "shape" of the universe. Cornish and Weeks have published a very nice survey article about this rather unlikely collaboration of cosmologists and topologists in [5]. These ideas were among those presented at a very unusual conference on connections between cosmology and topology at Case Western Reserve University in the Fall of 1998. One fallout of the conference was the publication of the popular science book *How the Universe Got its Spots,* by Janna Levin ([11]), a Cambridge astrophysicist who spoke at the meeting[4]. Since this conference, Weeks and several coauthors have released their preliminary findings on the shape of the universe ([14]), but their work seems at odds with recent measurements of the Hubble Telescope that seem to indicate an "open" universe. Stay tuned for further developments.

Analyzing the "shape" of the universe (or of Flatland) in this manner can establish, up to homeomorphism, what surface or higher-dimensional analog we live in. However, it can't explain how the object is embedded in the surrounding "space."

[4]Levin's book contains a lot of information about sorting out the structure of the universe from a topological point of view, written in the context of letters to her mother. This is a truly unique scientific document, well worth reading.

For example, the following two renditions of a surface are homemorphic, but are embedded in $\mathbb{R}^3$ in different ways:

Studying how a particular curve, surface or higher-dimensional analog is embedded in $\mathbb{R}^n$ is an enormously complex endeavor. For example, the entire field called *knot theory* is devoted to how objects homeomorphic to the circle can live in $\mathbb{R}^3$ (and higher dimensions). Surprisingly, knot theory has become enormously important in mathematical physics in recent years, as a way to study models for fundamental particles based on strings and superstrings.

10.2 SURFACES AND HIGHER-DIMENSIONAL MANIFOLDS

We begin with the long-promised precise definition of surface.

Definition 10.2.1. *A surface is a Hausdorff space which is locally homeomorphic to the open unit disk $B_1((0,0)) \subset \mathbb{R}^2_{\mathcal{U}^2}$. That is, a space $X_\mathcal{T}$ is a surface if X is Hausdorff and if for every $x \in X$, there exists an open neighborhood U_x of x with a homeomorphism $f_x : U_x \to B_1((0,0)) \subset \mathbb{R}^2_{\mathcal{U}^2}$.*

Of course, since any two open disks in $\mathbb{R}^2_{\mathcal{U}^2}$ are homeomorphic, we needn't have been quite so specific in our definition. Perhaps surprisingly, the Hausdorff requirement is necessary – there are examples of spaces which are locally homeomorphic to open real disks, but which are not Hausdorff. Such examples are too pathological, however, to be considered surfaces in the usual sense[5].

Surfaces are special cases of a more general "animal":

Definition 10.2.2. *An n-dimensional manifold (or n-manifold) is a Hausdorff space locally homeomorphic to the usual open unit ball $B_1(\vec{0}) \subset \mathbb{R}^n_{\mathcal{U}^n}$.*

So a surface is a 2-manifold, and a 1-manifold should be, at least intuitively, a curve. The following examples illustrate these concepts.

[5]For examples of this particular pathology, among many others, see *Counterexamples in Topology* by Seebach and Steen ([16]).

EXAMPLE 10.1

Consider the subspace $C = \{(\cos t, \sin t, t) : t \in \mathbb{R}\} \subset \mathbb{R}^3_{\mathcal{U}^3}$, pictured here:

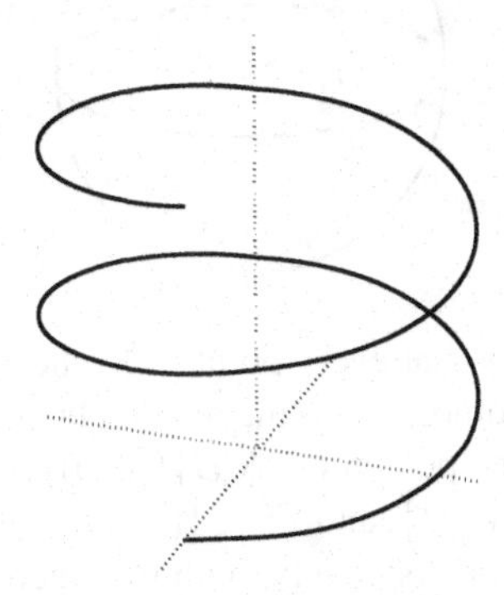

Note that C is also the image of the vector-valued function $\vec{r} : \mathbb{R} \to \mathbb{R}^3$ given by $\vec{r}(t) = (\cos t, \sin t, t)$. Then the parametrised curve C is a 1-manifold, since for every point $(\cos(t_0), \sin(t_0), t_0) \in C$, the arc $\{(\cos t, \sin t, t) : t \in (t_0 - \epsilon, t_0 + \epsilon)\}$ is homeomorphic to the open interval $(-\epsilon, \epsilon)$, by the inverse of the exponential map e of Example 5.3.

Of course, the most obvious example of a surface is just the usual Euclidean plane:

EXAMPLE 10.2

The usual plane $\mathbb{R}^2_{\mathcal{U}^2}$ is a surface. Given $(x_0, y_0) \in \mathbb{R}^2$, the neighborhood $B_1((x_0, y_0))$ is homeomorphic to the "standard" unit disk $B_1((0,0))$ by a simple translation (add (x_0, y_0) to every vector in $B_1(0,0)$), which is obviously a homeomorphism.

Somewhat less trivial is the following example:

EXAMPLE 10.3

The unit sphere $S^2 = \{(x, y, z) : x^2 + y^2 + z^2 = 1\} \subset \mathbb{R}^3_{\mathcal{U}^3}$ is a surface.

One can picture the neighborhoods around points in the sphere as in the following picture.

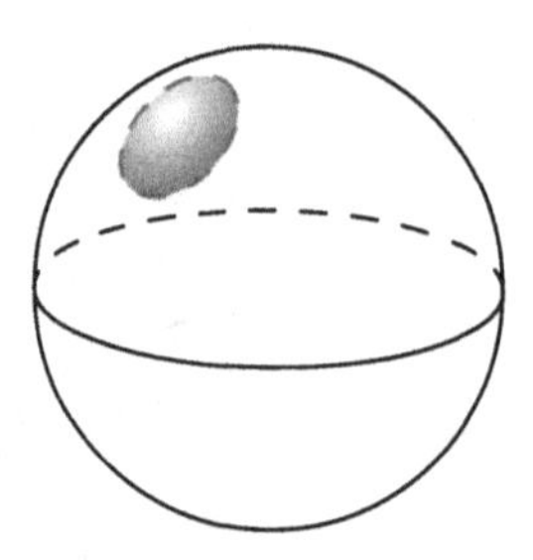

More precisely, we can show that every point in S^2 lies in at least one of two open sets, each of which is homeomorphic to the whole plane. Since the entire real plane $\mathbb{R}^2_{\mathcal{U}^2}$ is homeomorphic to the open unit disk $B_1((0,0))$, this will finish our example. The open sets are $S^2 \setminus \{(0,0,1)\}$, and $S^2 \setminus \{(0,0,-1)\}$,, the sphere with the north pole and south pole deleted, respectively, both of which are clearly open since the sphere is a T_1-space. We'll show that $S^2 \setminus \{(0,0,1)\}$ is homeomorphic to $\mathbb{R}^2$ by a function known as *stereographic projection*:

$$s : S^2 \setminus \{(0,0,1)\} \to \mathbb{R}^2$$

defined as in the following picture:

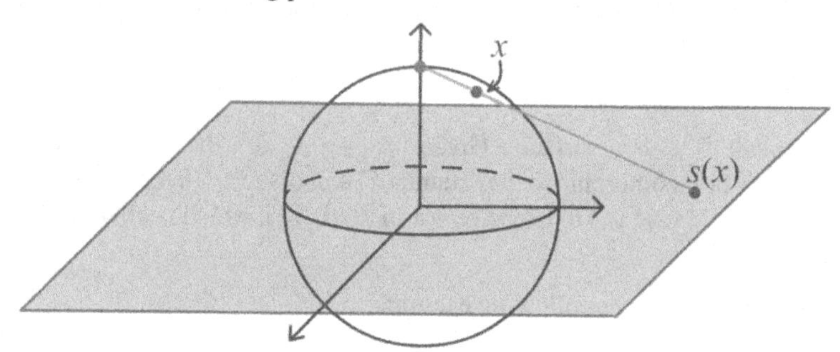

The function is defined by $s((x,y,z)) = \frac{1}{1-z}(x,y)$, which sends the point (x,y,z) in the sphere to a point in the plane in the same direction as $(x,y,0)$, with magnitude determined by the height z of the original point. One can check easily that the northern hemisphere of S^2 (except the north pole $(0,0,1)$) gets sent to points in the plane outside the unit circle, while the southern hemisphere maps to the interior of the unit disk. The equator of S^2 stays "fixed", mapping to the unit circle in the plane. It's easy to check that s is one-to-one and onto, and that s and its inverse are both continuous. Similarly, $S^2 \setminus \{south\,pole = (0,0,-1)\}$ is homeomorphic to the plane, and hence to an open disk, completing the proof.

■ EXAMPLE 10.4

Consider the unit square $[0,1]^2_{\mathcal{U}^2}$ with the equivalence relation $(x,0) \sim (x,1)$ and $(0,y) \sim (1,y)$, for all $x \in [0,1], y \in [0,1]$. Then the quotient space $([0,1]^2_{/\sim})_{\mathcal{U}^2_{/\sim}} = T^2$ is the usual torus:

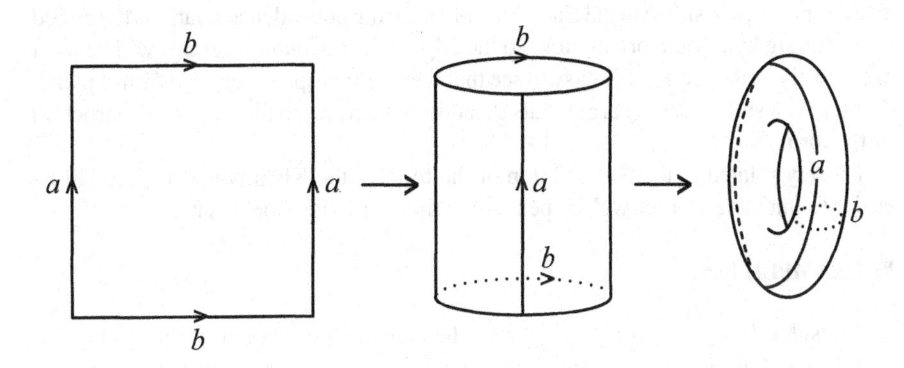

To see that the torus T^2 is actually a surface, we need to exhibit around each of its points a neighborhood homeomorphic to the usual planar disk. The easiest way to see this is to use the quotient map $p : [0,1]^2 \to T^2$. For a point $\vec{z} \in T^2$, we examine the inverse image $p^{-1}(\{\vec{z}\})$, which is a *subset* of $[0,1]^2$, not a point. If $p^{-1}(\{\vec{z}_0\})$ is a singleton, then it's easy to take a small enough open disk U around $p^{-1}(\{\vec{z}_0\})$ in $[0,1]^2$ so that $p(U)$ is the desired neighborhood around $\vec{z}$, with the restriction $p|_U$ the required homeomorphism. If $p^{-1}(\{\vec{z}_1\})$ is a doubleton set, the diagram below shows how to glue together two half-open disks to get the desired neighborhood. The worst case is if $\vec{z}_2$ is the image of the corner points of the square, where we need to glue together 4 "quarter-open" disks.

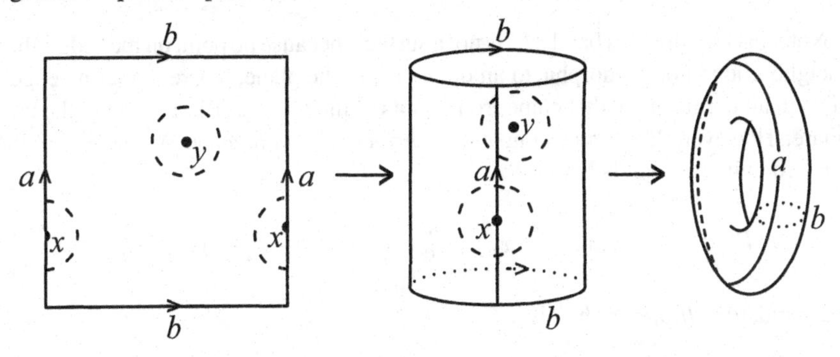

One can also see that T^2 has such neighborhoods around each of its points by showing that $T^2 \cong S^1 \times S^1$, the product of two circles (each given the $\mathcal{U}^2$-subspace topology), then using the quotient space construction of S^1 as $[0,1]_{/\sim}$ from Example 5.3.

Since S^1 is constructed from an interval, one can get neighborhoods of each of its points homeomorphic to open intervals. Thus, for each point of T^2, we can get a neighborhood homeomorphic to a product of two open intervals. One then shows that the open unit square $(0,1)^2$ is homeomorphic to the usual open disk to complete the proof.

More generally, we can construct surfaces by identifying edges of polygons (with an even number of sides) in the Euclidean plane, using equivalence relations described by assigning letters and orientations to the edges. Such *planar diagrams* will be used extensively in the sequel. It's easy to see that the quotient space obtained from a planar diagram is a surface, using arguments essentially identical to that for the construction of the torus.

Closely related to the construction of the torus is the Klein bottle, which will be easier to understand after we've spent some time with the Moebius band.

■ EXAMPLE 10.5

Consider the unit square $[0,1]^2_{\mathcal{U}^2}$ with the equivalence relation $(0,y) \sim (1,y)$ for all $y \in [0,1]$. Then the quotient space $([0,1]^2_{/\sim})_{\mathcal{U}^2_{/\sim}} = M^2$ is the Moebius band:

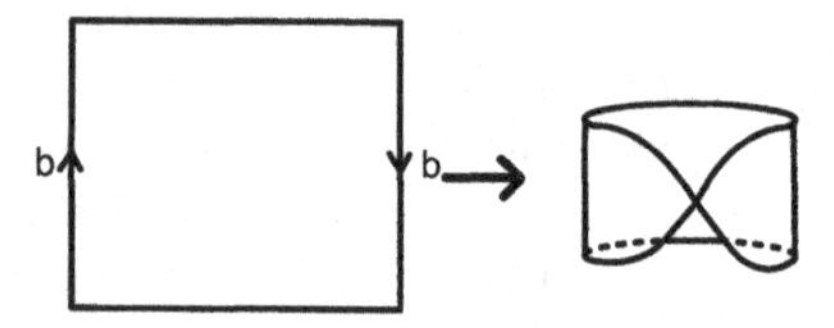

Note that the Moebius band M^2 is not a surface, because no point on the "edge" has a neighborhood homeomorphic to an open disk in the plane, where the term "edge" refers to all points of the that come from points of the form $(x,0)$ or $(x,1)$ in the unit square. However, M^2 is an example of a *surface with boundary*. We recall that the standard unit open ball is $\mathbb{R}^n$ is

$$B_1(\vec{0}) = \{(x_1, x_2, \ldots, x_n) : x_1^2 + x_2^2 + \cdots + x_n^2 < 1\} \subset \mathbb{R}^n_{\mathcal{U}^n}.$$

The standard *half-open n-ball* is

$$H_1(\vec{0}) = \{(x_1, x_2, \ldots, x_n) : x_1^2 + x_2^2 + \cdots + x_n^2 < 1, x_1 \geq 0\} \subset \mathbb{R}^n_{\mathcal{U}^n}.$$

So the standard half open 2-ball looks like

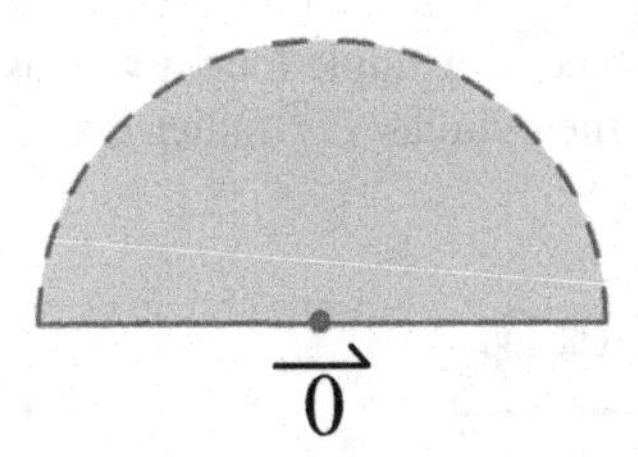

Definition 10.2.3. *An n-dimensional manifold with boundary is a Hausdorff space locally homeomorphic to the usual open unit ball $B_1(\vec{0})$ or the standard half-open n-ball $H_1(\vec{0})$ in $\mathbb{R}^n_{\mathcal{U}^n}$.*

Of course, we'll refer to a 2-manifold with boundary as a surface with boundary. The term "boundary" here is possibly a source of confusion, so we'll belabor the point somewhat. For a surface with boundary $X_\mathcal{T}$, we'll use the term *surface boundary* to mean the set of all points in X which have only neighborhoods homeomorphic to $H_1(\vec{0})$, not $B_1(\vec{0})$. Thus the surface boundary of X differs greatly from our familiar set $Bdy(X)$, as defined in Definition 4.4.1, even though $Bdy(X)$ might be given two different interpretations here. If one thinks of X as a subspace of itself, $Bdy(X) = \emptyset$. If one thinks of the surface X as a subspace of some Euclidean space like $\mathbb{R}^3$, when appropriate, then $Bdy(X) = X$, as one can easily check by looking at $S^2 \subset \mathbb{R}^3$. Instead, the *surface boundary* of X is generally a curve or union of curves. For example, the cylinder C^2 given by

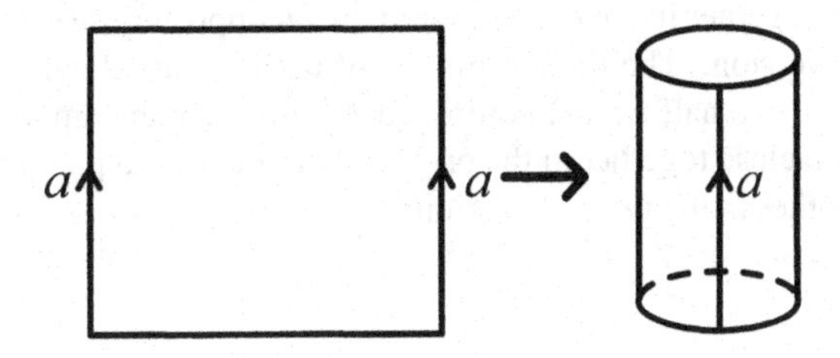

and the Moebius band are both surfaces with boundary. The surface boundary of C^2 is the disjoint union of two circles. The surface boundary of M^2 is a single circle, as one can check by building a physical Moebius band and marking the (single) edge.

It's easy to see that the Moebius band is a one-sided object (a *nonorientable surface with boundary*), as one can illustrate by constructing a physical representation and trying to draw a line down the center of the strip – to get the line to close up (to meet itself) one has to draw on "both sides" of the band. In other words, there is no "outward" direction that can be specified consistently along this simple closed curve. Another interesting property shows up if one cuts the band along this line (the "equator"). The resulting object is a single loop, twice as long as the original band, which is homeomorphic to the cylinder C^2 rather than M^2. The following picture illustrates this:

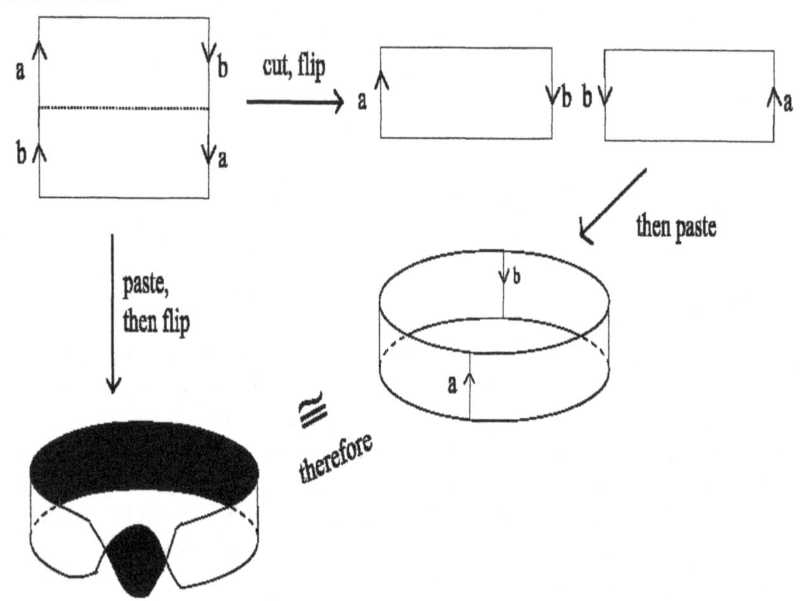

It may not be immediately obvious that the resulting band is in fact homeomorphic to the cylinder, but one can assure oneself of this by illustrating with a belt: the transformation of unbuckling the belt (a cylinder, in its usual state) and adding a single half-twist before rebuckling (a Moebius belt!) isn't a homeomorphism, because points which were close together in the cylinder are now on opposite edges from each other in the rebuckled version. The transformation of taking a usual cylindrical belt and adding a full twist (two half twists) before rebuckling *is* a homeomorphism, because points which were close together in the original belt are sent to points which are just as close to each other in the twisted version:

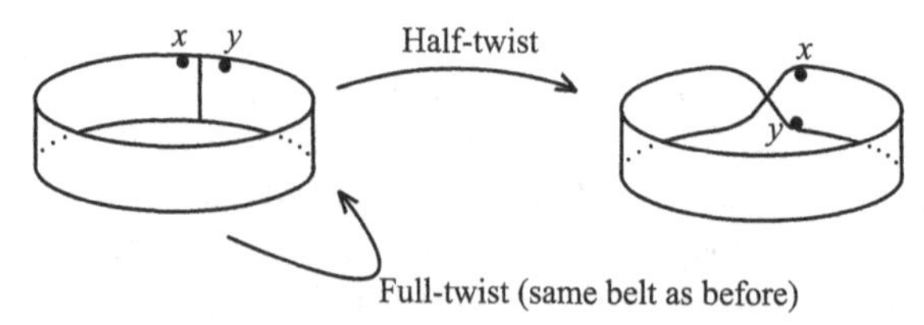

Now we're in a better position to understand what goes into the construction of a Klein bottle.

Definition 10.2.4. *Consider the unit square $[0,1]^2_{\mathcal{U}^2}$ with the equivalence relation $(x,0) \sim (1-x,1)$ and $(0,y) \sim (1,y)$, for all $x \in [0,1], y \in [0,1]$. Then the quotient space $([0,1]^2_{/\sim})_{\mathcal{U}^2_{/\sim}} = K^2$ is the Klein bottle:*

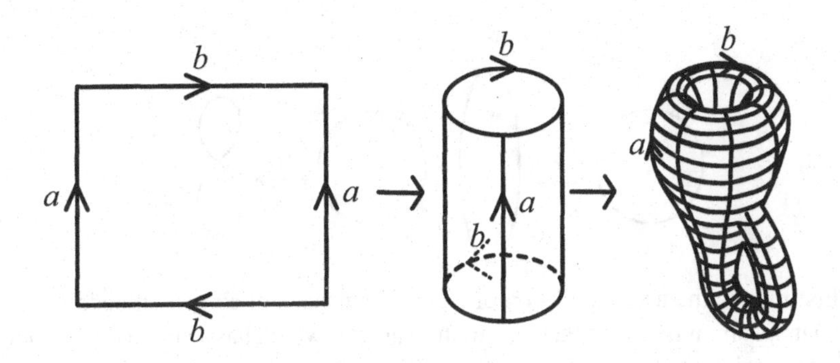

The circle worth of points where the Klein bottle *appears* to intersect itself is due to our tendency to try to see objects in three dimensions, rather than the four dimensions that this object lives in naturally. To make this clearer, we need a bit more terminology:

Definition 10.2.5. *A topological space $X_\mathcal{T}$ is said to be embedded in $\mathbb{R}^n_{\mathcal{U}^n}$ if there is a continuous and open injection $i : X \to \mathbb{R}^n$. That is, $X_\mathcal{T}$ is homeomorphic to a subspace of Euclidean $\mathbb{R}^n$. The homeomorphism i is referred to as an embedding of X in $\mathbb{R}^n$.*

When a space X can be embedded in $\mathbb{R}^n$, we'll use the shorthand notation $X \hookrightarrow \mathbb{R}^n$.

Sometimes, as in the case of the Klein bottle, a space X can be nearly embedded in, say, $\mathbb{R}^3$, except for some points of self intersection. In other words, the object is "locally-embedded" in the Euclidean space. The following definition makes this idea precise:

Definition 10.2.6. *A topological space $X_\mathcal{T}$ is said to be immersed in $\mathbb{R}^n_{\mathcal{U}^n}$ if there is a continuous and open function $j : X \to \mathbb{R}^n$ such that for every point $j(x_0)$ in $Im(j)$, there is a $\mathcal{T}$-open neighborhood U_{x_0} of x_0 such that the restriction $j|_{U_{x_0}}$ is an embedding. The "global" map $j : X \to \mathbb{R}^n$ is referred to as an immersion of X into $\mathbb{R}^n$.*

We'll use the notation $X \propto\!\!\rightarrow \mathbb{R}^n$ to indicate an immersion of X into Euclidean n-space. Note that every embedding of a space into $\mathbb{R}^n$ is also an immersion, but that the images of immersions can have self-intersections. To clarify the difference between embeddings and immersions, consider the following two functions from the usual circle to $\mathbb{R}^2$, whose images are given here:

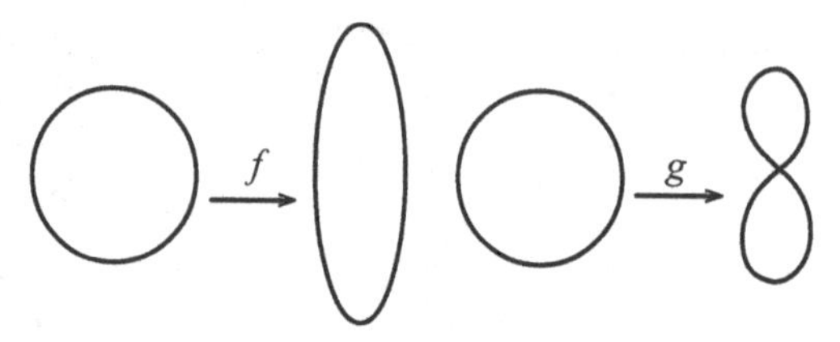

The first map is the usual inclusion of S^1 as a subspace of $\mathbb{R}^2$, an embedding. The second map, an immersion, sends S^1 to the figure 8, which has a single double point. Given any point on the image of g, we can find a neighborhood around any *one* of its preimages that maps homeomorphically by g. However, any neighborhood of a point in $Im(g)$ need not be the homeomorphic image of a neighborhood in S^1.

Some surfaces embed in $\mathbb{R}^2$, such as the plane itself and (less obviously), the cylinder (which is homeomorphic to an annular ring in the plane). Indeed, this is an easy way to see that the cylinder is not homoemorphic to the Moebius band, since the Moebius band cannot be embedded in $\mathbb{R}^2$. It's easy to see that many surfaces can be embedded in $\mathbb{R}^3$: the usual sphere S^2 is defined as a subspace of $\mathbb{R}^3$ and the usual torus T^2 can be shown to be homeomorphic to a subspace of $\mathbb{R}^3$ (see the exercises). However, there is no embedding of the Klein bottle K^2 in $\mathbb{R}^3$. It can be shown, in fact, that any immersion of K^3 in $\mathbb{R}^3$ has at least a circle worth of double points. The Klein bottle (and any other surface) can be embedded in $\mathbb{R}^4$ and immersed in $\mathbb{R}^3$.[6]

The torus is a fairly simple surface, with a readily understood inside and outside (which is why innertubes hold air). Its construction is, in essence, building a cylinder in two perpendicular directions. The Klein bottle, on the other hand, is a pretty complicated surface, because it is *nonorientable* or one-sided, like the Moebius band. Its construction is a cylinder in one direction, and a Moebius band in the perpendicular direction. One might ask what sort of surface results from doing a Moebius band in both directions?

[6]Hassler Whitney proved that any n-manifold can be embedded in $\mathbb{R}^{2n}$ and immersed in $\mathbb{R}^{2n-1}$ in 1939. That every surface immerses in $\mathbb{R}^3$ is a special case of the well-known *Immersion Conjecture,* due to Whitney and Hirsch (given in the 1950s), that predicted that any n-manifold can be immersed in $\mathbb{R}^{2n-\alpha(n)}$, where $\alpha(n)$ is the number of ones in the binary notation for n. The conjecture was proved by Ralph Cohen in the early 1980s.

Definition 10.2.7. *Consider the unit square* $[0,1]^2_{\mathcal{U}^2}$ *with the equivalence relation* $(x,0) \sim (1-x,1)$ *and* $(0,y) \sim (1,1-y)$, *for all* $x \in [0,1], y \in [0,1]$. *Then the quotient space* $([0,1]^2_{/\sim})_{\mathcal{U}^2_{/\sim}} = P^2$ *is the real projective plane:*

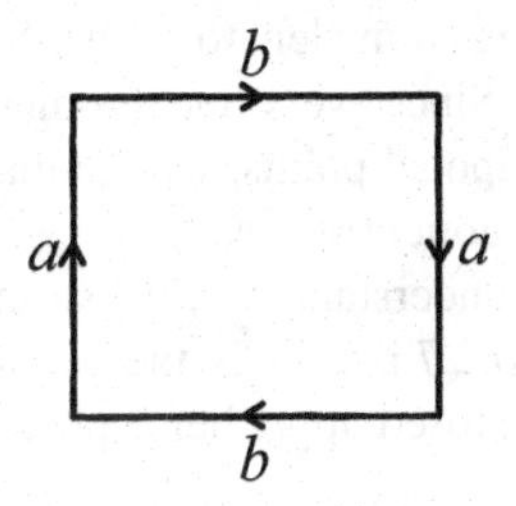

This picture shows only how to *construct* the projective plane – it does not show any embedding or immersion of the resulting surface into $\mathbb{R}^3$, because the projective plane is such a difficult space to picture or to "construct" a model for. The most commonly used immersion of the projective plane is originally due to W. Boy in 1902, called Boy's surface. A giant stainless steel model of Boy's surface is the focal point of the grounds of the Mathematicshes Forschungsinstitut Oberwolfach in southern Germany, dramatically displaying the amazing twisting necessary to immerse P^2 in three-space.[7] It can be shown, with some difficulty, that any immersion of P^2 into $\mathbb{R}^3$ has at least one triple point (and lots of double points). The existence of this triple point is the first case of an extremely important open problem in topology, known as the Kervaire/Arf Invariant One problem.[8]The projective plane is nonorientable, as one can see by its construction: it reverses orientation in either the north-south or east-west directions in the square.

To see why the name "projective plane" is apt for the last example, one needs to think about projective geometry. In this particular type of geometry, a point is specified by "homogeneous coordinates" (x_0, x_1, x_2), where the x_is are real numbers, at least

[7]For details, see the article *Die Boysche Fläche in Oberwolfach* (3 pages, in German) by H. Karcher and U. Pinkall in *Mitteilungen der DMV,* issue 1 (1997). The article can be downloaded from the Oberwolfach website at `www.mfo.de`.

[8]The Kervaire/Arf Invariant One problem looks at the existence of n-manifolds which have certain *framings* which can be classified by certain symmetric bilinear forms. It has been shown by Bill Browder that framed manifolds with Kervaire invariant one can exist only with dimensions $2^k - 2$, for $k = 2, 3, 4, \ldots$. Such manifolds have been shown to exist in dimensions 2,6,14,30 and 62 (the last by Barratt, Jones and Mahowald in the late 1970s). The problem of whether or not there exists such a manifold in higher cases is an open (and very hard) problem that has occupied homotopy theorists for quite a while.

one of which is nonzero. We use the term homogeneous because (x_0, x_1, x_2) and (y_0, y_1, y_2) represent the same point if and only if one is a scalar multiple of the other: $x_i = \lambda y_i$ for $i = 1, 2, 3$ and $\lambda \neq 0$. Thus, in usual $\mathbb{R}^3$, all the points which lie on the same line through the origin are identified into a single point on the projective plane. Equivalently, consider the following equivalence relation of the sphere S^2: $w \sim -w$ (in addition to every point being equivalent to itself). So we're just identifying all the antipodal points in the sphere. Since every line through the origin in $\mathbb{R}^3$ hits the usual unit sphere in exactly two antipodal points, we see that the geometrically-described projective plane is really the quotient space $S^2_{/\sim}$.

Now we're in a position to understand why the surface given by identifying edges of a square as in Definition 10.2.7 is exactly the geometrically-described projective plane $S^2_{/\sim}$: first, consider the closed upper hemisphere of S^2:

$$H = \{(x, y, z) \in S^2 : z \geq 0\}.$$

Under the quotient space construction of the projective plane, we identify all antipodal points in S^2. So when we look at the hemisphere $H \subset S^2$, we see that for points w in H above the $x-y$-plane (nonnegative z coordinate), the antipodal point $-w$ is not in H. The only points in H whose antipodes are also in H are the points on the $z = 0$ circle, the equator. Hence, we could also define the projective plane as $H_{/2}$, where the equivalence relation identifies all points on the boundary circle with their diametrical opposites. Equivalently, we could take any space homeomorphic to the upper hemisphere H, like the closed unit disk $B_1(\vec{0}) \subset \mathbb{R}^2_{\mathcal{U}^2}$, and perform the same identification. Now it's easy to see why identifying the edges of a "2-gon" as below yields the same surface as our original Definition 10.2.7:

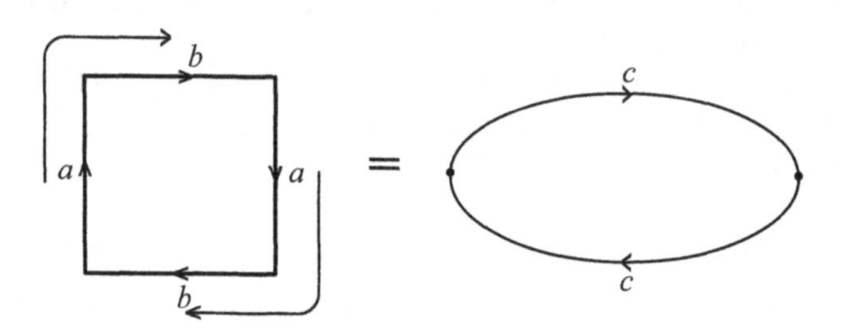

Exercises

1. Show that the entire real plane $\mathbb{R}^2_{\mathcal{U}^2}$ is homeomorphic to the open unit disk $B_1((0, 0))$, exhibiting a precise homeomorphism.

2. Prove that the stereographic projection $s : S^2 \setminus \{(0, 0, 1)\} \to \mathbb{R}^2$ is a homeomorphism. Hint: check that s is one-to-one and onto using ideas from polar

coordinates, and that s and its inverse are both continuous using calculus results about rational functions.

3. Prove that $T^2 \cong S^1_{\mathcal{U}^2} \times S^1_{\mathcal{U}^2}$. (Hint: First, think of the usual description of the torus given by identifying edges of a square.Next, note that one can write down an explicit function $I \times I \to S^1 \times S^1$, using the quotient map that constructs S^1 from the interval. Now notice that these two activities are actually doing the same thing!)

4. Prove that $T^2 \cong \{(x, y, z) : (\sqrt{x^2 + y^2} - 2)^2 + z^2 = 1\} \subset \mathbb{R}^3_{\mathcal{U}^3}$.

5. Build a physical Moebius band and cut along the meridian, as discussed in the paragraph following Definition 10.2.3. Verify that the resulting band is homeomorphic to a cylinder.

6. Cut a physical Moebius band along a line one-third of the distance from the bottom edge, continuing all the way around until the cut meets itself. The resulting object should we two linked bands. What surface with boundary is each homeomorphic to? Why?

7. What surface results from the following quotient space: take two Moebius bands and identify the edge of one with the edge of the other? Hint: consider the following limerick:
A mathematician named Klein
thought the Moebius band divine.
He said "If you glue
the edges of two,
you'll get a weird bottle like mine."

8. What surface results from forming the quotient space $M^2_{/\sim}$, where all the points on the edge of the Moebius band are identified together?

9. Show that removing an open disk from a projective plane yields a Moebius band.

10.3 CONNECTED SUMS OF SURFACES

We've seen several examples of surfaces so far, but no method to use these examples to produce new surfaces. In just about every area of mathematics, one has operations on the objects of study that allow one to produce new examples of these objects from old. In studying groups, for example, one can produce new groups by taking direct sums of existing groups. In topological spaces, the analog of the direct sum is the product of spaces. However, this will not allow us to produce new surfaces from existing surfaces, because the Cartesian product of manifolds increases the dimension. For example, the circle S^1 is a 1-manifold (a curve), but $S^1 \times S^1 \cong T^2$, the torus, a 2-manifold. Similarly, the product of two surfaces is a 4-manifold (an interesting object, but not a surface). Alternatively, one might also consider the union of two surfaces, which can indeed be a surface (if the component surfaces are disjoint, for example), but which is not connected. A more satisfactory operation on surfaces is the *connected sum*, defined below.

An informal description of the connected sum of two surfaces will help to make the formal definition clearer. Given two connected surfaces X_1 and X_2 (without boundary, of course, unless we explicitly say so), we will build a new surface, the connected sum of X_1 and X_2, denoted $X_1\#X_2$, as follows. First, choose any point $x_1 \in X_1$, and let U_1 denote a neighborhood of x_1 which is homeomorphic to the usual open planar disk. (Such a neighborhood always exists, since X_1 is a surface.) Then the boundary of the neighborhood in X_1 is homeomorphic to the unit circle S^1. Similarly, for any point $x_2 \in X_2$, let U_2 denote a neighborhood of x_2 homemorphic to the open planar disk. We then "cut" the neighborhoods U_1 and U_2 out of the surfaces and "sew" the remainder of X_1 to X_2, identifying $Bdy(U_1)$ with $Bdy(U_2)$. The picture below illustrates the construction.

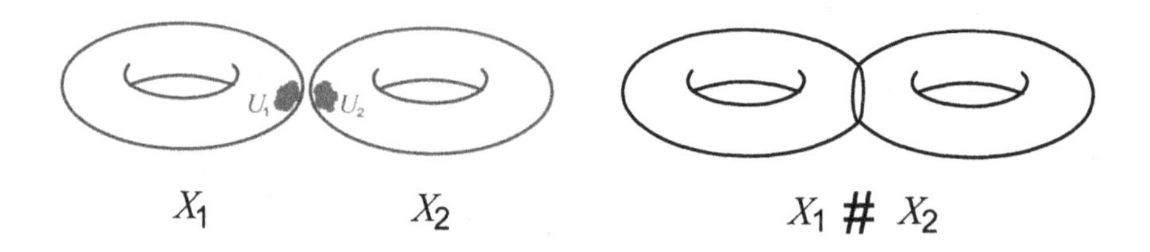

To make this precise, we need the following lemma.

Lemma 10.3.1. *Any continuous and open injection of the open unit disk into a subspace of* $\mathbb{R}^n$ *extends over the circle. That is, if* $f : B = B_1(\vec{0})_{\mathcal{U}^2} \to X_\tau$, *with* $X_\tau \subset \mathbb{R}^n_{\mathcal{U}^n}$ *is a continuous and open injection, then* f *extends to a continuous injection of the closed unit disk* $\overline{f} : Cl(B) = D^2 \to X$.

The proof is left as an exercise (with hints). Now we're in a position to make a precise definition of connected sum.

Definition 10.3.1. *Let X_τ and Y_ν be any connected surfaces. For any $x \in X$ and $y \in Y$, let U be a τ-neighborhood of x with $f : B_1(\vec{0}) \to U$ a homeomorphism and let V be a ν-neighborhood of y with $g : B_1(\vec{0}) \to V$ a homeomorphism. Then the connected sum of X with Y is defined as the quotient space*

$$X\#Y = \left[(X \setminus U)_\tau \coprod (Y \setminus V)_\nu\right]_{/\sim},$$

where the equivalence relation is given by $\overline{f}(t) \sim \overline{g}(t)$ for all $t \in S^1$.

What we haven't yet shown is that this definition "makes sense": that is, we should show that $X\#Y$ is well-defined, up to homeomorphism, regardless of the choices one makes in the construction. Precisely, we should show that we get homeomorphic spaces if we choose different points in X and in Y to center our neighborhoods at, or if we choose different neighborhoods of these points. We also should show that the resulting space $X\#Y$ is indeed a surface. We'll be a bit less than rigorous here, argue that it's plausible that this construction is well-defined since *every* point in a surface has a neighborhood homeomorphic the the usual open disk. Thus, regradless of the choice of points in X or Y, the connected sum $X\#Y$ can be built. It takes a bit more subtlety to see that all such choices result in spaces which are homeomorphic. We can also gain some graphical insight as to why $X\#Y$ is a surface if X and Y are. If one chooses a point $z \in X\#Y$ which is not on the identified "rims", then we can find a neighborhood of z homeomorphic to the open disk wholly within X or wholly within Y, as the following picture shows. If z is chosen on the (now common) boundary of U and of V, then we paste together the resulting half-neighborhoods from X and Y to get the desired neighborhood of z.

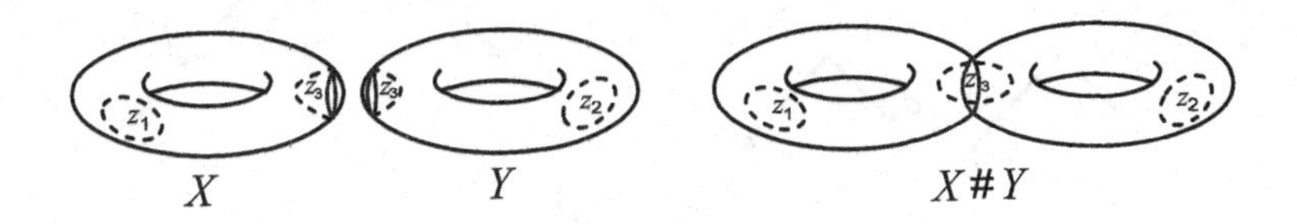

As a nice example of how the connected sum is a surface, we work out the *planar diagram* for the connected sum of two tori, $T^2\#T^2$.

■ **EXAMPLE 10.6**

Recall that the torus is given by identifying edges of a square, so our two tori are given by:

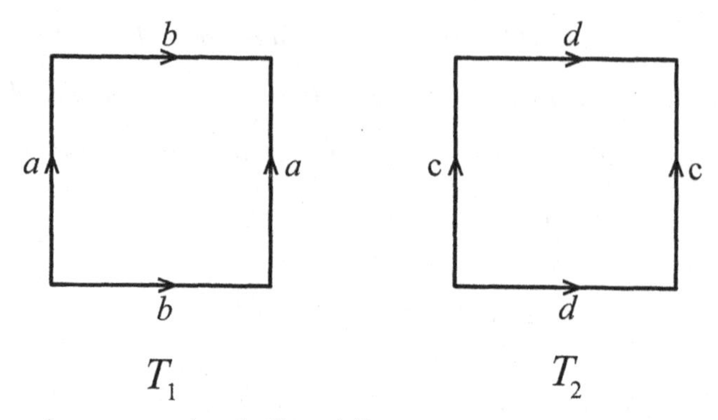

We can choose any points in T_1 and T_2 to determine our neighborhoods, so we'll choose the vertex in each, for simplicity:

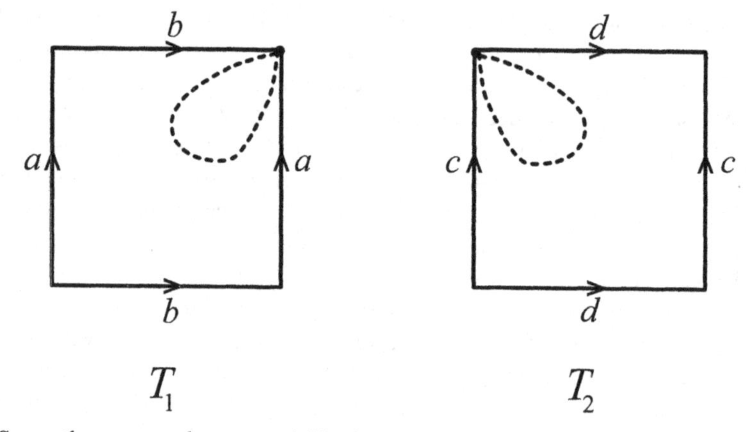

So we throw away the open neighborhoods and identify the boundaries:

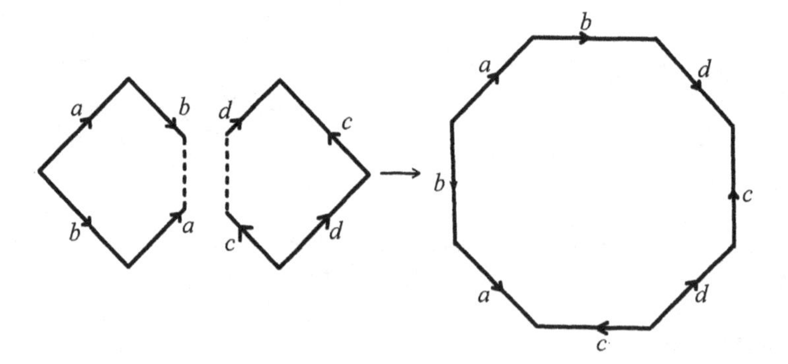

This example also points that a surface can be determined by a *planar diagram* for a polygon other than a square. Recall, for example, the projective plane P^2 can be given by identifying the edges of 2-sided polygon:

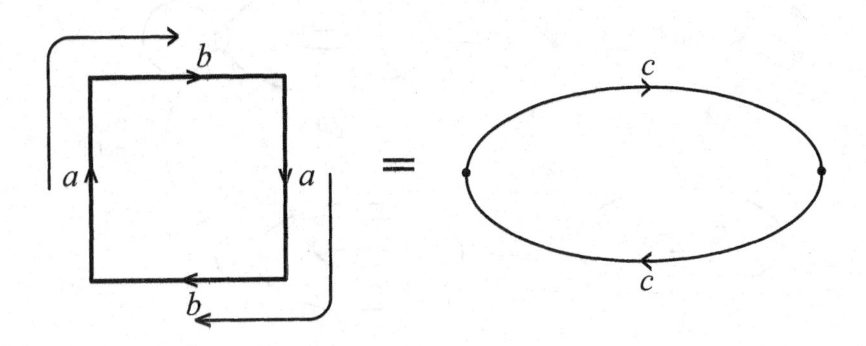

Similarly, the following 2-gon shows how to construct the sphere S^2:

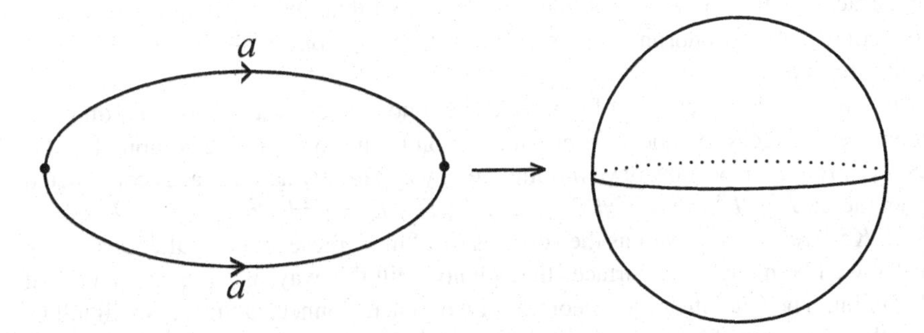

Note that we can perform a connected sum of more than two surfaces at a time, and that associativity holds, up to homeomorphism:

$$X_1\#(X_2\#X_3) \cong (X_1\#X_2)\#X_3.$$

This allows us to be a bit more relaxed about notation and to write such connected sums without parentheses, without ambiguity.

One can see intuitively that the following Theorem is plausible:

Theorem 10.3.2. *For any surface X, the connected sum $X\#S^2$ is homeomorphic to X.*

The following pictures are convincing evidence that the theorem is true.

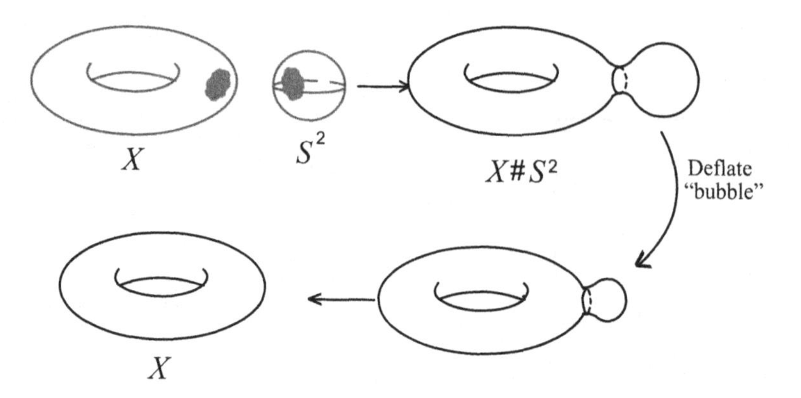

So the sphere S^2 acts, at least up to homeomorphism, as an identity for the operation #, connected sum. If one considers the collection of all homeomorphism classes of surfaces, then # is a well-defined operation, so that one might well ask: is the collection of homeomorphism classes of surfaces a *group* under the operation of connected sum?

In any case, the operation of connected sum now gives us a plethora of compact, connected surfaces to choose from. In addition to the roster we've compiled so far (S^2, T^2, K^2, P^2), we now have infinitely many apparently new surfaces constructed from these: $T^2\#T^2, T^2\#T^2\#T^2, \ldots, P^2\#P^2, P^2\#P^2\#P^2, \ldots, K^2\#K^2$, $K^2\#K^2\#K^2, \ldots$, as well as the surfaces that "mix" these spaces. Of course, there might well be many other surfaces that are not built this way, as far as we can tell at this point. The Classification Theorem for Compact, Connected Surfaces will tell us that these connected sums are, in fact, *all* possible such surfaces.

Note that $X\#K^2$ and $X\#P^2$ are nonorientable, regardless of whether or not the surface X is orientable. To see why, imagine oneself as a Flatlander, traversing the "world" $X\#K^2$. If one circumnavigates just the K^2 end of the world, one's sense of inward and outward is reversed. Thus, one could start on the "inside" of X, proceed into the K^2 region, then return to one's starting point, facing "outward," rather than "inward."

Exercises

1. Prove Lemma 10.3.1: If $f : B = B_1(\vec{0})_{\mathcal{U}^2} \to X_\tau$ is a continuous and open injection, with $X_\tau \subset \mathbb{R}^n$, then f extends to a continuous injection of the closed unit disk $\overline{f} : Cl(B) \to X$. Hint: If $x \in S^1 = Bdy(B)$, then there exists a sequence of points $\{x_n\}_{n\in\mathbb{N}} \subset B$ such that $x_n \to x$, by Theorem 9.4.4. Define $\overline{f}(x)$ as $\lim_{n\to+\infty} f(x_n)$. Show f is continuous and 1-1.

2. Prove precisely that Theorem 10.3.2 is true: for any surface X, $X \# S^2 \cong X$. Hint: The pictures following the statement of the theorem give the intuitive idea. To be precise, show that for any neighborhood U in S^2 which is homeomorphic to the planar disk, $S^2 \setminus U \cong B_1(\vec{0})$. Hence "sewing" $S^2 \setminus U$ onto $X \setminus V$ is the same, up to homeomorphism, as sewing a disk onto $X \setminus V$.

3. Show precisely that $P^2 \# P^2 \cong K^2$. (This will be proved in the next section, if you're stuck, but it's a nice exercise to work it out on your own.)

4. Start with the surface S^2 and remove two disjoint open disks. Identify the boundaries of these disks with each other. What is the resulting surface? Why?

5. Answer the question posed above: is the collection of homeomorphism classes of surfaces a *group* under the operation of connected sum?

10.4 THE CLASSIFICATION THEOREM

In this section, we'll state the Classification Theorem for Compact, Connected Surfaces and see where our known examples fit into the list that the theorem provides. We'll defer the proof until after we've added one more tool to our arsenal in Section 10.5.

Theorem 10.4.1. *(Classification Theorem for Compact, Connected Surfaces) Every compact, connected surface is homeomorphic to one of the following:*

- *the sphere, S^2;*
- *a connected sum of n tori, $T^2 \# T^2 \# \cdots \# T^2 = nT^2$, for some $n \in \mathbb{N}$;*
- *a connected sum of n projective planes, $P^2 \# P^2 \# \cdots \# P^2 = kP^2$, for some $k \in \mathbb{N}$.*

Some of our familiar examples of compact, connected surfaces show up in obvious ways in the list: S^2, T^2 and P^2 are the prime examples of each of the three types of surfaces, respectively. However, there are some examples that we're aware of, that, at first glance, do not fit into the classification scheme. This means either that our Classification Theorem is wrong or that we need to look more closely at these examples to see that they are indeed homeomorphic to spaces on the list.

The first such example is the Klein bottle, K^2, which is quite easy to construct and understand, yet lies in apparent contradiction to our theorem! We know that K^2 is nonorientable, so it seems unlikely to show up in the list as a sphere or a connected sum of tori, which are obviously oriented. Our only chance, then, is that the Klein bottle is a connected sum of projective planes.

Theorem 10.4.2. $K^2 \cong P^2 \# P^2 = 2P^2$.

Proof: We show how to build a simple planar diagram for the connected sum of two projective planes. Note that we're not really "cutting" the space – we're just forming a planar representation for it, where any "cuts" will be "sewn up" when we follow the diagram to form the actual surface. We start with two projective planes and "play with" the diagram to get:

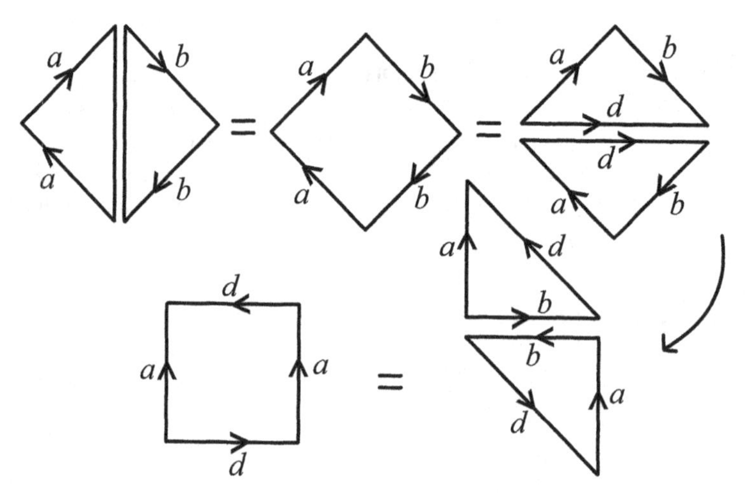

which is exactly the planar diagram for a Klein bottle. □

The next example of a surface that appears to violate the classification system shows up when one thinks critically about the statement of the theorem. We see that connected sums consisting of all tori or of all projective planes appear on the list, but there's no mention of surfaces built as connected sums of both types of spaces. The simplest case of this is $T^2 \# P^2$: where does this fit into the list?

Theorem 10.4.3. $T^2 \# P^2 \cong 3P^2$.

Proof: We'll present two proofs of this result. The first is a rather elaborate exercise involving planar diagrams:

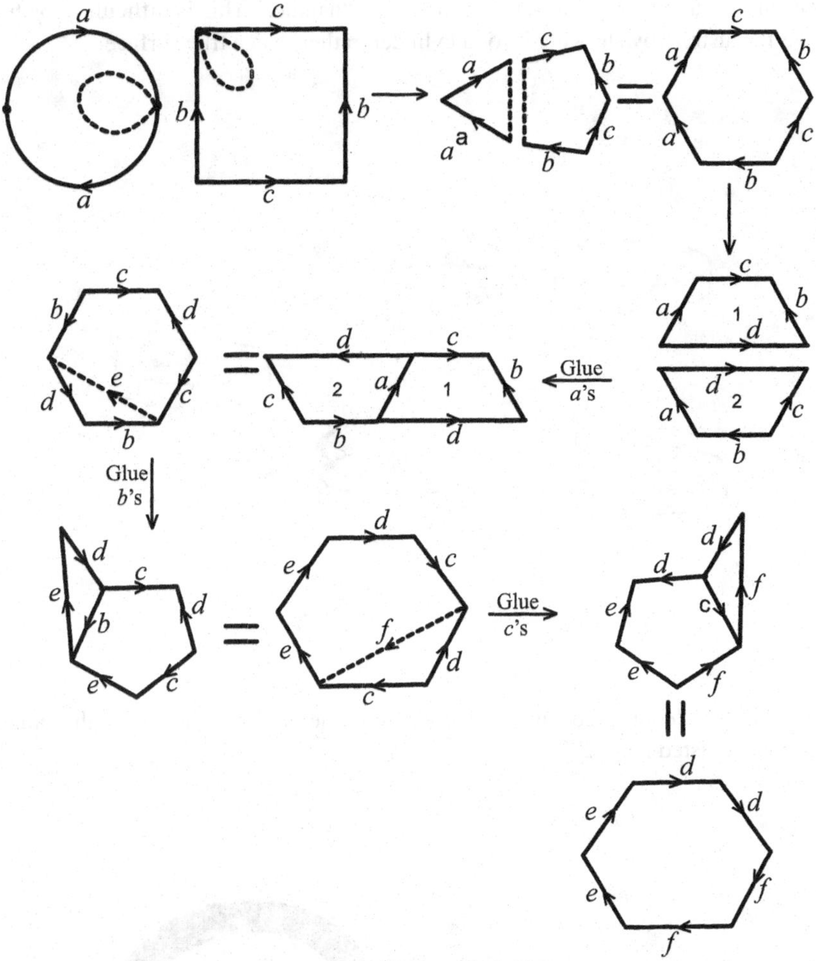

which is easily recognizable as the planar diagram for the connected sum of three projective planes, especially after seeing how to get the diagram for $2P^2$ in the proof of Theorem 10.4.2. However, this proof is actually much more complicated than is necessary. The second proof of the result is adapted from Massey's classic text ([12]). The idea is quite simple. First, recall that the projective plane is constructed by doing the "Moebius half-twist" to a rectangle in two perpendicular directions. We'll present what happens when we form the connected sum of a torus with a Moebius band, as a simpler way to see what happens with $T^2\#P^2$. I've come to call this demonstration the "Slinky trick," because it's easy to present in a classroom with a very large Moebius band and the vintage toy. (The old metal springs work better than the newer plastic versions for this.) First, we observe that "connected summing" a torus onto a surface X is the same as adding a “handle.” This is difficult to picture, so we'll illustrate by showing it for a cylinder, rather than a true surface:

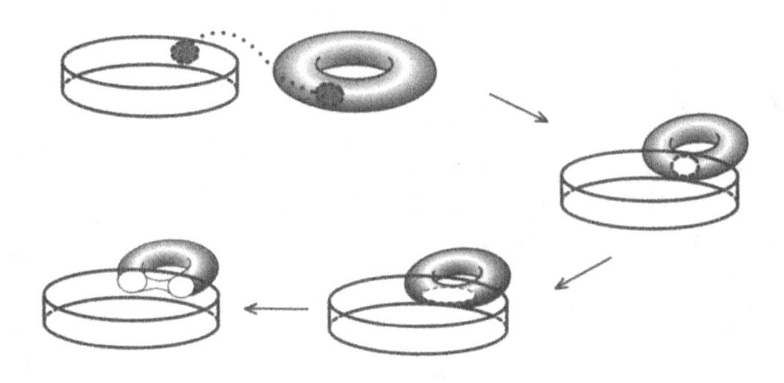

Note also that “connected summing” a Klein bottle onto a surface X is the same as adding a “twisted handle”:

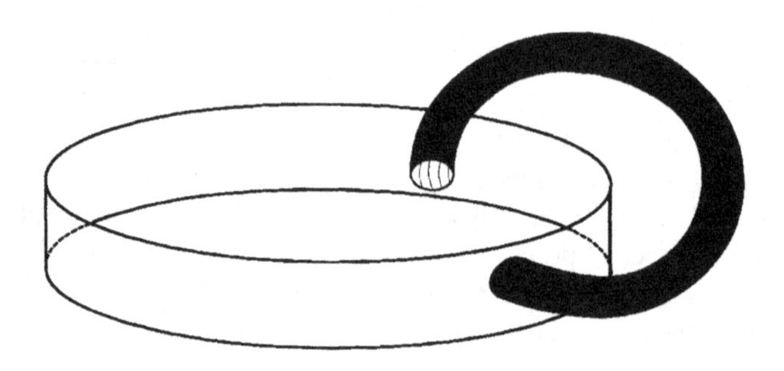

So when we add a torus onto the Moebius band, this is homeomorphic to the spaces shown below:

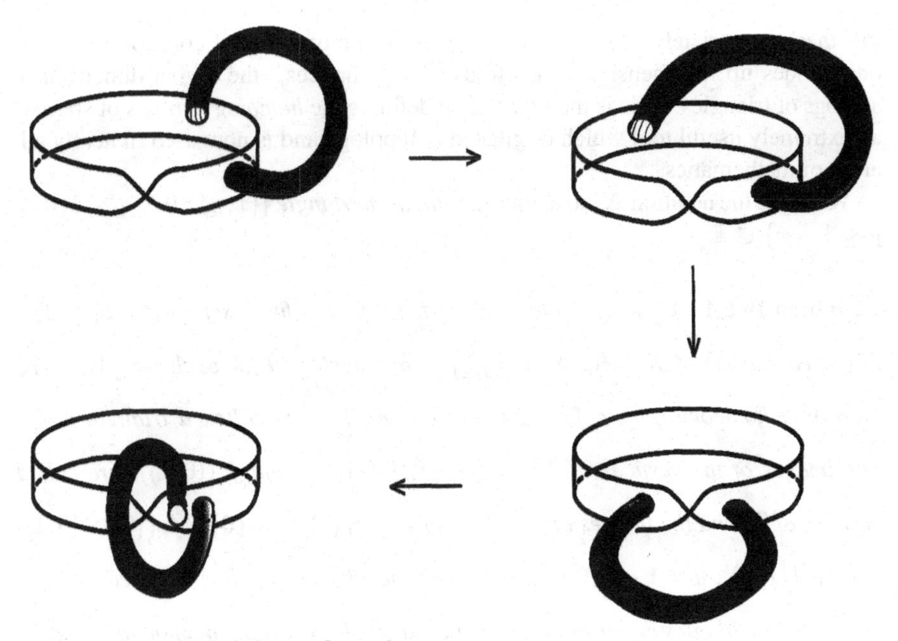

So we see that $T^2\#M^2 \cong K^2\#M^2$, so that

$$T^2\#P^2 \cong K^2\#P^2 \cong (P^2\#P^2)\#P^2 = 3P^2.$$

Slick, eh? □

Exercises

1. Where in the Classification Theorem list does $T^2\#K^2$ fit?
2. Where in the Classification Theorem list does $K^2\#K^2$ fit?
3. Let $X = T^2\#T^2$ and $Y = T^2\#T^2$. Remove *two* open disks from X and *two* open disks from Y and identify the boundary circles from X with those from Y. What is the resulting surface?
4. Let $X = T^2\#T^2$ and $Y = P^2\#P^2$. Remove *two* open disks from X and *two* open disks from Y and identify the boundary circles from X with those from Y. What is the resulting surface?

10.5 TRIANGULATIONS OF SURFACES

The easiest proof of the Classification Theorem requires that we be able to *triangulate* any given surface: to divide it up into regions homeomorphic to triangles which fit together appropriately. This is a special case of a more general concept, in which one divides up n-dimensional manifolds into "simplices," the higher-dimensional analogs of triangles. This is the first step in defining the *homology groups* of spaces, an extremely useful tool which originated in topology and is now used in nearly all areas of mathematics.

We'll use the notation Δ^2 to denote the *standard triangle* $\{(x,y) : 0 \leq x \leq 1, 0 \leq y \leq 1 - x\} \subset \mathbb{R}^2_{\mathcal{U}^2}$.

Definition 10.5.1. *A triangulation of a surface X is a finite set $\{T_1, T_2, \ldots, T_k\}$ of closed subsets of X, with $X = \bigcup_{i=1}^n T_i$ and such that for each $i = 1, \ldots, n$, there is a homeomorphism $f_i : \Delta^2 \to T_i$. Each T_i is called a triangle in X. The images of the vertices of Δ^2, $f_i((0,0))$, $f_i((1,0))$ and $f_i((0,1))$, are called vertices of T_i and the images of the edges of Δ^2, $f_i([0,1] \times \{0\})$, $f_i(\{0\} \times [0,1])$ and $f_i(\{(x,y) : y = 1 - x, 0 \leq x \leq 1\})$ are the edges of T_i. It is required that any two distinct triangles either be disjoint, have a single vertex in common, or have an entire edge in common. That is,*

$$T_i \cap T_j = \begin{cases} \emptyset \\ \textit{vertex} \\ \textit{edge} \end{cases}$$

The following pictures demonstrate some intersections which are not permitted in triangulations:

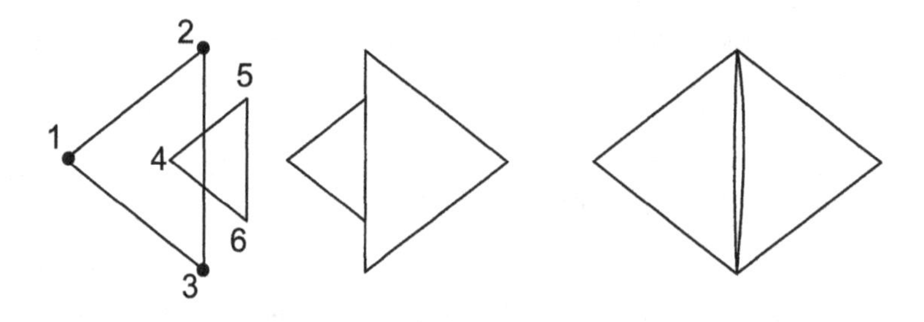

EXAMPLE 10.7

The easiest example of a "legal" triangulation of a surface is the following:

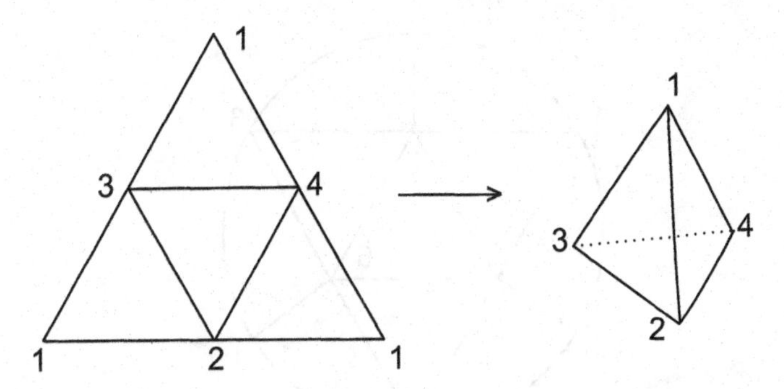

This presents a polygon in the plane which forms a surface homeomorphic to S^2 when we identify the edges as indicated by the numbering. Further, when viewed as S^2, this shows the four closed subsets, each homeomorphic to Δ^2 in an obvious way, with the intersections of the distinct triangles being vertices and edges, as required.

EXAMPLE 10.8

The following figure shows a triangulation of the torus T^2:

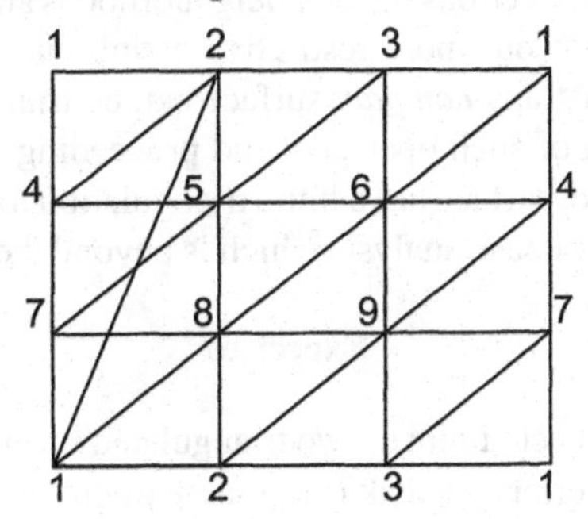

Note that we could specify this example without a picture, just by listing the 18 triangles:

124	245	235
356	361	146
457	578	658
689	649	479
187	128	289
239	379	137

■ **EXAMPLE 10.9**

A triangulation of the projective plane P^2 is given by:

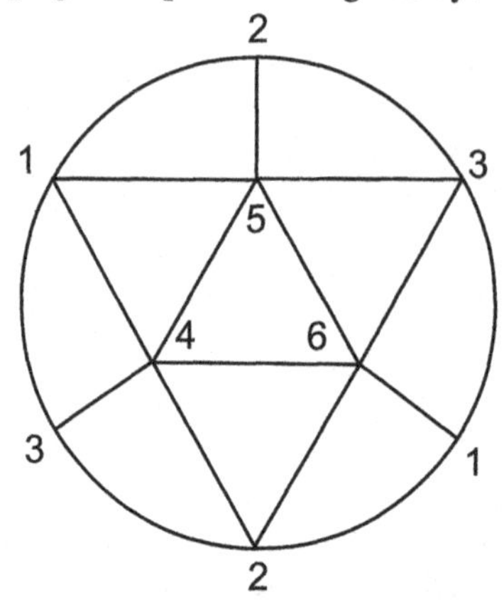

Note that we have required that our triangulations consist of *finitely* many triangular subsets of the surface. Thus, noncompact surfaces like $\mathbb{R}^2$ cannot be triangulated by our definition. If we allowed infinite sets of triangular subsets, one could prove that any surface can be triangulated, by starting with the fact that each point of a surface has a neighborhood homeomorphic to an open planar disk (and hence, homeomorphic to the interior of Δ^2). One then tries to arrange that the intersections of such neighborhoods are replaced with edges of triangles, etc. With our more restrictive definition, one can use a similar argument to prove that any *compact* surface can be triangulated, by choosing a finite subcollection of such open sets and proceeding as above. A rigorous proof that any compact surface has a finite triangulation follows this outline, but requires some pretty intense analysis which is beyond the scope of this course.

Exercises

1. Prove that the connected sum of two triangulated surfaces is triangulated. Hint: cutting out the interior of a disk is the same as cutting out the interior of a ....

10.6 PROOF OF THE CLASSIFICATION THEOREM

Our goal in this section is to present a proof of Theorem 10.4.1: Every compact, connected surface is homeomorphic to one of the following:

- the sphere, S^2;

- a connected sum of n tori, $T^2\#T^2\#\cdots\#T^2 = nT^2$, for some $n \in \mathbb{N}$;
- a connected sum of n projective planes, $P^2\#P^2\#\cdots\#P^2 = kP^2$, for some $k \in \mathbb{N}$.

The proof we present is algorithmic, in that we show a program that, given a compact, connected surface X, allows one to recognize which of the surfaces listed in the theorem that X is homeomorphic to. This approach to the proof first appeared in Siefert and Threfall's work ([17]), and is presented in Massey's book ([12]), and most published proofs of this theorem follow similar lines. A wholly different proof, due to John Conway, is called the ZIP proof (for "Zero Irrelevancy Proof") and can be found in an appendix to Week's text ([19]).
Proof: Let X be a compact, connected surface. We recognize which surface X is homeomorphic to by the following steps.
Step 1: Obtain a planar model for X. As we discussed in the previous section, every compact, connected surface X has a (finite) triangulation. That is, $X = \bigcup_{i=1}^{n} T_i$, where each triangle subset T_i is homeomorphic to the standard triangle $\Delta^2 \subset \mathbb{R}^2_{\mathcal{U}^2}$ and distinct triangles intersect only at a single vertex or an edge. We use this triangulation to produce a planar model for X.

We need to put a particular numbering on these triangles in X, in order to make the construction of the planar model precise. Let any of the triangles be labeled T_1. Choose for T_2 any triangle sharing an edge e_2 with T_1; T_3 is any triangle with an edge e_3 in common with T_1 or T_2, etc., up through T_n. If this process broke down before exhausting the set of triangles, we would have two disjoint sets of triangles $\{T_1, \ldots, T_k\}$ and $\{T_{k+1}, \ldots, T_n\}$ such that all f the triangles in the first set have at most a vertex in common with any triangle from the second set, so that X would be either disconnected (if the sets of triangles are pairwise disjoint and triangles are closed in X) or not a surface (since one can't find an appropriate neighborhood around the common vertex, as the picture below illustrates).

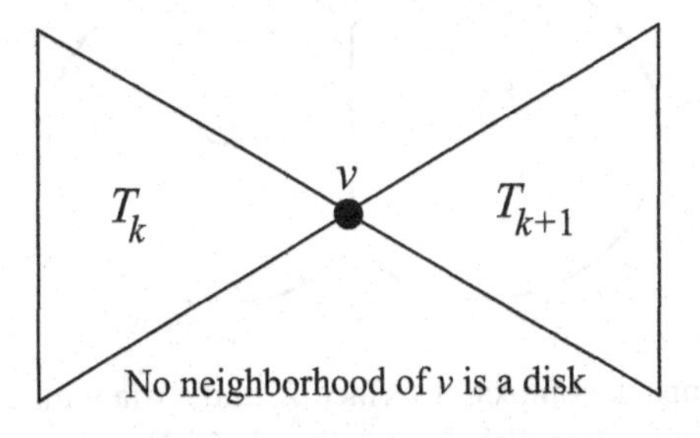

Now that we have an ordering to the triangles and a sequence of edges $e_2, e_3, \ldots, e_n$, we'll build the planar model by assuming we have n disjoint copies of the standard triangle Δ^2, and letting $T = \coprod_{i=1}^{n} \Delta_i^2 \subset \mathbb{R}^2_{\mathcal{U}^2}$. For each i, we have a continuous and open injection $f_i : \Delta_i^2 \to X$, which sends Δ^2 homeomorphically onto T_i. Thus we have a continuous and open map $f : \coprod_{i=1}^n \Delta_i^2 \to X$, which maps in all of the separate triangles at once. It's easy to see that $X = (\bigcup_{i=1}^n T_i)_{/\sim}$, the quotient space obtained by identifying the appropriate edges of the triangles: the edge e_i shows up in T_i and one of the lower triangles $T_1, \ldots, T_{i-1}$. If we identify the corresponding preimages of the edges under f, we obtain a *connected* polygon $D = \left(\coprod_{i=1}^n \Delta_i^2\right)_{/\sim}$. The following example makes this clearer:

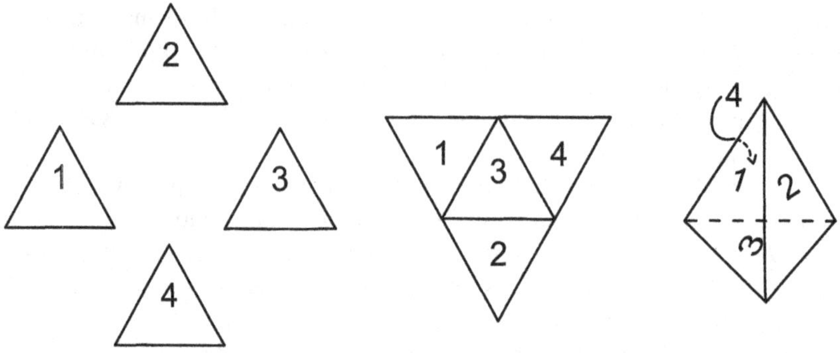

Here we see the separate planar triangles whose images glue together to form S^2, and the corresponding identifications of the planar triangles that form D. Note that the edges e_2, e_3 and e_4 are in the interior of D. We now show that for a general triangulation of a compact, connected surface $X = \bigcup_{i=1}^n T_i$, the quotient space $D = \left(\coprod_{i=1}^n \Delta_i^2\right)_{/\sim}$ is homeomorphic to a closed planar disk. This is easy to see by induction, since each triangle Δ_i^2 is homeomorphic to the closed planar disk $D^2 = Cl(B_1(\vec{0})) \subset \mathbb{R}^2_{\mathcal{U}^2}$ and since the quotient space obtained by gluing together two closed disks along a segment of their boundaries is also homeomorphic to D^2:

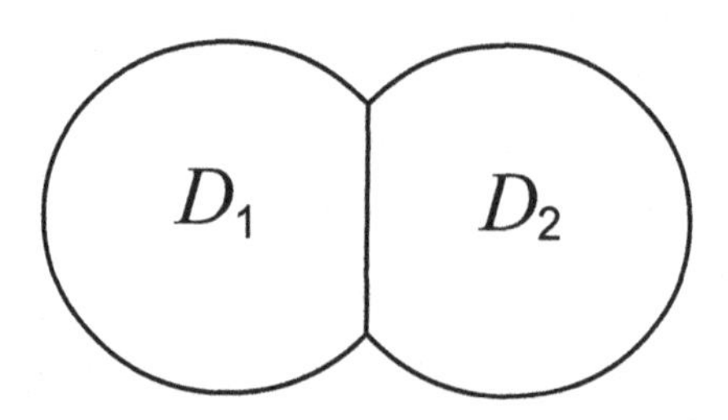

So given a compact, connected surface X and a triangulation of it, we have constructed a closed planar disk D such that $X \cong D_{/\sim}$, where the equivalence relation $\sim$ identifies certain pairs of edges on the boundary of D, as we wished. We continue with the triangulation of the sphere S^2 illustrated above, showing how the planar model can be simplified:

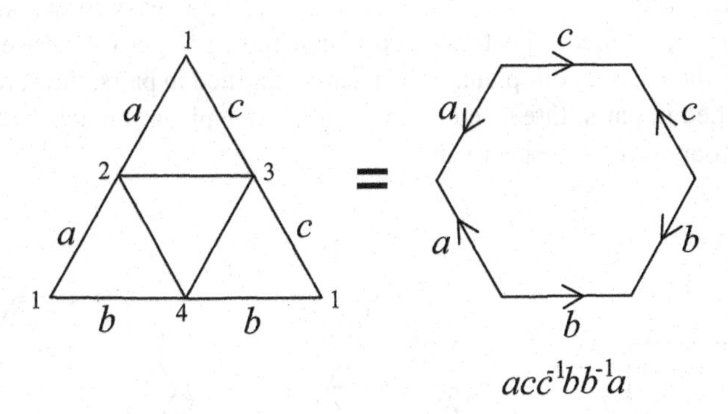

Step 2: Elimination of spheres. For our compact, connected surface X, we have our planar model $X \cong D_{/\sim}$, with the equivalence relation $\sim$ identifying certain pairs of edges on the boundary of D. Any adjacent pairs of edges of the form xx^{-1} may be eliminated from the planar model, if there are at least 4 edges in the model, as the following sequence of pictures illustrates:

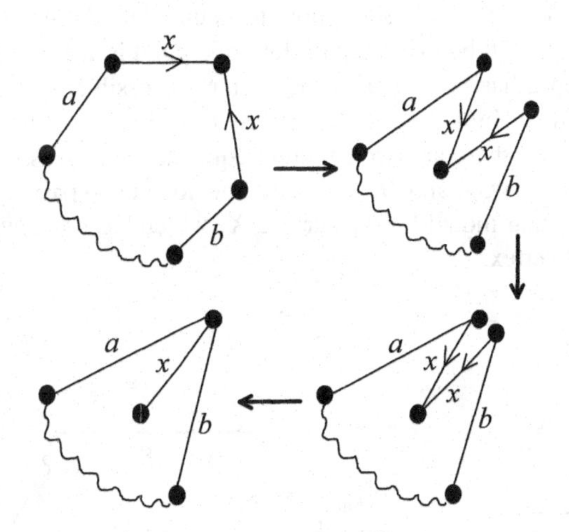

Note that this is a "combinatorial" interpretation of the homeomorphism $X\#S^2 \cong X$.

We eliminate *all* such pairs of edges from our planar model D. If the resulting planar model has fewer than 4 edges, it must be of the form aa^{-1} (a sphere) or bb (a projective plane), and we're finished. If the planar model has 4 or more edges, we continue to Step 3.

Step 3: Elimination of all but one vertex. This step is easy to explain, but it often must be applied repeatedly to a planar model to get the desired result. Although the edges in our planar models are identified in pairs, the vertices can be identified in pairs, threes, etc. For example, the planar models below have one and four vertices, respectively:

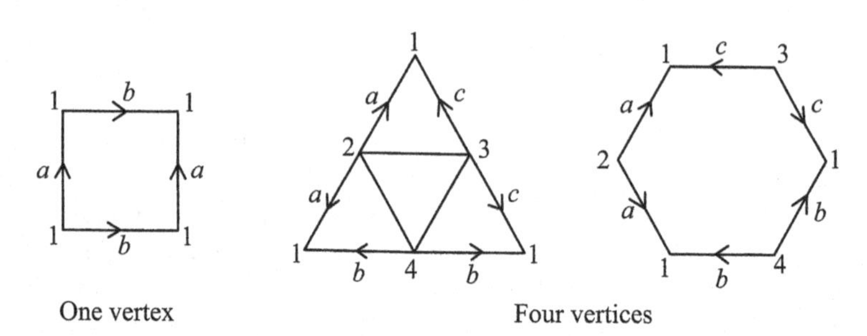

One vertex Four vertices

These illustrate the planar models for the torus (usual model) and the sphere (tetrahedron model). We now show how to manipulate the planar model so that all the vertices will be identified to the same point in the finished surface. Assume that step 2 has been carried out as far as possible, so that there are no redundant spheres in the model. therefore, the model must have at least 4 edges. Suppose that there are two different equivalence classes of vertices in the model, labeled, say, A and B. We show how to cut and paste the diagram, yielding a new planar model for the surface X, which has one fewer A vertex and one more B vertex.

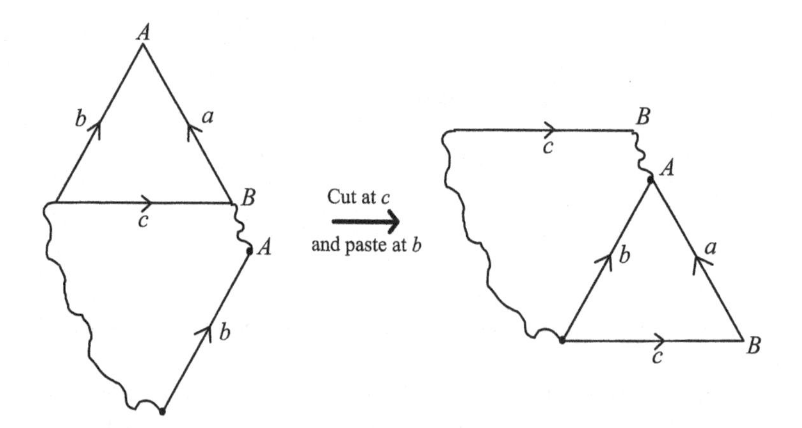

After reducing the occurrence of the unwanted vertex by one, we must go back and eliminate any redundant spheres that may have cropped up (i.e. do Step 2 again). Then we continue with Step 3, as needed. A full example follows.

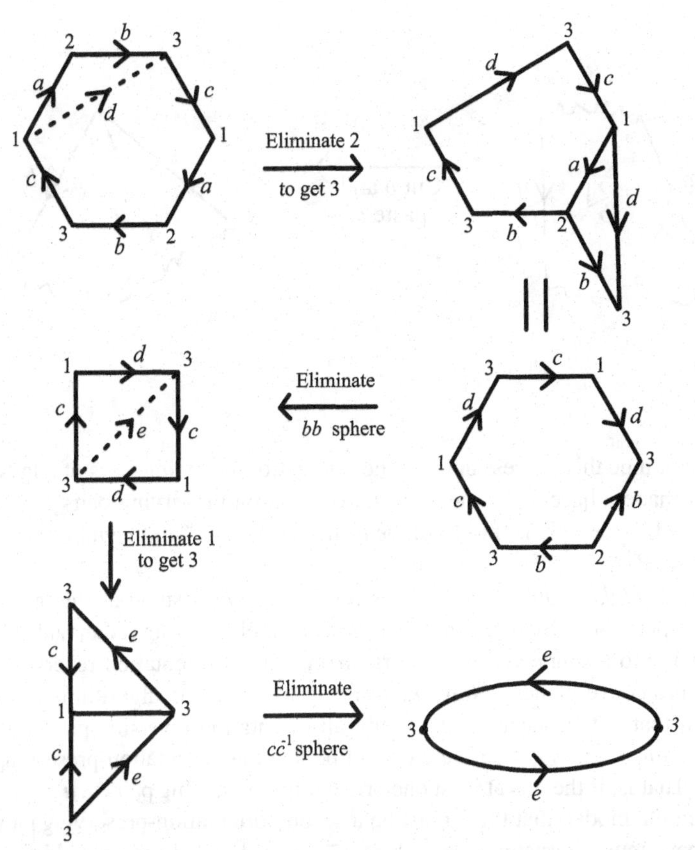

Note that in the example, we recognize the surface at the fourth stage. However, being true to the algorithm, we perform Step 2 and get ee, a projective plane.

Step 4: Make any orientation-reversing pairs adjacent. We now show that any pair of orientation-reversing pairs of edges can be made adjacent (although under different names). Thus, a planar model that looks like $\ldots a \ldots a \ldots$ can be cut and pasted into one that looks like $\ldots bb \ldots$, which is a projective plane P^2 connected summed with the rest of the model. Here's how:

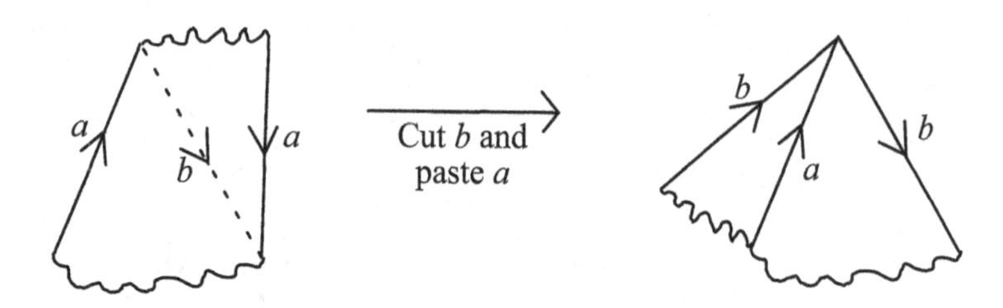

We continue this process until all the orientation-reversing pairs of edges have been made adjacent. If there are no orientation-preserving pairs left, we're finished, since a planar model of the form $a_1a_1a_2a_2\ldots a_ka_k$ yields the surface $kP^2 = P^2\#\cdots\#P^2$.

Step 5: Making any leftover pairs into tori. We assume at this point that we've performed Steps 2 through 4 as needed, eliminating redundant spheres, reducing to a single vertex, and organizing all the orientation-reversing pairs into projective planes. Now, we're ready to finish the algorithm. First, we claim that if the planar model is not all orientation-reversing pairs (such as $a_1a_1a_2a_2\ldots a_ka_k$), then there cannot be just one orientation-preserving pair left. That is, if there is at least one orientation-preserving pair $\ldots a\ldots a^{-1}\ldots$ left in the model, that there must be a second orientation-preserving pair, and the pairs must alternate: $\ldots a\ldots b\ldots a^{-1}\ldots b^{-1}\ldots$. To see why this is so, we ask, what if we had just one orientation-preserving pair, without an intervening second orientation-preserving pair? Then the planar model would look like this:

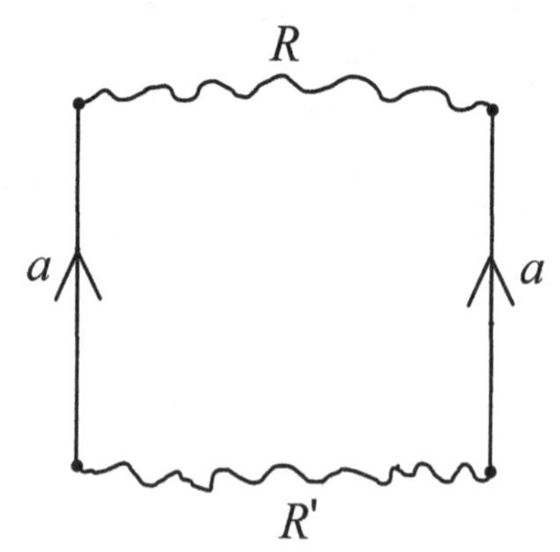

Note that the edges a and a^{-1} separate the remainder of edges in the model into two regions, R and R', with the property that all of the edges in R must be identified with edges back in R, and all of the edges in R' are paired with other edges in R'. But this can't happen, since we've already reduced the model down to one vertex, so that the initial and terminal points of a *must* be identified in the end.

Now we know that if there are any orientation-preserving pairs, there must be at least two such pairs, distributed as $\ldots a \ldots b \ldots a^{-1} \ldots b^{-1} \ldots$. Here's how to get these edges adjacent, using the following two cut-and-paste moves:

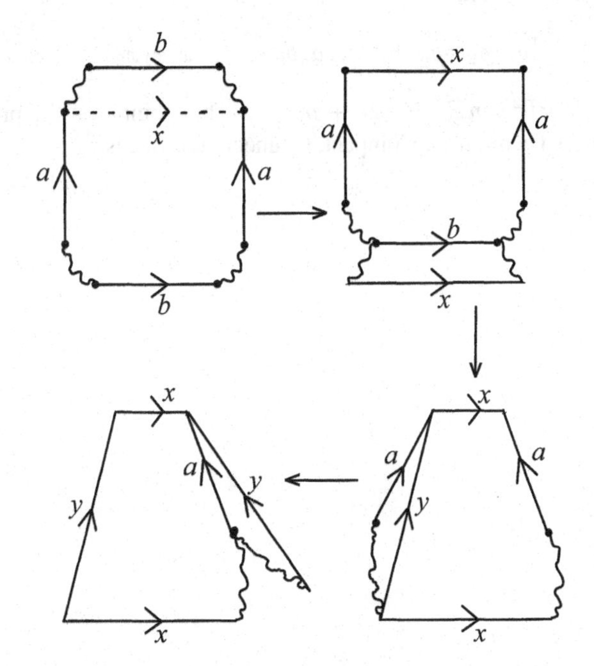

Continue this process until all orientation-preserving pairs are in adjacent groups of four, like $xyx^{-1}y^{-1}$. If there are no remaining orientation-reversing pairs, then our planar model is of the form

$$a_1 b_1 a_1^{-1} b_1^{-1} a_2 b_2 a_2^{-1} b_2^{-1} \ldots a_k b_k a_k^{-1} b_k^{-1},$$

which identifies to form a sum of k tori, and we're done. If there are remaining orientation-reversing pairs, then we're of the form

$$a_1 b_1 a_1^{-1} b_1^{-1} a_2 b_2 a_2^{-1} b_2^{-1} \ldots a_k b_k a_k^{-1} b_k^{-1} x_1 x_1 x_2 x_2 \ldots x_m x_m,$$

which is easily seen to produce $kT^2 \# mP^2$. This isn't on the list of surfaces given in the theorem, but we recall that Theorem 10.4.3 states that $T^2 \# P^2 \cong 3P^2$, so that $kT^2 \# mP^2 \cong (2k + m)P^2$.

To summarize, given a surface X, a triangulation of X produces a planar model (Step 1). We first eliminate redundant spheres, "canceling out" adjacent edges of the form aa^{-1} (Step 2). If the diagram reduces to S^2 or P^2 at this point, we're done. If not, we reduce to a single vertex (Step 3) and make any orientation-reversing pairs of edges adjacent (Step 4). If the diagram reduces to $a_1a_1a_2a_2 \ldots a_ka_k$ at this point, we're done, since this yields the surface $kP^2 = P^2\# \cdots \#P^2$. If not, we make any remaining sets of orientation-preserving pairs adjacent. If the model reduces to $a_1b_1a_1^{-1}b_1^{-1}a_2b_2a_2^{-1}b_2^{-1} \ldots a_kb_ka_k^{-1}b_k^{-1}$ at this point, we're done, since this is kT^2. If not, we must have

$$a_1b_1a_1^{-1}b_1^{-1}a_2b_2a_2^{-1}b_2^{-1} \ldots a_kb_ka_k^{-1}b_k^{-1}x_1x_1x_2x_2 \ldots x_mx_m,$$

which yields $kT^2\#mP^2 \cong (2k+m)P^2$. This completes the proof of the Classification Theorem for Compact, Connected Surfaces. □

To see how this algorithm works in practice, we'll proceed with an example from beginning to end. For this example, we start with the "serial number" $abcdec^{-1}da^{-1}b^{-1}e^{-1}$ and work out the details below.

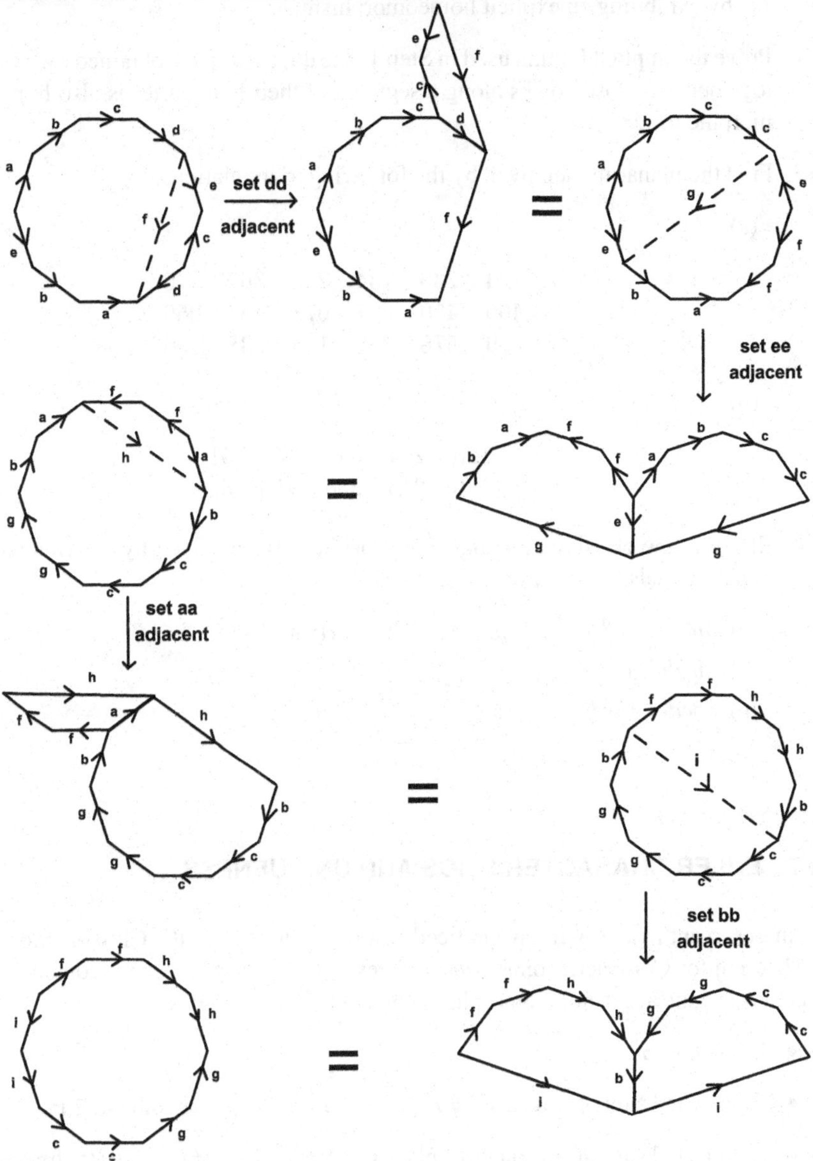

set dd adjacent

=

set ee adjacent

=

set aa adjacent

=

set bb adjacent

=

Thus we end with $ffhhg^{-1}g^{-1}c^{-1}c^{-1}i^{-1}i^{-1}$, which is the connected sum of 5 projective planes.

Exercises

1. Prove that the standard triangle Δ^2 is homeomorphic to the closed planar disk D^2 by exhibiting an explicit homeomorphism.

2. Prove the implicit lemma used in Step 1: the quotient space obtained by gluing together two closed disks along a segment of their boundaries is also homeomorphic to D^2.

3. Find the planar model given by the following triangulations:

   (a)

   $$\begin{array}{cccccc} 124 & 236 & 134 & 246 & 367 & 346 \\ 469 & 459 & 698 & 678 & 457 & 259 \\ 289 & 578 & 358 & 125 & 238 & 135 \end{array}$$

   (b)

   $$\begin{array}{ccccc} 123 & 234 & 345 & 451 & 512 \\ 136 & 246 & 356 & 416 & 526 \end{array}$$

4. Illustrate the classification algorithm for the surfaces given by the following planar models:

   (a) $aa^{-1}fbb^{-1}f^{-1}e^{-1}gcc^{-1}g^{-1}dd^{-1}e$ (Hint: Think "box.")

   (b) $abcdabcd$

   (c) $abcda^{-1}bcd$

10.7 EULER CHARACTERISTICS AND UNIQUENESS

An observant reader will have noticed that our statement of the Classification Theorem for Compact, Connected Surfaces was a bit vague: Every compact, connected surface is homeomorphic to one of the following:

- the sphere, S^2;
- a connected sum of n tori, $T^2 \# T^2 \# \ldots \# T^2 = nT^2$, for some $n \in \mathbb{N}$;
- a connected sum of n projective planes, $P^2 \# P^2 \# \ldots \# P^2 = kP^2$, for some $k \in \mathbb{N}$.

What we've really proved is that every compact connected surface $X_\mathcal{T}$ is homeomorphic to *at least* one surface from this list. We've yet to show that this is a list of *distinct* homeomorphsm classes. That is, we have yet to eliminate the possibility, say, that a connected sum of n tori is homeomorphic to a connected sum of m tori, if $n \neq m$. We already know that spheres and connected sums of tori cannot be homeomorphic to connected sums of projective planes, because projective planes are nonorientable, while spheres and tori are oriented. However, we need another tool to show that the Classification Theorem does indeed provide a list with no overlap.

One tool that would work well for this purpose is the *fundamental group* of a surface, which will be defined in detail in Chapter 11. We'll prove there that one can associate with every topological space X_τ a group, denoted $\pi_1(X_\tau)$, called the fundamental group of X. We'll also show that homeomorphic spaces have isomorphic fundamental groups (but not conversely), so that one could show the classification scheme above has no overlaps if the fundamental groups of these surfaces are not isomorphic. Unfortunately, calculating the fundamental groups of connected sums is pretty difficult, so we'll use a simpler tool, instead.

The tool we'll use is called the *Euler characteristic* of a surface,[9] which is a much simpler topological invariant. The definition relies on being able to find a triangulation of a space, however, so that we'll restrict our discussion of Euler characteristics to compact surfaces. However, there is a version of Euler characteristic that can be defined for nearly all topological spaces, using the *singular homology groups*[10].

We'll begin by dealing with compact surfaces with a given triangulation.

[9]Leonhard Euler (1707-1783) was one of the most prolific mathematicians in history. He first used the combinatorial idea now known as the Euler characteristic to solve the problem of whether or not it is possible to find a path across the bridges of Koenigsberg that crosses each bridge exactly once. Ask your instructor for a diagram of the bridges and islands in the Pregel River and see for yourself. Euler's work in analysis, both real and complex, laid the foundations for our modern approaches to calculus.

[10]Homology theory is a branch of algebra which was invented (discovered?) by topologists in an attempt to "count the n-dimensional holes" is spaces. It dates back to work of Henri Poincare (c1890), who looked at a triangulation of a given manifolds and used this to define a sequence of abelian groups, which encoded much of the structure of the manifold. This work was continued by many topologists (notably Cech and Alexander), until being subsumed in the "singular" homology theory of Eilenberg, MacLane and Steenrod. Nowadays, homology theory is used in many areas of mathematics, from geometry to hard-core analysis.

Definition 10.7.1. *Let X be a compact surface with a triangulation $\{T_1, T_2, \ldots, T_n\}$. Then the Euler characteristic of X is*

$$\chi(X) = V - E + F,$$

where

- *V = the total number of vertices of X,*
- *E = the total number of edges of X, and*
- *F = the total number of faces or triangles of X.*

From this definition, it is not at all apparent that the Euler characteristic depends just on X. At first glance, $\chi(X)$ appears to depend on the *choice* of triangulation of X. However, we've seen different triangulations of the same surface already. For example, here are two different triangulations of S^2:

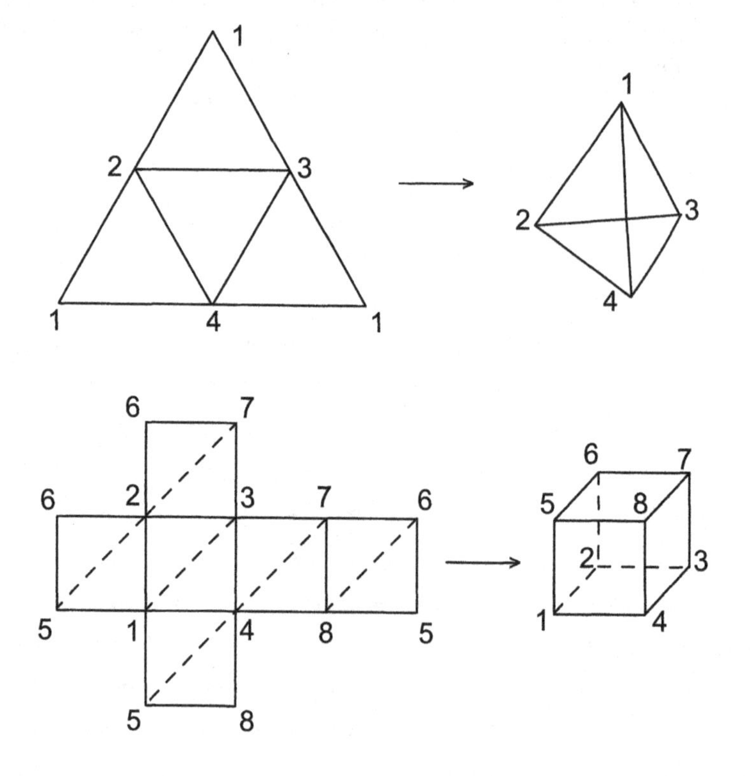

In the tetrahedron triangulation of the sphere, we see that $V = 4, E = 6$ and $F = 4$, so that $\chi(S^2) = 2$ *with the tetrahedron triangulation*. In the box triangulation, we see that $V = 8, E = 18$ and $F = 12$, so that $\chi(S^2) = 2$ with this triangulation, as well.

Of course, we would hope that $\chi(X)$ depends only on the space X, and is independent of the triangulation chosen for X. To see why this is so, we'll look not just at triangulations of surfaces, but at subdivisions of surfaces into "polygonal" regions with n sides for any natural number n. Here are some pictures of such polygonal regions:

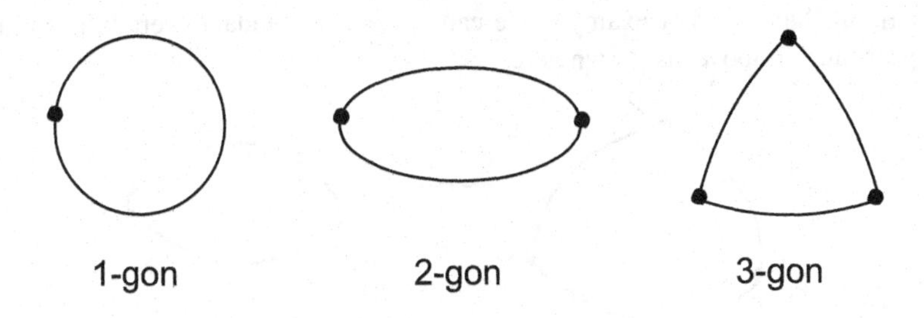

It's easy to take our arguments on triangulations of surfaces in Section 10.5 and translate them into polygonal subdivisions of surfaces. Of course, then, any compact surface can be given a finite polygonal subdivision (with a triangulation as one example).

Given any subdivision of a compact, connected surface X, we define the Euler characteristic as before:

$$\chi(X) = V - E + F,$$

where V is the number of vertices in the polygonal subdivision, E is the number of edges, and F is the number of faces (or polygonal regions). Note that even edges of this sort

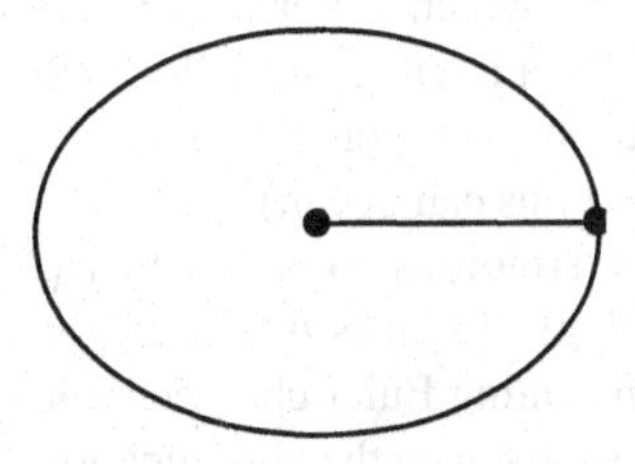

do not alter the Euler characteristic $\chi(X)$, since they add one edge and one vertex to the calculation $V - E + F$, canceling each other out. Indeed, we can easily see that $\chi(X)$ is invariant under the following changes to a polygonal subdivision:

1. Adding a vertex in the interior of an edge (since this adds one vertex and adds a new edge by dividing the original edge into two, canceling out in $V - E + F$).

2. Adding a face by connecting two existing vertices by an edge (since this adds one edge and one face, canceling out).

3. Adding an edge connecting an existing vertex to a new vertex in the interior of a region (as in the picture above).

Note that the inverses of these procedures will also leave the Euler characteristic unchanged. For example, we can erase a "redundant" vertex, reversing procedure 1 above, as shown here:

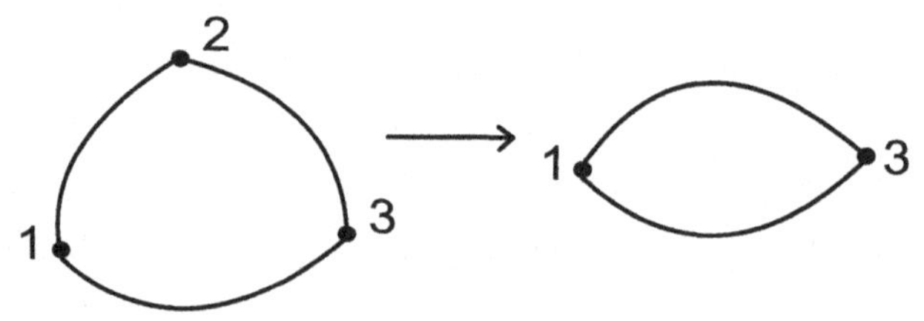

Similarly, one can reverse procedure 2 by amalgamating two adjacent regions by erasing the edge separating them, and can reverse procedure 3 by eliminating "redundant" edges and vertices.

With these procedures in mind, it seems intuitively clear that the Euler characteristic of a compact, connected surface is independent of the triangulation one chooses to compute it. Given two triangulations on such a surface X, it seems quite reasonable that one should be able to perform the three procedures above (and their inverses) repeatedly to one triangulation to transform it into the other. Since these procedures leave the calculation of χ unchanged, this would show that we get the same answer for $\chi(X)$ no matter what triangulation we use. However, this argument works only when the two triangulations of X do not intersect in "pathological" ways. It's possible to get two triangulations of a simple surface like S^2 which are each nice and finite in their own right, but where triangles in one set intersect triangles in the other in infinitely many points. (See the "Topologist's sine curve" in the exercises of Section 6.4 for an idea of how such intersections can occur.)

So we will not present a rigorous proof here that $\chi(X)$ is well-defined, since a proof for general topological spaces is well beyond the scope of this book. However, our goal in presenting Euler characteristics for compact surfaces is to show that the categories given by the classification theorem of disjoint. This will be accomplished with fairly complete rigor in Chapter 11, so we can avoid feeling too guilty about this omission.

The Euler characteristic of a connected sum of surfaces is quite easy to compute:

Theorem 10.7.1. *If X and Y are compact, connected spaces, then*

$$\chi(X\#Y) = \chi(X) + \chi(Y) - 2.$$

Proof: Assume we have triangulations of X and Y. To form $X\#Y$, we remove a disk (homeomorphic to the interior of a triangle/face) from each and indentify the boundaries on the two surfaces. Thus, we remove from $X \cup Y$ exactly 2 faces, 3 edges and 3 vertices, so that $V - E + F$ is lowered by two from adding those quantities from X and Y. □

Now we can easily compute the Euler characteristics of our fundamental surfaces:

- $\chi(S^2) = 2$.
- $\chi(T^2) = 0$.
- $\chi(P^2) = 1$.

Using induction, we can see the Euler characteristic of each of the surfaces in the classification theorem:

- $\chi(S^2) = 2$.
- $\chi(nT^2) = 2 - 2n$.
- $\chi(mP^2) = 2 - m$.

If we assume that the Euler characteristic is a topological invariant, then the classification theorem and the discussion above prove the following:

Theorem 10.7.2. *Two compact, connected surfaces X and Y are homeomorphic if and only if $\chi(X) = \chi(Y)$ and both are orientable or both are nonorientable.*

This lovely theorem shows just how easy the "fundamental problem of topology" is in the case of compact, connected surfaces: telling such spaces apart reduces to the easily-calculated Euler characteristic and determining whether or not the spaces are orientable! Such theorems are quite rare in any area of mathematics, particularly in a rather venerable field like topology.

Exercises

1. Explicitly calculate the Euler characteristic of T^2 from the usual "square" presentation.

2. Explicitly calculate the Euler characteristic of $2T^2$ from the usual "octagonal" presentation.

3. Explicitly calculate the Euler characteristic of P^2 from the usual 2-gon presentation.

4. Explicitly calculate the Euler characteristic of K^2 from the usual "square" presentation.

CHAPTER 11

FUNDAMENTAL GROUPS AND COVERING SPACES

11.1 INTRODUCTION

The content of this chapter is, to a great extent, just an extended illustration of the mathematical concept of a *functor*. A functor is a precise way of taking a problem in one mathematical setting (such as the problem of telling whether two topological spaces $X_\mathcal{T}$ and $Y_\mathcal{V}$ are homeomorphic) and "translating" it into a problem in a different setting (such as determining whether two groups are isomorphic), where one hopes that the new problem is easier to solve than the original. One of the first examples of a functor is the idea of the "fundamental group" of a topological space.

To be precise about what we mean by a functor, we'll need to think in terms of *categories*. Intuitively, a category is a "setting" in which one does mathematics. For example, one may be working on a problem about topological spaces, as opposed to the entirely different setting of abelian groups. Precisely, a category is a collection of *objects*, together with a collection of *morphisms*, functions between objects that preserve the structure inherent to the objects. For example, the category $\mathcal{G}p$ of groups has as its objects the set of all groups, where the morphisms are all possible group homomorphisms. Recall that a group homomorphism from a group G to a group H is a function from the set G to the set H that respects the group operations: $f(g_1 g_2) = f(g_1)f(g_2)$. In other words, the morphism must preserve the group structure. An example closer to home for us is the category $\mathcal{T}op$, whose collection of objects consists of all topological spaces and whose morphisms include all continuous functions of spaces. Again, recall that a continuous function $f : X_\tau \to Y_\nu$ is a function between the underlying sets X and Y, which must respect the topological structure: f^{-1}(any ν – open set) must be τ-open. (On occasion, we might need to work in the category of "pointed spaces" $\mathcal{P}t\mathcal{T}op$ – that is, we specify particular "basepoints" $x_0 \in X_\tau$ and $y_0 \in Y_\nu$ and require that morphisms in the category be continuous maps that send the basepoint in X to the basepoint in Y.) Other examples of categories include $\mathcal{R}g$ of rings (morphisms are ring homomorphisms) and $\mathcal{S}ets$ of sets (where the objects are all sets and the morphisms include *all* functions).

A functor from a category $\mathcal{C}$ to a category $\mathcal{C}'$ is a function

$$F : \{objects\, in\, \mathcal{C}\} \to \{objects\, in\, \mathcal{C}'\}$$

that assigns to every object X in $\mathcal{C}$ a unique object $F(X)$ in $\mathcal{C}'$, which is *natural*, meaning that F satisfies the following properties:

- If $f : X \to Y$ is a morphism in $\mathcal{C}$, then f induces a morphism $f_* : F(X) \to F(Y)$ in $\mathcal{C}'$.
- The identity morphism $i_X : X \to X$ for any object in $\mathcal{C}$ induces the identity morphism $i_* = i_{F(X)} : F(X) \to F(X)$.
- A composition of morphisms $X \xrightarrow{f} Y \xrightarrow{g} Z$ in $\mathcal{C}$ must induce the appropriate composition in $\mathcal{C}'$; That is, $(g \circ f)_* = g_* \circ f_* : F(X) \to F(Z)$.

You might reasonably ask why we need to be so precise about how functors treat the morphisms in our category. The answer is pretty straightforward, if considered carefully. We hope that our functor F, whatever it is, can accurately translate problems in category $\mathcal{C}$ to problems in category $\mathcal{C}'$. To do so, F must send objects that are equivalent in $\mathcal{C}$ to objects that are equivalent in $\mathcal{C}'$. Note that the properties above that characterize naturality do exactly this. As you've no doubt deduced already, two objects X and Y in a category $\mathcal{C}$ are equivalent (or isomorphic) if there are morphisms $f : X \to Y$ and $g : Y \to X$ that are inverses in $\mathcal{C}$; that is, if $g \circ f = i_X$ and $f \circ g = i_Y$. If so, our conditions guarantee that $F(X)$ and $F(Y)$ are isomorphic in $\mathcal{C}'$, with f_* and g_* being the isomorphisms.

Our goal in this chapter is to define precisely a functor from $\mathcal{P}t\mathcal{T}op$ to $\mathcal{G}p$, which takes problems in topological spaces (with basepoints) and continuous functions to problems in groups and group homomorphisms. This functor, known as the "fundamental group," assigns to each topological space X and each point $x_0 \in X$ a group, $\pi_1(X, x_0)$, in a way that is natural with respect to continuous functions between spaces. In particular, spaces that are homeomorphic (isomorphic in $\mathcal{T}op$) will have isomorphic fundamental groups. To define this functor, we'll have to look at the concept of a *homotopy* between two maps of spaces, $f, g : X \to Y$, which continuously deforms f into g, and at the related concept of a *homotopy equivalence* between two spaces.

One very important perk of working with fundamental groups is that it allows us to define a new topological invariant, called *simple-connectivity*. Intuitively, a space is simply-connected if every loop in the space can be shrunk to a point within the space. This invariant will allow us to show that $\mathbb{R}^2_{\mathcal{U}^2}$ is not homeomorphic to $\mathbb{R}^3_{\mathcal{U}^3}$, and will allow us to tell, for example, the torus T^2 apart from the sphere S^2 with little work.

11.2 HOMOTOPY OF FUNCTIONS AND PATHS

Recall that an "isomorphism" of topological spaces is a homeomorphism: A continuous function $j : X_\tau \to Y_\nu$ that has a continuous inverse $k : Y_\nu \to X_\tau$ (so that j must be one-to-one and onto, as well as continuous and open). The key idea here is that the compositions $k \circ j$ and $j \circ k$ are exactly *equal* to the identity maps i_X and i_Y, respectively. We will now generalize this concept, defining a weaker sort of equivalence between spaces, known as a *homotopy equivalence*, which is a continuous function $j : X_\tau \to Y_\nu$ that has an *inverse up to homotopy*. In other words, there is a continuous function $k : Y_\nu \to X_\tau$ so that the compositions $k \circ j$ and $j \circ k$ are "continuously deformable" into the identity maps i_X and i_Y, respectively. We make this notion precise here:

Definition 11.2.1. *Two continuous functions $f, g : X \to Y$ are homotopic (denoted $f \simeq g$) if there exists a continuous function*

$$F : X \times I \to Y,$$

where I is the unit interval $[0, 1]$ in the usual subspace topology, such that $F(x, 0) = f(x)$ for all $x \in X$ and $F(x, 1) = g(x)$ for all $x \in X$. We refer to the map F as a homotopy[1] *from f to g, and often use the shorthand notation $F : f \simeq g$.*

We'll frequently use the notation $F(x, 0) \equiv f(x)$ to indicate that this is an *identity*, valid for all $x \in X$. For example, in $\mathbb{R}$, $x^2 - 1 \equiv (x - 1)(x + 1)$ (since it's valid for all $x \in \mathbb{R}$), whereas the equation $x^2 = 1$ is true for only two such x's. Note that the definition above requires that F be continuous, so that $F|_{X \times \{t\}}$ must be continuous for each $t \in I$. Thus we have an infinite family of maps $X \to$Y, parametrized by t, which "interpolate" between f and g, which gives precise meaning to the intuitive idea that the homotopy F "continuously deforms" the map f into the map g. The diagram below illustrates this idea.

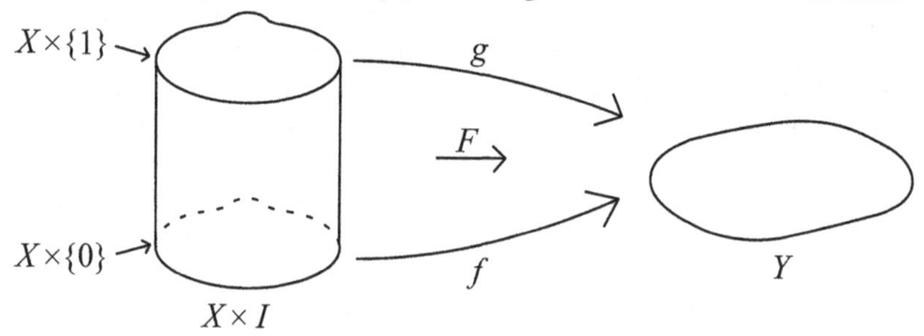

As a simple example, consider the following two maps from $\mathbb{R}_{\mathcal{U}}$ to $\mathbb{R}_{\mathcal{U}}$:

$$\begin{aligned} f(x) &= x \\ g(x) &= x^3 \end{aligned}$$

Then $f \simeq g$, as the following homotopy $F : \mathbb{R} \times I \to \mathbb{R}$ shows: $F(x, t) = (1 - t)x + tx^3$. First, we see that F is continuous, since it's a composition of continuous functions f and g with the operations of +,- and multiplication, all of which are continuous maps from $\mathbb{R} \times \mathbb{R}$ to $\mathbb{R}$. Next, we note that $F(x, 0) \equiv x$ and $F(x, 1) \equiv x^3$, so that we have the required homotopy. However, there's nothing specific about the functions $f(x) = x$ and $g(x) = x^3$ that comes into play in defining the homotopy F. Indeed, *any* two functions $f, g : \mathbb{R}_{\mathcal{U}} \to \mathbb{R}_{\mathcal{U}}$ are

[1]The concept of a homotopy between two maps $f, g : X \to Y$ is due to L.E.J. Brouwer (1881–1966), who published this idea, along with many other seminal concepts, in a series of papers appearing from 1910-1912. In these papers, he made precise many of the ideas of Henri Poincaré, whose 1895 paper*Analysis Situs* is often cited as the beginning of the subject of topology. Brouwer's well-known fixed-point theorem will be proved in Section 11.6.

homotopic, using the homotopy $F(x,t) = (1-t)f(x) + tg(x)$. This property of $\mathbb{R}_{\mathcal{U}}$ is quite important, as we'll see later.

Definition 11.2.2. *A homotopy equivalence from a space X to a space Y is a map $f : X \to Y$ that has a homotopy inverse; that is, there is a map $g : Y \to X$ such that $g \circ f \simeq i_X$ and $f \circ g \simeq i_Y$. We use the notation $X \simeq Y$ to indicate that X is homotopy equivalent to Y.*

As an example, consider the subspace $\{0\} \subset \mathbb{R}_{\mathcal{U}}$. Then the constant map $c_0 : \mathbb{R} \to \{0\}$ is a homotopy inverse to the inclusion $j_0 : \{0\} \hookrightarrow \mathbb{R}$, as we can see easily: $c_0 \circ j_0 = i_{\{0\}}$ and $j_0 \circ c_0 \simeq i_{\mathbb{R}}$, using the homotopy F above, with

$$F(x,t) = (1-t)x + t0 = (1-t)x.$$

Thus the usual real line $\mathbb{R}_{\mathcal{U}} \simeq \{0\}$. In general, a space that is homotopy equivalent to a one-point space is called *contractible.*

We now specialize to the idea of a *path-homotopy*. We recall that a path in a space X is a continuous map $\alpha : I = [0,1]_{\mathcal{U}} \to X$, where $\alpha(0)$ is the initial point and $\alpha(1)$ is the terminal point of the path.

Definition 11.2.3. *Let α and β be two paths in a space X, having the same initial and terminal points. A path-homotopy between α and β is a continuous function $F : I \times I \to X$ such that*

$$\begin{aligned} F(s,0) &\equiv \alpha(s) \\ F(s,1) &\equiv \beta(s) \\ F(0,t) &\equiv \alpha(0) = \beta(0) \\ F(1,t) &\equiv \alpha(1) = \beta(1). \end{aligned}$$

We'll use the notation $\alpha \simeq_p \beta$ to indicate that α is path-homotopic to β, and use the notation $F : \alpha \simeq_p \beta$ to show that F is the path-homotopy.[2] Note that

[2] The idea of path-homotopy actually predates the more general idea of homotopic maps. Certainly Poincaé used the idea of path-homotopy extensively in his 1895 work, although the concept may be even older.

this is a stronger property than just saying that the two maps are homotopic, because the last two conditions of the definition require that for each $t \in I$, $F(*, t) : I \to X$ be a path *from* $x_0 = \alpha(0) = \beta(0)$ *to* $x_1 = \alpha(1) = \beta(1)$. In other words, the path-homotopy F is a continuous, parametrized family of paths, each of them starting at x_0 and ending at x_1. The figure below illustrates the idea:

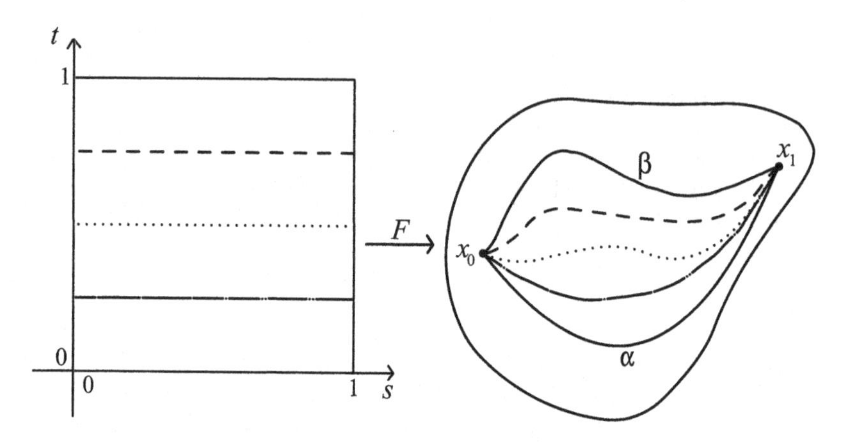

As a simple example, consider the following two paths in $\mathbb{R}^2_{\mathcal{U}^2}$:

$$\begin{aligned} \alpha(t) &= (\cos \pi t, \sin \pi t), \\ \beta(t) &= (1 - 2t, 0). \end{aligned}$$

Both are paths starting at the point (1,0) and ending at (-1,0). (One should sketch the images of α and β to see this clearly.) We can show that α and β are path-homotopic by using the same sort of homotopy that we use for maps into $\mathbb{R}^1$ above. Define $F : I \times I \to \mathbb{R}^2$ by

$$F(s, t) = (1 - t)\alpha(s) + t\beta(s).$$

First, we see that F is continuous, by the continuity of the trigonometric functions (proved in calculus) and that of the usual arithmetic operations. Further, $F(0, t) = (1, 0)$ for all $t \in I$ and $F(1, t) \equiv (-1, 0)$, so that for each t, $F(*, t)$ is a path from $(1, 0)$ to $(-1, 0)$. Finally, $F(s, 0) = \alpha(s)$ for all $t \in I$, and $F(s, 1) \equiv \beta(s)$, so that F is indeed the desired path-homotopy. The following diagrams show why F is referred to as a "straight line" homotopy from α to β.

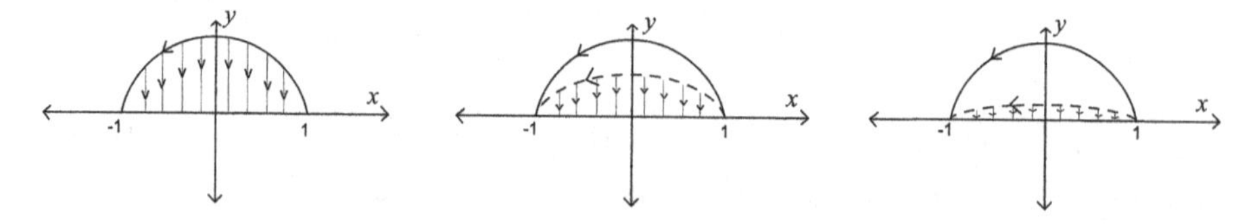

Note that there's nothing particularly special about the paths α and β in this example. In fact, *any* two paths in $\mathbb{R}^2$ with the same initial and terminal points are path-homotopic, with the straight-line homotopy between them being the easiest to write down. However, if we delete a point from $\mathbb{R}^2$, the resulting space no longer has this property. Consider the following two paths in $\mathbb{R}^2 \smallsetminus \{(0,0)\}$:

$$\begin{aligned} \alpha(t) &= (\cos \pi t, \sin \pi t), \\ \beta(t) &= (\cos \pi t, -\sin \pi t). \end{aligned}$$

Then, as before, α and β are both paths from $(1,0)$ to $(-1,0)$, but the straight-line homotopy F between them is no longer well-defined and no longer continuous. In fact, it will turn out that the paths α and β are *not* path-homotopic in $\mathbb{R}^2 \smallsetminus \{(0,0)\}$, although we need results from Section 11.6 to prove this.

The following two theorems show that the relations $\simeq$ and $\simeq_p$ are, as one might expect, equivalence relations.

Theorem 11.2.1. *For any spaces X and Y, homotopy is an equivalence relation on the set of continuous functions from X to Y.*

Theorem 11.2.2. *For any space X, path-homotopy is an equivalence relation on the set of paths from x_0 to x_1 in X.*

We'll prove the second result, leaving the first as an exercise.

Proof: First, we show that $\simeq_p$ is reflexive. If α is any path from x_0 to x_1, then $\alpha \simeq_p \alpha$, as the path-homotopy $F(s,t) = \alpha(s)$ easily demonstrates. Next, we'll show that $\simeq_p$ is symmetric. If $\alpha \simeq_p \beta$, then there must be a path-homotopy $F : I \times I \to X$ with $F(s,0) \equiv \alpha(s)$ and $F(s,1) \equiv \beta(s)$. Define a new path-homotopy G by $G(s,t) = F(s,1-t)$. Then G is continuous, because it's F composed with a continuous arithmetic operation on I, and $G(s,0) \equiv \beta(s)$ while $G(s,1) \equiv \alpha(s)$. Since the endpoints remain at x_0 and x_1 for each t, we see that $\alpha \simeq_p \beta$ implies $\beta \simeq_p \alpha$. Finally, we'll show that $\simeq_p$ is transitive. Let $F : \alpha \simeq_p \beta$ and $G : \beta \simeq_p \gamma$ be two path-homotopies deforming α to β and β to γ, respectively. Then we'll produce the desired path-homotopy from α to γ by concatenating F and G. Precisely, define $H : I \times I \to X$ by

$$H(s,t) = \begin{cases} F(s,2t) & \text{if } t \in [0, \frac{1}{2}], \\ F(s,2t-1) & \text{if } t \in [\frac{1}{2}, 1]. \end{cases}$$

The map H is well-defined, since

$$H(s, \frac{1}{2}) = F(s,1) = G(s,0) = \beta(s).$$

Further, H is continuous by the Pasting Lemma, Theorem 5.5.1. Finally, it's easy to see that $H(s,0) \equiv \alpha(s)$ and $H(s,1) \equiv \gamma(s)$, and that H preserves the endpoints. The figure below shows graphically how H is constructed.

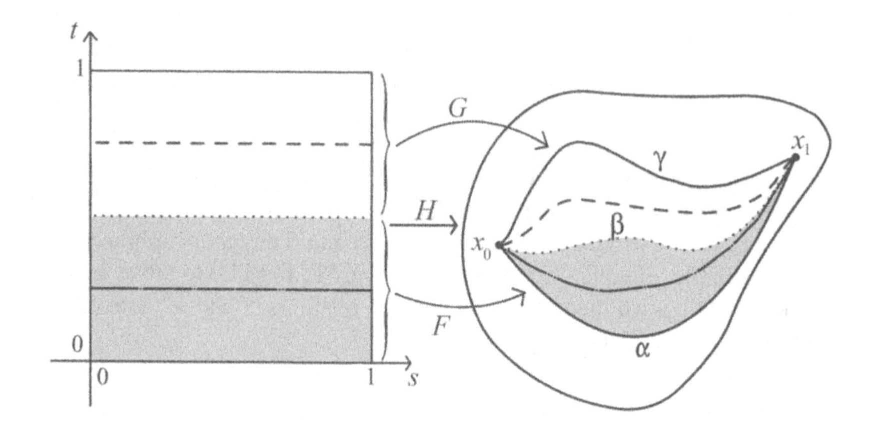

Exercises

1. Prove that if $f, f' : X \to Y$ are homotopic and $g, g' : Y \to Z$ are homotopic, then the compositions $g \circ f$ and $g' \circ f'$ are homotopic.

2. Let $[X,Y]$ denote the set of *homotopy classes of maps* from X to Y. Show that $[X,I]$ has exactly one element. (*Hint*: What's the easiest map to think about?) Show that $[I,Y]$ has exactly one element *if* the space Y is path-connected. Give a counterexample showing that the path-connected hypothesis on Y is necessary.

3. A space X is said to be *contractible* if the identity map on X is homotopic to a constant map.

   (a) Prove that $\mathbb{R}_{\mathcal{U}}$ is contractible.

   (b) Prove that a contractible space is path-connected. (*Hint*: Use the constant map in the definition to help construct the paths that you need.)

   (c) Show that $[X,Y]$ has exactly one element if Y is contractible.

   (d) Show that $[X,Y]$ has exactly one element if X is contractible and Y is path-connected.

11.3 AN OPERATION ON PATHS

Our goal in this section is to introduce an operation on the set of paths in any space X, with an eye toward defining a group to be associated to each space X. Unfortunately, the operation on paths works *only* when the terminal point of the first path agrees with the initial point of the second path. Even then, the operation fails to have most of the properties we want: associativity, identity (or, more precisely, identities), and inverses. However, if we look at *path-homotopy classes* of paths, rather than the paths themselves, everything can be made to work nicely. We begin by defining the operation.

Definition 11.3.1. *Let α and β be two paths in a space X, with $\alpha(1) = \beta(0)$. We define the concatenation $\alpha * \beta : I \to X$ as follows:*

$$\alpha * \beta(s) = \begin{cases} \alpha(2s) & \text{if } s \in [0, \frac{1}{2}], \\ \beta(2s-1) & \text{if } s \in [\frac{1}{2}, 1]. \end{cases}$$

The new path $\alpha * \beta$ is well defined, since $\alpha * \beta(\frac{1}{2}) = \alpha(1) = \beta(0)$, and $\alpha * \beta$ is continuous, by the Pasting Lemma. (Note that the two subspaces on which α and β are defined are $[0, \frac{1}{2}]$ and $[\frac{1}{2}, 0]$, both of which are closed as subspaces of $[0, 1]_{\mathcal{U}}$, so the Pasting Lemma applies nicely.) The best way to think about the operation is that one first does the path α at twice the usual speed, followed immediately by doing the path β at twice the usual speed.

One might expect that the *constant path* at x_0, defined by $c_{x_0}(s) \equiv x_0$, should act as a left identity for the paths from x_0 to x_1. Certainly the picture one draws looks correct. However, the path $c_{x_0} * \alpha$ and the path α are very different, despite the fact that their images are equal (which explains why the sketch you've no doubt drawn by now looks correct). Why? Well,

$$c_{x_0} * \alpha = \begin{cases} x_0 & \text{if } s \in [0, \frac{1}{2}], \\ \alpha(2s-1) & \text{if } s \in [\frac{1}{2}, 1], \end{cases}$$

so that it hits the same points as α does, but for very different values of s, in general. To get around this, we will work with the equivalence classes given by path-homotopy, which we proved to be an equivalence relation at the end of the last section.

We will use the notation $[\alpha]$ to denote the path-homotopy class of the path α, where α runs from x_0 to x_1 in X:

$$[\alpha] = \{\mu : I \to X \mid \mu \text{ is a path, } \mu(0) = x_0,\ \mu(1) = x_1, \text{ and } \mu \simeq_p \alpha\}.$$

We now define the operation $*$ on path-homotopy classes, by mandating that

$$[\alpha] * [\beta] = [\alpha * \beta],$$

for paths α from x_0 to x_1 and paths β from x_1 to x_2. Again, this operation is defined on path-homotopy classes *only* when the terminal point of all paths in the first path-homotopy class agrees with the initial point of all paths in the second path-homotopy class. As it stands, however, there's no reason to believe that this definition makes sense. The following result proves that the operation $*$ and path-homotopy are indeed compatible.

Theorem 11.3.1. *The concatenation operation is well defined on path-homotopy classes.*

Proof: Let α be any path in X from x_0 to x_1, and β any path from x_1 to x_2. If $\alpha \simeq_p \alpha'$, via a path-homotopy F, and $G : \beta \simeq_p \beta'$, we need to show that $\alpha * \beta \simeq_p \alpha' * \beta'$. We can do this by concatenating the path-homotopies F and G as follows. Define $H : I \times I \to X$ by

$$H(s,t) = \begin{cases} F(2s,t) & \text{if } s \in [0, \frac{1}{2}], \\ G(2s-1,t) & \text{if } s \in [\frac{1}{2}, 1]. \end{cases}$$

The H is continuous by the Pasting Lemma, since F and G are continuous and $F(1,t) \equiv x_1 \equiv G(0,t)$. Further, $H|_{I\times\{0\}} = \alpha * \beta$, and $H|_{I\times\{1\}} = \alpha' * \beta'$, as we wished. The figure below illustrates the construction of H. □

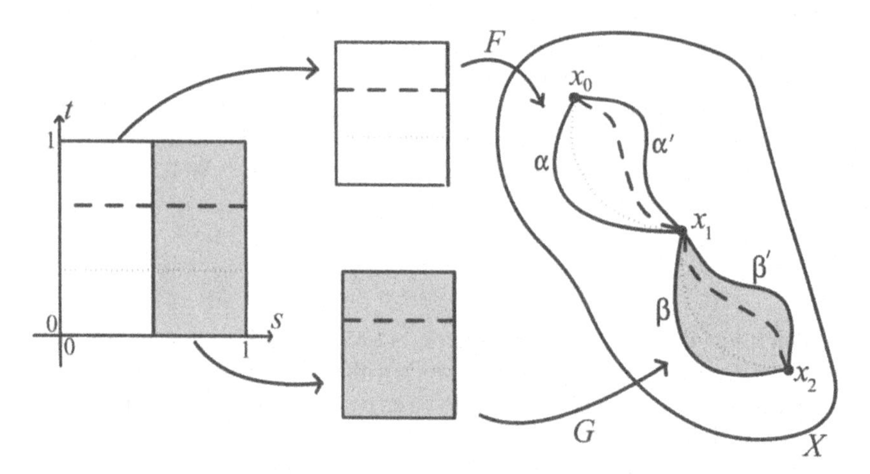

We now have a well-defined operation on path-homotopy classes of paths, whenever the endpoints match up appropriately. We now show that this operation is the one we're looking for.

Theorem 11.3.2. *For all paths in a space X,*

1. *The operation $*$ is associative; whenever $(\alpha * \beta) * \gamma$ is defined, so is $\alpha * (\beta * \gamma)$, and*
$$[(\alpha * \beta) * \gamma] = [\alpha * (\beta * \gamma)].$$
2. *The operation $*$ has right and left identities; if α is a path from x_0 to x_1, then*
$$[c_{x_0}] * [\alpha] = [\alpha] = [\alpha] * [c_{x_1}].$$
3. *Every path in X has a right and left inverse; if α is a path from x_0 to x_1, then the path $\overline{\alpha} : I \to X$ defined by $\overline{\alpha}(s) = \alpha(1 - s)$ has*
$$[\alpha] * [\overline{\alpha}] = [c_{x_0}] \text{ and } [\overline{\alpha}] * [\alpha] = [c_{x_1}].$$

Note that the inverse path $\overline{\alpha}$ is simply the original path α traversed backward. The proof of the theorem involves constructing path-homotopies using fairly elementary geometric ideas. The figures will go a long way toward making the path-homotopies clear.
Proof:

1. We need to show that $(\alpha * \beta) * \gamma \simeq_p \alpha * (\beta * \gamma)$, whenever α, β and γ are paths from x_0 to x_1, x_1 to x_2 and x_2 to x_3, respectively. Recall that $(\alpha * \beta) * \gamma$ is defined as follows:
$$(\alpha * \beta) * \gamma(s) = \begin{cases} \alpha(4s) & \text{if } s \in [0, \frac{1}{4}], \\ \beta(4s - 2) & \text{if } s \in [\frac{1}{4}, \frac{1}{2}], \\ \gamma(2s - 1) & \text{if } s \in [\frac{1}{2}, 1]. \end{cases}$$
So $(\alpha * \beta) * \gamma$ does α at 4 times the usual speed, followed by β at $4\times$ speed, then γ at $2\times$ speed. The path $\alpha * (\beta * \gamma)$ has the same image in the space X, but does α at twice speed, followed by β and γ at $4\times$ speed. We need a path-homotopy that continuously adjusts these speeds. The easiest is this "linear" version:
$$F(s,t) = \begin{cases} \alpha(\frac{4s}{t+1}) & \text{if } s \in [0, \frac{t+1}{4}], \\ \beta(4s - t - 1) & \text{if } s \in [\frac{t+1}{4}, \frac{t+2}{4}], \\ \gamma(\frac{4s-t-2}{2-t}) & \text{if } s \in [\frac{t+2}{4}, 1]. \end{cases}$$

The figure below illustrates how the path-homotopy F is put together:

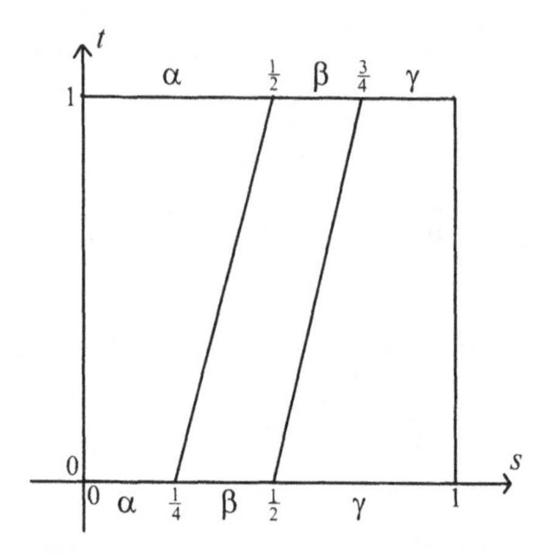

To see that this does the trick, let's examine the first case more closely. We're defining F to take on values of the path α for s between 0 and the line $t = 4s - 1$, alias $s = \frac{4s}{t+1}$. When $t = 0$, we're doing α at $4\times$ speed, as we wish. When $t = 1$, we've slowed down to doing α at twice speed. In between, the speed at which α is traversed slows as t increases. Finally, when (s, t) lies on the line $t = 4s - 1$, alias $s = \frac{4s}{t+1}$, we get $F(s, t) = \alpha(\frac{4s}{4s-1+1}) = \alpha(1) = x_1$, as we wish. Similarly, we get the desired endpoints of the paths for all (s, t) on the lines indicated, so that F is continuous by the Pasting Lemma.

2. We'll prove that $c_{x_0} * \alpha \simeq_p \alpha$. The other case is quite similar. The homotopy G that does this is suggested by the following diagram:

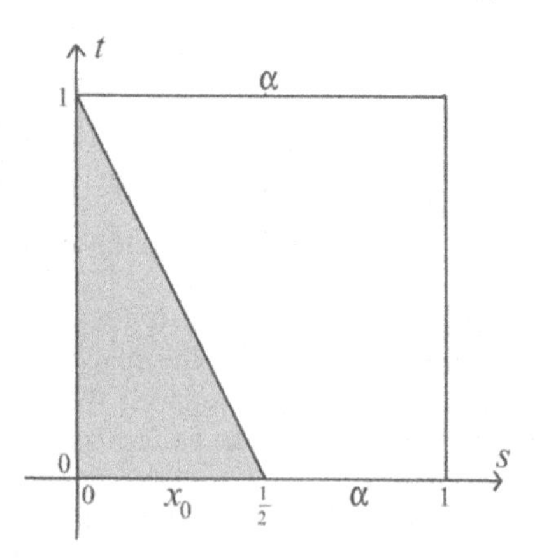

Note that when $t = 0$, we want to "sit" at x_0 from $s = 0$ to $s = \frac{1}{2}$, then do α at $2\times$ speed. At $t = 1$, we want just α at unit speed. The dividing line is given by $t = -2s + 1$, alias $s = \frac{1-t}{2}$. So the homotopy must be

$$G(s,t) = \begin{cases} x_0 & \text{if } s \in [0, \frac{1-t}{2}], \\ \alpha(\frac{2s}{t+1}) & \text{if } s \in [\frac{1-t}{2}, 1]. \end{cases}$$

Again, we check to see that when (s, t) is on the dividing line, we get $G(s, t) = x_0$, so that G is continuous.

3. For α any path from x_0 to x_1, its inverse (both right and left) is the backward path $\overline{\alpha}$, given by $\overline{\alpha}(s) = \alpha(1 - s)$. We need to show that $\alpha * \overline{\alpha} \simeq_p c_{x_0}$. The other case is similar. The idea is to set up a path homotopy that does all of $\alpha * \overline{\alpha}$ at the bottom, and sits still at x_0 at the top. To do this, the easiest way is to do less and less of $\alpha * \overline{\alpha}$ as you go farther up the t-axis in the path homotopy, as the following diagram suggests:

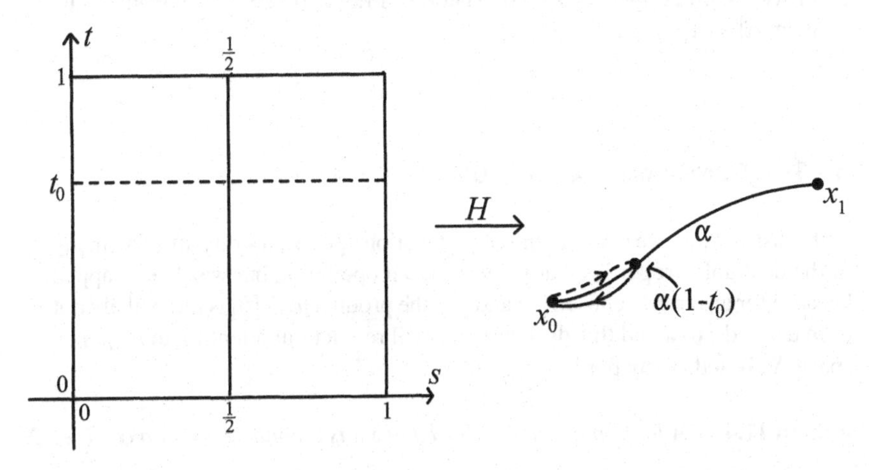

This suggests that we define the path-homotopy H as follows:

$$H(s,t) = \begin{cases} \alpha(2(1-t)s) & \text{if } s \in [0, \frac{1}{2}], \\ \alpha(2(1-t)(1-s)) & \text{if } s \in [\frac{1}{2}, 1]. \end{cases}$$

So when $t = 0$, $H(s, 0)$ does all of α and $\overline{\alpha}$, both at twice speed. When $t = 1$, one sits quietly at x_0 for all s. In between, at height t we do α from $\alpha(0)$ out to $\alpha(1 - t)$ (when $s = \frac{1}{2}$), then turn around and go back via $\overline{\alpha}$. Again, we get continuity by the Pasting Lemma, since the paths agree at the endpoints. □

So we have a nice operation on paths, but it doesn't have the properties that we generally want an operation to have, at least on the *paths* themselves. When we work with *path-homotopy classes* of paths, we do indeed get the properties we want: associativity, left and right identities and inverses. However, the operation is still not very satisfactory: $\alpha * \beta$ is defined **only** when we're lucky enough that the terminal point of α is equal to the initial point of β. So we definitely do not have a group structure on the set of path-homotopy classes of paths in a space X, because for arbitrarily chosen $[\alpha]$ and $[\beta]$, the product $[\alpha] * [\beta]$ is most likely not even defined. However, this operation does give

$$\{[\alpha] \mid \alpha : I \to X \text{ is a path}\}$$

the structure of a *groupoid*. We'll get around this difficulty in the next section.

Exercises

1. Prove the other half of part 2 of Theorem 11.3.2: If α is a path from x_0 to x_1, then $[\alpha] = [\alpha] * [c_{x_1}]$.
2. Prove the other half of part 3 of Theorem 11.3.2: If α is a path from x_0 to x_1, then $[\overline{\alpha}] * [\alpha] = [c_{x_1}]$.

11.4 THE FUNDAMENTAL GROUP

In the last section, we saw that the concatenation operation $*$ on paths had many of the desirable properties that we want in an operation, but was handicapped because for arbitrarily chosen $[\alpha]$ and $[\beta]$, the product $[\alpha] * [\beta]$ is most likely not even defined. To avoid this difficulty, we will restrict our attention to *loops*in a space X, based at any particular point x_0.

Definition 11.4.1. *A loop in a space X based at x_0 is a continuous map $\alpha : I \to X$ such that $\alpha(0) = \alpha(1) = x_0$; that is, a loop based at x_0 is a path whose initial and terminal points are both x_0.*

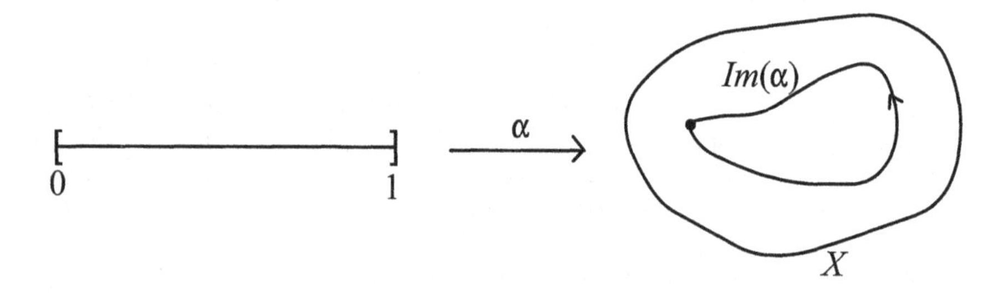

With this in mind, we make the main definition of this chapter:

Definition 11.4.2. *For X any space and any $x_0 \in X$, we define the fundamental group*[3] *of X based at x_0, denoted by $\pi_1(X, x_0)$, to be the set of path-homotopy classes of loops in X based at x_0, with concatenation operation $*$; that is,*

$$\pi_1(X, x_0) = \{[\alpha] \mid \alpha : I \to X \text{ is a path, } \alpha(0) = \alpha(1) = x_0\}$$

under the concatenation operation $*$.

The operation is defined for any two path-homotopy classes of loops at x_0, since the initial points and terminal points of both loops must be x_0, so we've avoided the problem that's plagued us so far. Note that Theorem 11.3.2 shows that the operation does indeed make the set $\pi_1(X, x_0)$[4] into a group, with identity $[c_{x_0}]$. First, we need an easy example or two.

■ EXAMPLE 11.1

Let x_0 be any point in $\mathbb{R}^2_{\mathcal{U}^2}$. Then $\pi_1(\mathbb{R}^2, x_0) = 0$, the group with only one element (namely, the identity). To see why, consider the straight-line homotopy

$$F(s,t) = (1-t)\alpha(s) + tx_0,$$

which shrinks any loop α based at x_0 to the constant loop c_{x_0}. Thus, for any $[\alpha] \in \pi_1(\mathbb{R}^2, x_0)$, we see that $[\alpha] = [c_{x_0}]$.

Of course there's nothing particularly special about $n = 2$ here: $\pi_1(\mathbb{R}^n, x_0) = 0$ for any $n \in \mathbb{N}$, using the same homotopy. In fact, any convex subset[5] of $\mathbb{R}^n$ has 0 for its fundamental group.

[3] Henri Poincaré defined this concept in his groundbreaking 1895 paper, using the notation $\alpha \vee \beta$ to denote the concatenation product.

[4] The notation used here is quite suggestive – are there groups $\pi_2(X, x_0), \pi_3(X, x_0)$, and so on ? The answer is yes, there is, in fact, an infinite sequence of these "higher homotopy groups" for any space X. Roughly speaking, these higher homotopy groups completely determine a space X up to homotopy equivalence (but not up to homeomorphism).

[5] A subset $A \subset \mathbb{R}^n$ is convex if for every pair of points x_0 and $x_1 \in A$, the entire line segment from x_0 to x_1 is contained in A. For example, $\mathbb{R}^n$ is convex, as is $B_1(\vec{0})$, the solid ball of radius 1 (in the usual metric).

Since the examples above all have trivial fundamental groups, one might ask whether there's an easy example of a space X with $\pi_1(X, x_0) \neq 0$, for some $x_0 \in X$. The answer to this depends on your definition of "easy." It will turn out, for example, that the unit circle $S^1 \subset \mathbb{R}^2$ has a nontrivial fundamental group. Intuitively, the loop $\alpha(s) = (\cos(2\pi s), \sin(2\pi s))$, which wraps around the circle once counterclockwise, cannot be shrunk to the constant loop at $(1, 0)$. However, the easiest way to prove this precisely will require the machinery of covering spaces (Section 11.5).

So far, it appears that the fundamental group $\pi_1(X, x_0)$ depends strongly on the choice of the basepoint x_0. For path-connected spaces X, this dependence is only up to isomorphism, as the following construction will show (after some work).

Let $\alpha : I \to X$ be any path from x_0 to x_1. Given any loop β based at x_0, we can use α to build a loop based at x_1 as follows. Define $\tilde{\alpha}(\beta)$ to be the loop $\overline{\alpha} * \beta * \alpha$, which is based at x_1:

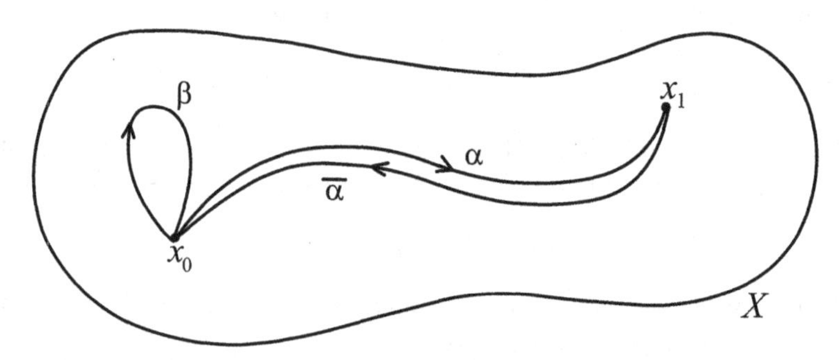

In fact, we can use this idea to define such a construction on the path-homotopy classes of loops.

Theorem 11.4.1. *Let $\alpha : I \to X$ be any path from x_0 to x_1. Then the construction*

$$\tilde{\alpha}([\beta]) = [\overline{\alpha}] * [\beta] * [\alpha]$$

defines an isomorphism

$$\tilde{\alpha} : \pi_1(X, x_0) \to \pi_1(X, x_1).$$

Proof: First, we'll have to prove that $\tilde{\alpha}$ is well-defined. Since we know that the concatenation operation $*$ is well-defined on path-homotopy classes, $\tilde{\alpha}$ must be well-defined. Next, we'll show that $\tilde{\alpha}$ is a homomorphism. Let β and γ be any two loops based at x_0. Then

$$\begin{aligned}\tilde{\alpha}([\beta] * \tilde{\alpha}[\gamma]) &= ([\overline{\alpha}] * [\beta] * [\alpha]) * ([\overline{\alpha}] * [\gamma] * [\alpha]) \\ &= [\overline{\alpha}] * [\beta] * [c_{x_0}] * [\gamma] * [\alpha] \\ &= [\overline{\alpha}] * ([\beta] * [\gamma]) * [\alpha] \\ &= \tilde{\alpha}([\beta] * [\gamma]).\end{aligned}$$

Now we'll show that $\tilde{\alpha}$ is an isomorphism by constructing an inverse. Note that $\overline{\alpha}$ is the inverse path for α, and it defines a function

$$\tilde{\overline{\alpha}} : \pi_1(X, x_1) \to \pi_1(X, x_0).$$

Then an easy[6] algebraic calculation (really!) shows that $\tilde{\overline{\alpha}}$ and $\tilde{\alpha}$ undo each other, proving the theorem. (Yes, you should verify that the computation is, indeed, easy.) □

So for path-connected spaces, the fundamental group depends on the choice of basepoint only up to isomorphism:

Corollary 11.4.2. *If X is path-connected and x_0 and x_1 are two points in X, then $\pi_1(X, x_0) \cong \pi_1(X, x_1)$.*

This corollary is an easy consequence of Theorem 11.4.1.

Note, however, that isomorphisms between groups are not necessarily the identity homomorphism. Different paths from x_0 to x_1 might result in very different isomorphisms between the fundamental groups.

Since the fundamental group of a path-connected space is unique, up to isomorphism, we can make the following definition without fear of ambiguity:

Definition 11.4.3. *A space X is simply-connected if X is path-connected and if $\pi_1(X, x) = 0$ for any $x \in X$.*

So the examples above (viz., $\mathbb{R}^n$ and its convex subsets) are all simply-connected. We shall see soon what a powerful property this is.

So far, then, we have shown how to assign a group to each space X, which is unique up to isomorphism if the space is path-connected. Now we need to show that this construction can be used effectively to tell when spaces are not homeomorphic. In other words, we need to show that the fundamental group is a functor from the category of topological spaces and continuous functions

[6]Beware of authors who use terminology like this! In many cases, the phrase "it is easily shown that ...." should be translated as "I don't know how to write down the details of ...."

to the category of groups and group homomorphisms. The first step is to show that a continuous map between spaces will induce a homomorphism between the fundamental groups of those spaces.

Theorem 11.4.3. *Let $f : X \to Y$ be a continuous function. Then the function*

$$f_* : \pi_1(X, x_0) \to \pi_1(Y, f(x_0))$$

given by $f_*([\alpha]) = [f \circ \alpha]$ is a homomorphism of groups.

We will refer to f_* as the homomorphism induced by f. The illustration below helps explain how f_* is defined:

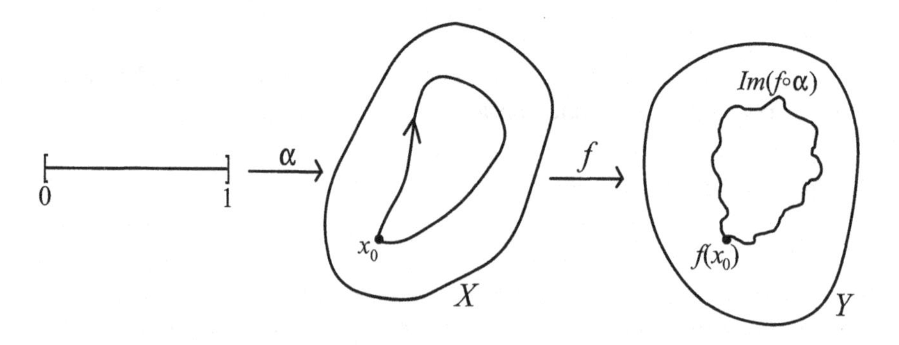

Proof: Let α be any loop in X based at x_0. Then $f \circ \alpha : I \to Y$ is a loop based at $f(x_0)$, since it's the composite of two continuous functions. We need to show that f_* is well-defined on path-homotopy classes of loops, and that it's a homomorphism of groups. If $\alpha \simeq_p \alpha'$ via a path-homotopy F, then $f \circ F : I \times I \to Y$ is a path-homotopy showing that $f \circ \alpha \simeq_p \alpha'$, so f_* is well-defined. To see that f_* is a homomorphism, let α and β be any two loops in X based at x_0. Recall that the product of these loops is given by

$$\alpha * \beta(s) = \begin{cases} \alpha(2s) & \text{if } s \in [0, \frac{1}{2}], \\ \beta(2s-1) & \text{if } s \in [\frac{1}{2}, 1]. \end{cases}$$

Composing with the map $f : X \to Y$ yields

$$f \circ (\alpha * \beta)(s) = \begin{cases} f \circ \alpha(2s) & \text{if } s \in [0, \frac{1}{2}], \\ f \circ \beta(2s-1) & \text{if } s \in [\frac{1}{2}, 1], \end{cases}$$

which is exactly $(f \circ \alpha) * (f \circ \beta)$. Thus $f_*[\alpha * \beta] = f_*[\alpha] * f_*[\beta]$, as we wished. □

Note that choosing a different basepoint in X, say, x_1, may well result in a different homomorphism:

$$f_* : \pi_1(X, x_1) \to \pi_1(Y, f(x_1)).$$

So our notation f_* is a bit misleading, because it depends not only on the map $f : X \to Y$ but also on the choice of basepoint in X. We'll try to avoid any confusion that this might cause in the sequel.

Notice also that the following result is essentially proved above:

Theorem 11.4.4. *Let f and g be continuous functions from X to Y with $f(x_0) = g(x_0)$. If $f \simeq g$, then the induced homomorphisms f_* and g_* are equal.*

Now we're ready to finish showing that the fundamental group construction yields a functor from spaces to groups:

Theorem 11.4.5. *The fundamental group construction is a functor from spaces (with basepoint) and continuous functions to groups and group homomorphisms. In other words,*

1. *If X is any space and $x_0 \in X$, then the identity map $i_X : X \to X$ induces the identity homomorphism*

$$i_{\pi_1(X,x_0)} : \pi_1(X, x_0) \to \pi_1(X, x_0);$$

that is, $(i_X)_* = i_{\pi_1(X,x_0)}$.

2. *If $f : X \to Y$ and $g : Y \to Z$ are continuous and $x_0 \in X$, then*

$$(g \circ f)_* = g_* \circ f_* : \pi_1(X, x_0) \to \pi_1(Z, (g \circ f)(x_0)).$$

Before we prove this result, we note that this yields exactly the property we wanted:

Corollary 11.4.6. *If $f : X \to Y$ is a homeomorphism, then for any $x_0 \in X$, $f_* : \pi_1(X, x_0) \to \pi_1(Y, f(x_0))$ is an isomorphism.*

Proof of Corollary: If $f : X \to Y$ is a homeomorphism, then there's an inverse map $g : Y \to X$ with $(g \circ f) = i_X$ and $(f \circ g) = i_Y$. Pick any point

$x_0 \in X$ and take fundamental groups based at x_0 and $f(x_0)$, respectively. Then $f_* : \pi_1(X, x_0) \to \pi_1(Y, f(x_0))$ has an inverse homomorphism, namely, g_*, so that f_* is an isomorphism. □

Proof of Theorem: Part 1 is trivial, since $i_X \circ \alpha = \alpha$ for any loop α. Part 2 is almost as easy:

$$\begin{aligned}(g \circ f)_*([\alpha]) &= [(g \circ f) \circ \alpha] \\ &= g_*([f \circ \alpha]) \\ &= g_*(f_*([\alpha])) \\ &= (g_* \circ f_*)([\alpha]).\end{aligned}$$

□

Exercises

1. For A a subspace of X, let $r : X \to A$ be a retraction. (i.e., the composition $A \xrightarrow{i} X \xrightarrow{r} A$ is id_A.) Prove that the induced homomorphism $r_* : \pi_1(X, a_0) \to \pi_1(A, a_0)$ is onto for all $a_0 \in A$.

2. Let $f : X \to Y$ be continuous, with X path-connected. Prove that the homomorphism induced by f is independent of the basepoint chosen in X, up to isomorphism. In other words, if α is a path in X from x_0 to x_1, then the following diagram commutes:

$$\begin{array}{ccc} \pi_1(X, x_0) & \xrightarrow{f_*} & \pi_1(Y, f(x_0)) \\ \Big\downarrow \hat{\alpha} & & \Big\downarrow \widehat{f \circ \alpha} \\ \pi_1(X, x_1) & \xrightarrow{f_*} & \pi_1(Y, f(x_1)). \end{array}$$

3. Given x_0 and x_1 in X, a path-connected space, prove that $\pi_1(X, x_0)$ is abelian if and only if for every pair of paths α and β from x_0 to x_1, the homomorphisms are identical: $\hat{\alpha} = \hat{\beta}$.

4. Prove Theorem 11.4.4: Let f and g be continuous functions from X to Y with $f(x_0) = g(x_0)$. If $f \simeq g$, then the induced homomorphisms f_* and g_* are equal.

11.5 COVERING SPACES

The idea of a *covering space* has been around for as long as that of the fundamental group. For our purposes, covering spaces will be treated primarily as computational devices. However, the topic of covering spaces is nearly a field of study in its own right, and we'll be merely skimming the surface of some very deep waters. The concept arises naturally in studying fundamental groups, but also shows up in parts of differential geometry, among other areas.[7]

Definition 11.5.1. *For B any topological space, a covering space for B is a space E, together with a continuous, onto map $p : E \to B$ such that for every point $b \in B$, there exists an open neighborhood U of b where*

$$p^{-1}(U) = \coprod_{\alpha \in \Lambda} V_\alpha,$$

a disjoint union of open sets in E, and for every $\alpha \in \Lambda$, the restriction $p|_{V_\alpha}$ is a homeomorphism from V_α to U. The map $p : E \to B$ is called a covering map, and we'll often refer to B as the "base space" and E as the "total space" of the covering space.

This definition may seem pretty complicated, but it's really just a very precise way of saying that every point in the base space has a neighborhood which is "evenly covered" by disjoint open sets in the total space. The first example and picture following should help clarify this idea. First, a *warning:* it's not enough that the inverse image $p^{-1}(U)$ be a disjoint union of open sets in E, each of which is homeomorphic to U; we further require that the restriction map $p|_{V_\beta}$ be a homeomorphism for each open set V_β. Even if V_β happens to be homeomorphic to U by some other map, we still require additionally that the restriction of the covering map be a homeomorphism.

Our first example is called a *trivial covering space*. Let X be any space. Give the set $\{1, 2, \ldots, n\}$ the discrete topology, (think of $\{1, 2, \ldots, n\}$ as a subspace of $\mathbb{R}_{\mathcal{U}}$). Then the projection function

$$p_X : X \times \{1, 2, \ldots, n\} \to X$$

[7] According to Jean Dieudonne, in his book *A History of Algebraic and Differential Topology, 1900–1960*, [6], the ideas of fundamental groups, classifying spaces and properly discontinuous actions of groups are so interconnected that each basically determines the other two. He points out that the more geometric idea of group actions actually arose first. See Ref. [6] for full details.

is a covering map. The following diagram makes this clear:

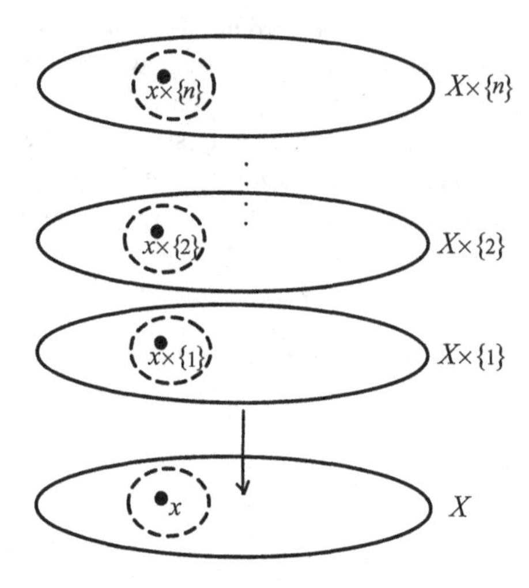

Note that the inverse image of every open set in the base space X has exactly n disjoint preimages in the total space. Whenever this occurs for all the evenly covered open sets in the base space, we'll say that the covering space is an *n-sheeted covering.*

The total space of the trivial covering space (often called the "trivial cover") is obviously disconnected whenever $n > 1$. The useful examples of covering spaces have total spaces that are path-connected. The following is such an example, and will be the key tool in calculating the fundamental groups of several spaces. Consider the map

$$exp : \mathbb{R} \to S^1$$

given by $exp(t) = e^{2\pi it} = (\cos(2\pi t), \sin(2\pi t))$, where the domain has the usual topology and S^1 is considered as a subspace of the usual plane. Then exp is onto (obviously, since exp "wraps" the line around S^1 infinitely many times) and continuous (since the basic trigonometric functions are continuous, as you proved in calculus). Further, for every point $b \in S^1$, there exists an open neighborhood of b in S^1 that is evenly covered by exp, as the following diagram illustrates:

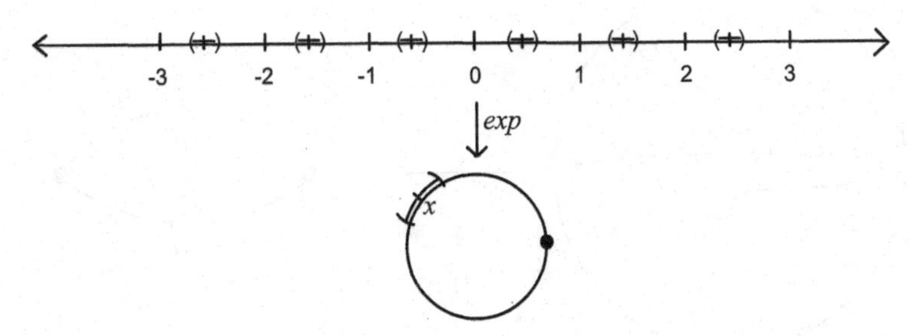

Note that $exp : \mathbb{R} \to S^1$ is an infinite-sheeted cover.

Recall (from the exercises of Section 10.2) that the usual torus T^2 is homeomorphic to the product space $S^1 \times S^1$. So the example p above can be exploited to construct a covering space for the torus:

$$exp \times exp : \mathbb{R}^2 \to T^2$$

given by $(exp \times exp)(s, t) = (e^{2\pi i s}, e^{2\pi i t})$ is an infinite-sheeted cover of $T^2 \cong S^1 \times S^1$:

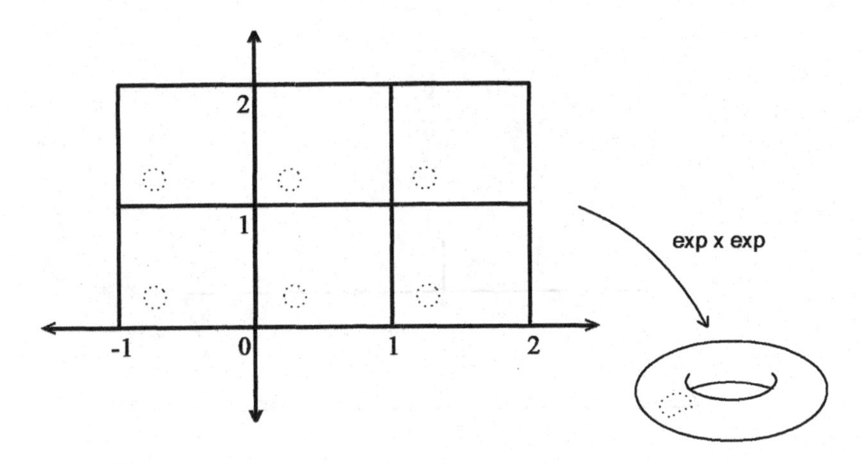

For our next example, recall that the projective plane P^2 (Definition 10.2.7) can be constructed as the quotient space $S^2_{/\sim}$, where $\sim$ is the equivalence relation that identifies antipodal points: $w \sim -w$. Then the projection $p : S^2 \to P^2$ that sends each $w \in S^2$ to its equivalence class in P^2 is certainly onto and continuous, by the definition of the quotient topology. Further, we assert that p is a covering map, as the following diagram illustrates:

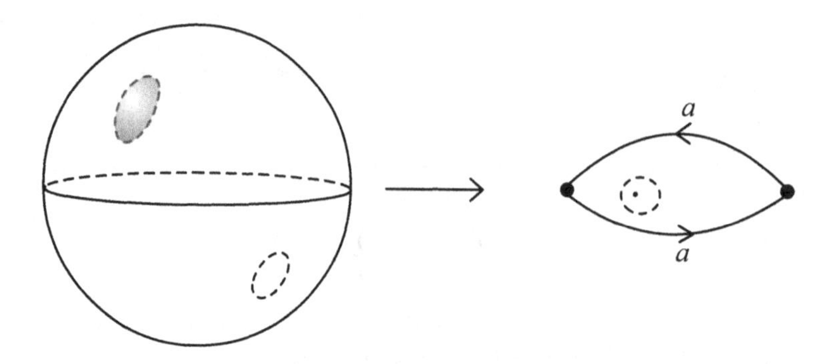

A precise proof that the map p is a covering map can be found in Ref. [13]. Note that $p : S^2 \to P^2$ is a 2-sheeted covering.

Our final example is one that is of importance in complex analysis and differential geometry. Consider the punctured plane $\mathbb{R}^2 \setminus \{\vec{0}\}$ in the usual subspace topology. The complex logarithm function maps a *Riemann surface* to $\mathbb{R}^2 \setminus \{\vec{0}\}$ as illustrated below:

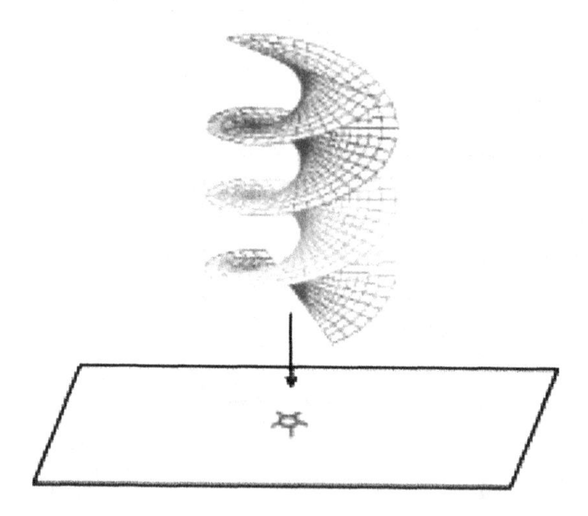

This is an infinite-sheeted cover of the punctured plane. One can write down the function precisely by thinking of the domain as $\mathbb{R} \times (0, +\infty)$ topologized as a subspace of the plane. Then the map is the composite $\mathbb{R} \times (0, +\infty) \overset{(exp,i)}{\longrightarrow} S^1 \times (0, +\infty) \overset{q}{\longrightarrow} \mathbb{R}^2 \setminus \{\vec{0}\}$ given by $q(x, t) = tx$. In this description, it's a "reasonable" conclusion that the composite is a covering map. Since we won't be using this example for any computations, there's no great moral obligation to work through the full details.

Two important properties make it possible to use covering spaces to compute fundamental groups: the Path Lifting Property and the Path-Homotopy Lifting Property.[8] These two concepts make very clear the connection between fundamental groups and covering spaces.

Definition 11.5.2. *Let $p : E \to B$ be any map of topological spaces. If $f : X \to B$ is any continuous map, a lifting of f is any map $\tilde{f} : X \to E$ such that $p \circ \tilde{f} = f$.*

The following diagram explains the term "lifting":

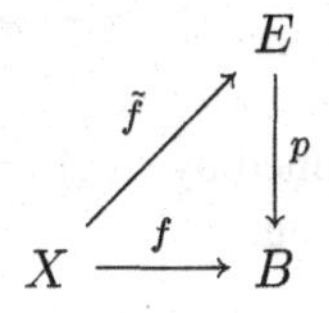

To see how this fits into the framework of covering spaces, recall the exponential covering: $exp : \mathbb{R} \to S^1$. We will see how certain paths in the base space S^1 lift to paths in the total space $\mathbb{R}$. Let $\alpha : I \to S^1$ be the loop that wraps around S^1 exactly once, counterclockwise, based at the east pole; that is, $\alpha(t) = e^{2\pi i t} = (\cos(2\pi t), \sin(2\pi t))$. Then the following diagram shows clearly how the loop α lifts to a path in $\mathbb{R}$:

[8] These two concepts appear to have been known to Poincaré and his successors, notably Tietze, although they did not formulate them explicitly or provide proofs. The first formal statement and proof of these ideas is due to Herman Weyl in a 1913 paper on Riemann surfaces. Reidemester's 1928 paper on fundamental groups and covering spaces presented these ideas for 3-manifolds, albeit rather informally. The classic 1934 textbook by Seifert and Threlfall [17] gives the first general (and precise) account of these notions.

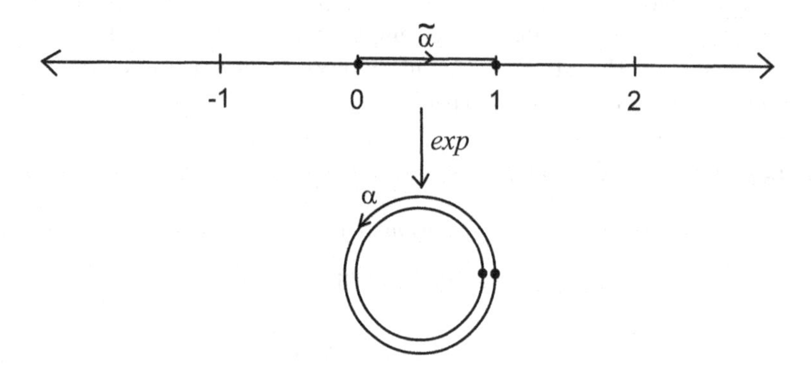

Note that the lifting $\tilde{\alpha}$ is defined by $\tilde{\alpha}(s) = s$. Note also that there are other possible liftings to $\mathbb{R}$, including:

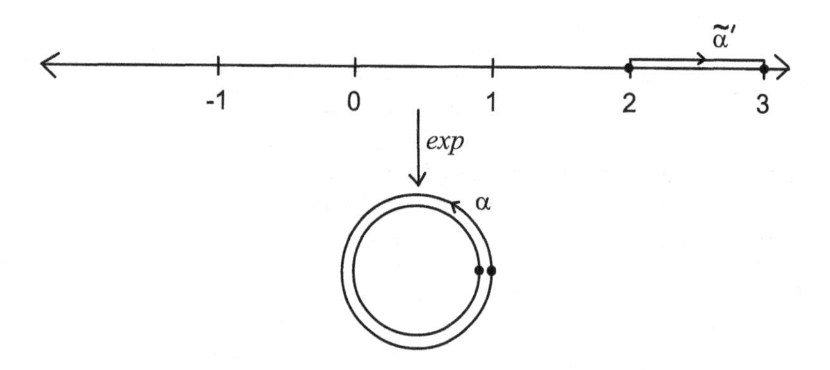

Here $\tilde{\alpha}'$ is defined by $\tilde{\alpha}'(s) = 2 + s$. Both $\tilde{\alpha}$ and $\tilde{\alpha}'$ are liftings of α, since composition with the covering map exp yields the original loop α.

Similarly, let $\beta : I \to S^1$ be the path given by $\beta(s) = (-\cos(\pi t), -\sin(\pi t))$. Then β starts at the east pole $(1, 0)$ and ends at the west pole $(-1, 0)$, traveling clockwise. A lifting of β is given as follows:

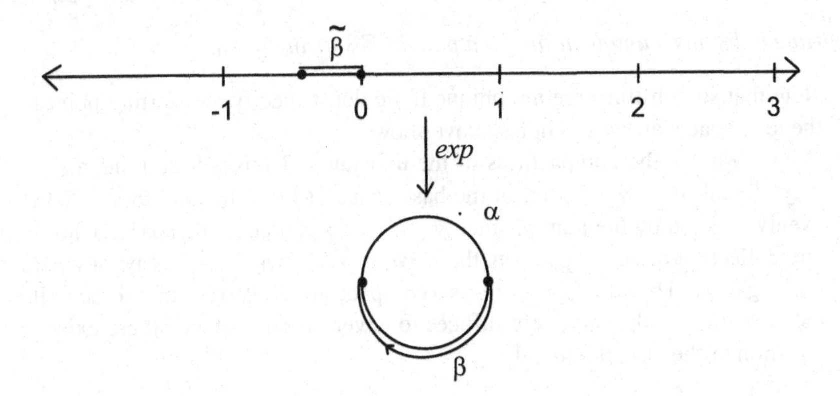

Here this particular lifting $\tilde{\beta}$ is defined by $\tilde{\beta}(s) = -\frac{s}{2}$. Again, one has other liftings, starting at other integer points in $\mathbb{R}$.

Finally, let $\gamma : I \to S^1$ be the loop that wraps around S^1 exactly twice, counterclockwise, based at the east pole; that is, $\gamma(t) = e^{4\pi i t} = (\cos(4\pi t), \sin(4\pi t))$. Then the following diagram shows clearly how the loop γ lifts to a path in $\mathbb{R}$:

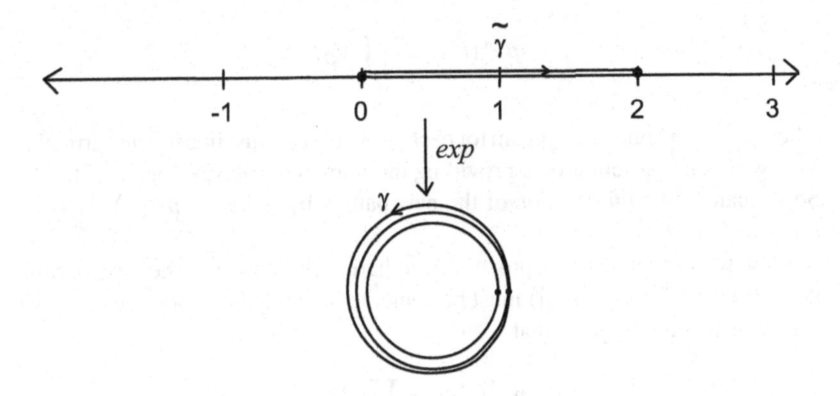

Note that the lifting $\tilde{\gamma}$ is defined by $\tilde{\gamma}(s) = 2s$. Again, this lifting is not at all unique. For example, $\tilde{\gamma}'$ defined by $\tilde{\gamma}'(s) = 17 + 2s$ is another perfectly good lifting of γ to $\mathbb{R}$.

The following theorem shows precisely what covering spaces have to do with loops.

Theorem 11.5.1. *(Path Lifting Property) Let $p : E \to B$ be a covering space. For any point $b_0 \in B$, let $e_0 \in p^{-1}(b_0)$, [i.e., let $p(e_0) = b_0$]. Then any path in B starting at b_0 has a unique lifting to a path in E starting at e_0.*

Note that such liftings are *not* unique if we don't specify the starting point in the total space, as the examples above show.

Proof: We use the compactness of the unit interval I in a very fundamental way. Recall that every point in the base space B has a neighborhood that is evenly covered by the map p, since $p : E \to B$ is a covering space. Choose any collection of such open sets that covers B. Let $\alpha : I \to B$ be any path starting at b_0. Then the image of α is a compact subset of B, so that some finite subcollection of the open sets suffices to cover $Im(\alpha)$. Hence there exists a partition of the closed interval

$$0 = s_0 < s_1 < s_2 < \cdots < s_{i-1} < s_i < \cdots < s_n = 1$$

such that the set $\alpha([s_{i-1}, s_i])$ is entirely contained in an open subset of B that is evenly covered by p. This enables us to construct the lifting $\tilde{\alpha}$ in finitely many steps.

To begin the new path, define $\tilde{\alpha}(0) = e_0$, the "basepoint" chosen in the total space E. To continue the path from here, recall that the initial point of the path α in B is b_0, and that b_0 (along with the entire portion of the path $\alpha([0, s_1])$) is contained in an open set U_0 which is evenly covered by p, so that

$$p^{-1}(U_0) = \coprod_{\beta \in \Lambda} V_\beta,$$

where $p|_{V_\beta}$ is a homeomorphism for each β. Thus, e_0 must live in *one* particular V_{β_0}, with the restriction of p providing the homeomorphism from V_{β_0} to U_0. So we can define this portion of the new path $\tilde{\alpha}$ by $\tilde{\alpha}(s) = (p|_{V_{\beta_0}})^{-1}(\alpha(s))$ for all $s \in [0, s_1]$.

Now we'll continue this process from here. The image of the next part of the partition of I, $\alpha([s_1, s_2])$ must be contained entirely in an open set U_1, that is evenly covered by p, so that

$$p^{-1}(U_1) = \coprod_{\beta \in \Lambda} W_\beta,$$

where $p|_{W_\beta}$ is a homeomorphism for each β. We've already defined how the lifting works at the point $s_1 \in I$: $\tilde{\alpha}(s_1) \in V_{\beta_1}$. But since $\alpha(s_1) \in (U_0 \cap U_1) \subset B$, we know that there must be some particular open set W_{β_1} in $p^{-1}(U_1)$ such that $\tilde{\alpha}(s_1) \in V_{\beta_0} \cap W_{\beta_1}$. This tells us which "sheet" in $p^{-1}(U_1)$ in which to define the continuation of the lifting: $\tilde{\alpha}(s) = (p|_{W_{\beta_1}})^{-1}(\alpha(s))$ for all $s \in [s_1, s_2]$. We continue in this manner for all of the *finitely many* pieces in the partition of I; if the lifting $\tilde{\alpha}$ has been defined for all $s \in [0, s_j]$, we note that $\alpha([s_j, s_{j+1}])$ is contained in a single open set U_j, which is evenly covered by p

$$p^{-1}(U_j) = \coprod_{\beta \in \Lambda} Z_\beta,$$

where $p|_{Z_\beta}$ is a homeomorphism for each β. Now we choose the unique sheet Z_{β_j} that contains the previously defined $\tilde{\alpha}(s_j)$ and define $\tilde{\alpha}(s) = (p|_{Z_{\beta_j}})^{-1}(\alpha(s))$ for all $s \in [s_j, s_{j+1}]$. Since the partition

$$0 = s_0 < s_1 < s_2 < \cdots < s_{i-1} < s_i < \cdots < s_n = 1$$

is finite, this inductive step allows us to define the new path $\tilde{\alpha}$ for all $s \in I$. The Pasting Lemma assures us that $\tilde{\alpha}$ is continuous, since it's defined so that the endpoints of the "subpaths" agree. Further, $\tilde{\alpha}$ is a lifting of α to the total space E by definition: $p \circ \tilde{\alpha}(s) = \alpha(s)$ for all $s \in I$, since each such s belongs to a subinterval $[s_j, s_{j+1}]$ for some j, and $\tilde{\alpha}(s) = (p|_{Z_{\beta_j}})^{-1}(\alpha(s))$ for such s.

The uniqueness of the lifting $\tilde{\alpha}$ is also proved a piece at a time. If one has another potential lifting of α, say, $\gamma : I \to E$ starting at e_0, then $\gamma(s)$ must agree with $\tilde{\alpha}(s)$ for all $s \in [0, s_1]$, since $p \circ \gamma(s)$ must equal $p \circ \tilde{\alpha}(s)$ for all such s, and since $p|_{V_{\beta_0}}$ is a homeomorphism. Hence, there are no other choices for the value of $\gamma(s)$. We continue in this manner; if $\gamma(s) = \tilde{\alpha}(s)$ for all $s \in [0, s_j]$, then the two paths must agree on $[s_j, s_{j+1}]$, since $p|_{Z_{\beta_j}}$ is a homeomorphism.

The following diagram illustrates the main idea of the proof: □

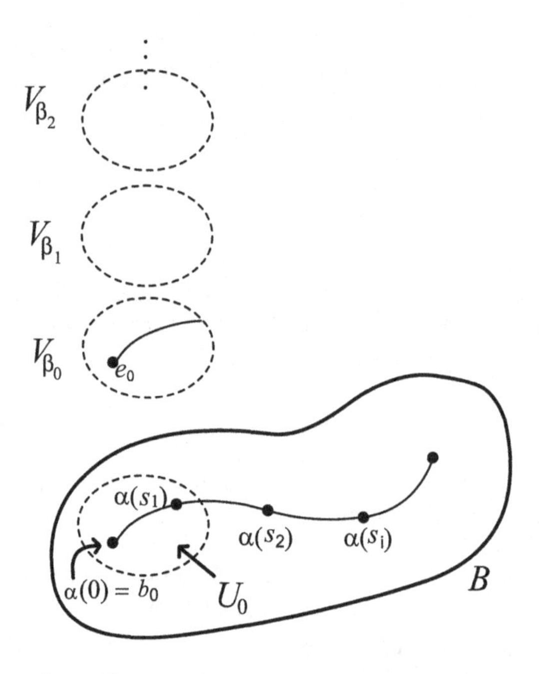

The next theorem shows that covering spaces work well with path-homotopies as well.

Theorem 11.5.2. *(Path-Homotopy Lifting Property) Let $p : E \to B$ be a covering space. For any point $b_0 \in B$, let $e_0 \in p^{-1}(b_0)$, [i.e., let $p(e_0) = b_0$]. Then any path-homotopy in $F : I \times I \to B$ starting at $F(0,0) = b_0$ has a unique lifting to a path-homotopy in $\tilde{F} : I \times I \to E$ starting at $\tilde{F}(0,0) = e_0$.*

Sketch of Proof: The proof is quite similar to the proof of the Path-Lifting Property, so we'll give just a sketch. Again, we will exploit the compactness of the unit square $I \times I$. We start the lifting $\tilde{F}$ by defining $\tilde{F}(0,0) = e_0$. Then we continue it using the Path Lifting Property: $\tilde{F}(0,t)$ is defined for all $t \in I$, so that the lifting is defined on the left-hand edge of $I \times I$. Now we proceed to define the lifting on the rest of the square. Since F is continuous, the image $F(I \times I)$ is a compact subspace of B. Since $p : E \to B$ is a covering space, every point of B is contained in a neighborhood that is evenly covered by p. The collection of such evenly covered neighborhoods is an open cover of $F(I \times I)$, so there must be a finite subcollection of evenly covered open sets whose union contains $F(I \times I)$. So there must be two partitions of the closed interval I,

$$0 = s_0 < s_1 < s_2 < \cdots < s_{i-1} < s_i < \cdots < s_n = 1$$

and

$$0 = t_0 < t_1 < t_2 < \cdots < t_{j-1} < t_j < \cdots < t_m = 1$$

such that the image of each rectangle $F([s_{i-1}, s_i] \times [t_{j-1}, t_j])$ is completely contained in *one* open set $U_{i,j}$ that is evenly covered by p. In particular, b_0 and all of $F(s,t)$ for $(s,t) \in [0, s_1] \times [0, t_1]$ must live in one open set $U_{0,0}$ that is evenly covered by p. Hence, we define the lifting for all $(s,t) \in [0, s_1] \times [0, t_1]$ by $\tilde{F}(s,t) = (p|_{V_{\beta_{0,0}}})^{-1} F(s,t)$, for whatever sheet $V_{\beta_{0,0}}$ contains e_0 and the appropriate left edge $F(0,s)$ for all $s \in [0, s_1]$. We continue by splicing together the lifting $\tilde{F}$ rectangle by rectangle, making sure to be on the same sheet of the cover as the next previously defined rectangle. Since there are only finitely many rectangles, an inductive step similar to that in the proof of the Path Lifting Property does the trick. Uniqueness is proved similarly, rectangle by rectangle. □

Exercises

1. Let $p : E \to B$ be a covering space, with the base space B connected. Prove that if $p^{-1}(b_0)$ consists of n points for one $b_0 \in B$, then $p^{-1}(b)$ has n elements for every $b \in B$. Such a covering space is called an *n-sheeted covering.*

2. Show that the function $f : S^1 \to S^1$ defined by $f(z) = z^2$ is a covering space of S^1 by itself. Note that it's a 2-sheeted cover.

3. Prove that the projection map $p_Y : X \times Y \to Y$ is a covering map if X is discrete.

11.6 FUNDAMENTAL GROUP OF THE CIRCLE AND RELATED SPACES

We apply the Path Lifting and Path-Homotopy Lifting Properties to compute the fundamental group of the circle S^1. The main result of this section is the following theorem:

Theorem 11.6.1. *The fundamental group of the circle is infinite cyclic; that is, $\pi_1(S^1, x_0) \cong \mathbb{Z}$, the additive group of the integers, for any point $x_0 \in S^1$.*

Proof: We will use the covering space for the circle discussed earlier: $exp : \mathbb{R}_{\mathcal{U}} \to S^1$ defined by $exp(t) = e^{2\pi i t} = (\cos(2\pi t), \sin(2\pi t))$. We use this to define a function that assigns to each loop in S^1 based at the east pole a unique integer, as follows. Let $\alpha : I \to S^1$ be any loop with $\alpha(0) = \alpha(1) = (1,0)$. Then for each point n in $exp^{-1}((1,0))$ there is a unique lifting

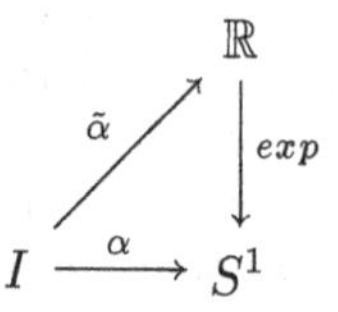

to a path in $\mathbb{R}$ that starts at n. Note that $exp^{-1}((1,0)) = \mathbb{Z}$, so that we can choose any integer for our starting point. We'll use the number 0, since it makes the rest of the proof easier. So for every loop α in S^1 based at the east pole, we look at the lifting $\tilde{\alpha}$ in $\mathbb{R}$ starting at 0. This (almost) defines the following function

$$f : \pi_1(S^1, (1,0)) \to \mathbb{Z}$$

by $f([\alpha]) = \tilde{\alpha}(1)$, the terminal point of the lifted path that starts at 0:

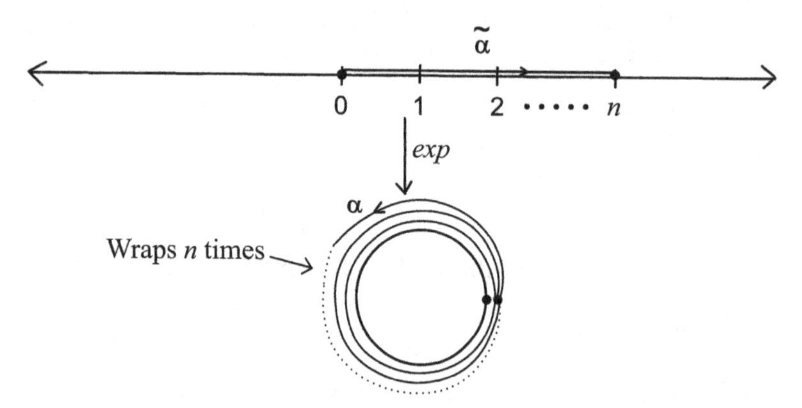

First, of course, we'll need to show that the function f is well-defined. In other words, we need to show that if α and β are loops in S^1 based at $(1,0)$ with $\alpha \simeq_p \beta$, then $\tilde{\alpha}(1) = \tilde{\beta}(1)$. This follows easily from the Path-Homotopy Lifting Property: If $\alpha \simeq_p \beta$ via the path-homotopy $F : I \times I \to S^1$, then F lifts to a path homotopy $\tilde{F} : I \times I \to \mathbb{R}$ with $\tilde{F}(0,0) = 0$, showing that $\tilde{\alpha} \simeq_p \tilde{\beta}$. (In fact, the lifting $\tilde{F}$ is unique, but we don't need this.) Hence the lifted paths end at the same point: $\tilde{\alpha}(1) = \tilde{\beta}(1)$. So we were, indeed, correct in writing the function f with $\pi_1(S^1, (1,0))$ as its domain, rather than the set of loops based at $(1,0)$.

Next, we'll show that the function f is onto. Consider any $n \in \mathbb{Z}$. We need a loop in S^1 that lifts to a path in $\mathbb{R}$ that starts at 0 and ends at n. Let's take the easiest route to this by defining $\gamma : I \to \mathbb{R}$ by $\gamma(s) = ns$. Then γ is continuous, running linearly from 0 to n. Then $exp \circ \gamma$ is a loop in S^1, based at $(1, 0)$, which lifts to γ by definition. Slick, eh? Alternatively, pick μ to be a loop in S^1 that "wraps" the interval I around S^1 counterclockwise exactly n times if $n > 0$; clockwise $|n|$ times if $n < 0$. The first construction is much easier to write down precisely.

Now we'll show that f is a homomorphism of groups, remembering that the operation on $\mathbb{Z}$ is usual addition. Let α and β be any two loops in S^1 based at $(0, 1)$. We need to show that

$$f([\alpha] * [\beta]) = f([\alpha * \beta]) = \widetilde{\alpha * \beta}(1)$$

is the same integer as $\tilde{\alpha}(1) + \tilde{\beta}(1)$. The difficulty here is that the operation $*$ on $\pi_1(S^1, (1, 0))$ involves first doing one loop (at twice speed) then the other. Intuitively, then, we hope to emulate this in $\mathbb{R}$ by first doing the first lifted path and then the second. Unfortunately, both $\tilde{\alpha}$ and $\tilde{\beta}$ are paths in $\mathbb{R}$ starting at 0, so we can't so them in succession. We'll have to produce a lifting for $\alpha * \beta$ that better emulates the operation $*$. Define $\delta : I \to \mathbb{R}$ by

$$\delta(s) = \begin{cases} \tilde{\alpha}(2s) & \text{if } s \in [0, \frac{1}{2}], \\ \tilde{\alpha}(1) + \tilde{\beta}(2s - 1) & \text{if } s \in [\frac{1}{2}, 1]. \end{cases}$$

In other words, if $f([\alpha]) = n$ and $f([\beta]) = m$, we define the path δ in $\mathbb{R}$ by first doing $\tilde{\alpha}$ from 0 to n, then doing the lift $\tilde{\beta}$ shifted to start at n, rather than 0. Thus δ is a path starting at 0 and ending at $n + m$, which is a lifting for $\alpha * \beta$, as we can see by applying exp to it. (Remember that exp applied to *any* integer is zero.) Further, $\delta(1) = \tilde{\alpha}(1) + \tilde{\beta}(1)$, by our definition of δ. Thus

$$f([\alpha] * [\beta]) = f([\alpha * \beta]) = \delta(1) = \tilde{\alpha}(1) + \tilde{\beta}(1).$$

Finally, we show that the homomorphism f is one-to-one. Recall that group homomorphisms are particularly nice in this regard; a homomorphism $j : G \to H$ is one-to-one if and only if the kernel of j is trivial, where $ker(j) = \{g \in G : j(g) = 1_H\}$. So we need to show that $ker(f) = [c_{(1,0)}]$, the identity of $\pi_1(S^1, (1, 0))$. Let α be any loop in S^1 based at $(1, 0)$ such that $f([\alpha]) = 0$. We need to show that $\alpha \simeq_p c_{(1,0)}$. Since $f([\alpha]) = 0$, we know that the lifting $\tilde{\alpha}$ ends at $0 = 1_{\mathbb{Z}}$. Hence, $\tilde{\alpha}$ is a loop in $\mathbb{R}$ based at 0. Since $\mathbb{R}$ is simply-connected (see Definition 11.4.3 and the discussion following it), we know that $\tilde{\alpha} \simeq_p c_0$, the constant loop at 0, by some path-homotopy, say, $G : I \times I \to \mathbb{R}$. Then $exp \circ G : I \times I \to S^1$ is the required path-homotopy deforming α into the constant loop at $(1, 0)$, as we wished. Alternatively, we could directly show that f is 1–1 by letting $f([\alpha]) = f([\beta])$, so that the liftings $\tilde{\alpha}$ and $\tilde{\beta}$ must end at the same point in $\mathbb{R}$. Since $\mathbb{R}$ is simply-connected, any two paths with the same initial and terminal points are path-homotopic in $\mathbb{R}$. Composing this path-homotopy with the covering map exp finishes the job. $\square$

This proof looks like it works only in the setting of the covering map $exp : \mathbb{R} \to S^1$. However, many of the ideas work in far more generality. We state the general theorem here:

Theorem 11.6.2. *Let $p : E \to B$ be a covering space, and let $e_0 \in p^{-1}(b_0)$. Define $f : \pi_1(B, b_0) \to p^{-1}(b_0)$ by $f([\alpha]) = \tilde{\alpha}(1)$, where $\tilde{\alpha}$ is the lifting of α that starts at e_0. Then f is onto. Further, if E is simply-connected, then f is a bijection.*

Note that the theorem does not claim that f is a homomorphism, since the set $p^{-1}(b_0)$ will not, in general, have a group operation. Because of this theorem, simply-connected covering spaces are particularly useful in determining the size of the fundamental group of the base space. In general, a covering space $p : E \to B$ where E is simply-connected is called a *universal covering space* for B.

Now we use the calculation of $\pi_1(S^1, x_0)$ to compute the fundamental group of the punctured plane $R^2 \setminus \{\vec{0}\}$.

Theorem 11.6.3. *The inclusion map $i : S^1 \hookrightarrow \mathbb{R}^2 \setminus \{\vec{0}\}$ induces an isomorphism of fundamental groups, for any $x_0 \in S^1$:*

$$i_* : \pi_1(S^1, x_0) \xrightarrow{\cong} \pi_1(\mathbb{R}^2 \setminus \{\vec{0}\}, x_0).$$

Thus $\pi_1(\mathbb{R}^2 \setminus \{\vec{0}\}, x_0) \cong \mathbb{Z}$.

Intuitively, the result seems quite plausible, since the fact that the nonpunctured plane $\mathbb{R}^2$ is simply-connected implies that any loop that doesn't "wrap around" the origin $\vec{0}$ is contractible to the constant loop. To prove this precisely, we exploit the functorial nature of the fundamental group.

Proof: Consider the map $r : \mathbb{R}^2 \setminus \{\vec{0}\} \to S^1$ given by $r(\vec{x}) = \frac{\vec{x}}{||\vec{x}||}$, which is continuous since $\vec{0}$ is not in the domain. In fact, r is a retraction of $\mathbb{R}^2 \setminus \{\vec{0}\}$ onto S^1, since any element $x \in S^1$ gets sent to itself by r. In other words, $r \circ i = i_{S^1}$, so it induces $(r \circ i)_* = i_{\pi_1(S^1, x_0)}$ for any $x_0 \in S^0$. We want to show that the other composite $i \circ r$ induces an isomorphism from $\pi_1(\mathbb{R}^2 \setminus \{\vec{0}\}, x_0)$ to itself.

Then following diagram makes the idea clear:

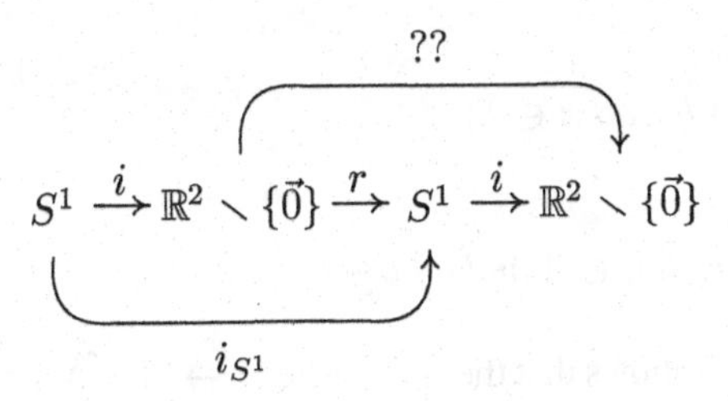

So we want to show that the map $i \circ r : \mathbb{R}^2 \setminus \{\vec{0}\} \to \mathbb{R}^2 \setminus \{\vec{0}\}$ induces an isomorphism of fundamental groups. It suffices to show that the composite is homotopic to the identity: $i \circ r \simeq i_{\mathbb{R}^2 \setminus \{\vec{0}\}}$, since homotopic maps that agree at the basepoint induce the same homomorphism on fundamental groups (Theorem 11.4.4). To see why these maps are homotopic, note that $i \circ r$ is basically just r, sending the entire punctured plane onto the unit circle S^1. It seems unlikely, at first glance, that this could be homotopic to the identity map $i_{\mathbb{R}^2 \setminus \{\vec{0}\}}$. We need to keep in mind, however, that the entire plane $\mathbb{R}^2$ has the property that *any* two maps into it are homotopic, so this should work out as long as we stay away from the (missing) origin. Precisely, define the following homotopy

$$F : \mathbb{R}^2 \setminus \{\vec{0}\} \times I \to \mathbb{R}^2 \setminus \{\vec{0}\}$$

by

$$F(\vec{x}, t) = tr(\vec{x}) + (1 - t)\vec{x} = t\frac{\vec{x}}{||\vec{x}||} + (1 - t)\vec{x}.$$

Note that $F(\vec{x}, t)$ is never $\vec{0}$, so that F is continuous by usual calculus considerations. Further, $F(\vec{x}, 1) = r(\vec{x})$ and $F(\vec{x}, 0) = \vec{x}$, for all $\vec{x} \in \mathbb{R}^2 \setminus \{\vec{0}\}$, so F is the desired homotopy from $i \circ r$ to $i_{\mathbb{R}^2 \setminus \{\vec{0}\}}$. Applying Theorem 11.4.4 now completes the proof. □

The main tool in the proof was the homotopy F, which showed that the map r was not just a run-of-the-mill retraction. In general, a retraction r from a space X to a subspace A has the property that $r \circ i = i_A$, where $i : A \hookrightarrow X$. The homotopy F above showed that the other composite $i \circ r : X \to X$ was, in the case of $X = \mathbb{R}^2 \setminus \{\vec{0}\}$, homotopic to the identity map i_X, keeping the subspace $A = S^1$ fixed throughout the homotopy. Such retractions are important enough to have a special name:

Definition 11.6.1. *A strong deformation retract from a space X to a subspace A is a map $r : X \to A$ such that there is a homotopy $F : X \times I \to X$ where*

1. *$F(x, 0) = x$ for all $x \in X$.*

2. *$F(x, 1) = r(x)$ for all $x \in X$.*

3. *$F(a, t) = a$ for all $a \in A$ and all $t \in I$.*

Note that property 3 shows that the composite $A \hookrightarrow X \xrightarrow{r} A$ must be i_A, so that r is indeed a retraction in the usual sense. Further, the homotopy F deforms $i \circ r$ into i_X while keeping the subspace A fixed.

The ideas in the proof of Theorem 11.6.3 can be used to prove the following:

Theorem 11.6.4. *If $r : X \to A$ is a strong deformation retract of X onto its subspace A, and if $a_0 \in A$, then both $i : A \hookrightarrow X$ and $r : X \to A$ induce isomorphisms:*

$$i_* : \pi_1(A, a_0) \xrightarrow{\cong} \pi_1(X, a_0)$$

and

$$r_* : \pi_1(X, a_0) \xrightarrow{\cong} \pi_1(A, a_0).$$

The proof is left as an exercise.

We close this section with an application, proving the classic fixed-point theorem of Brouwer:

Theorem 11.6.5. *Any continuous map from the usual two disk D^2 to itself has at least one fixed-point; that is, if $f : D^2 \to D^2$ is continuous, then there exists a point $x \in D^2$ such that $f(x) = x$.*

Proof: Let $f : D^2 \to D^2$ be continuous, and assume that f has no fixed points. Then we can use f to define a retraction from the disk to its boundary, $r : D^2 \to S^1$, as follows. For each $x \in D^2$, we've assumed that x is distinct from $f(x) \in D^2$. Thus we have a well defined ray, starting at $f(x)$ and continuing through x. Define $r(x)$ to be the point where this ray hits the boundary circle S^1:

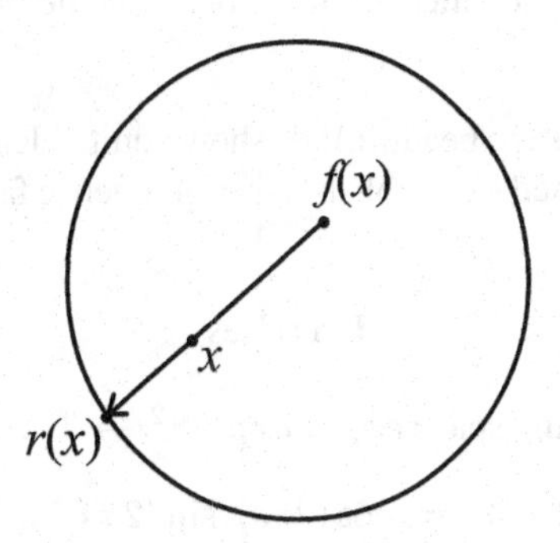

Since f is continuous, it takes points "near" x to points "near" $f(x)$, so that $r(x')$ will be "near" $r(x)$ if x' is "near" x, so that r is continuous.[9] Further, the construction of r shows that $r|_{S^1}$ is the identity map on S^1, so that r is a retraction; the composition

$$S^1 \xrightarrow{i} D^2 \xrightarrow{r} S^1$$

is the identity map on S^1. Now, apply the fundamental group functor, for any basepoint $x \in S^1$:

$$\begin{array}{ccccc} \pi_1(S^1, x) & \xrightarrow{i_*} & \pi_1(D^2, x) & \xrightarrow{r_*} & \pi_1(S^1, x) \\ \downarrow \cong & & \downarrow \cong & & \downarrow \cong \\ \mathbb{Z} & \longrightarrow & 0 & \longrightarrow & \mathbb{Z} \end{array}$$

The composite in the top row is the identity, since r is a strong deformation retract. But the bottom row composite can't be the identity on $\mathbb{Z}$, since it factors through the trivial group, so that the retraction r can't exist. We conclude, then, that our assumption must have been wrong, and that the map f must have at least one fixed point. □

We include the Brouwer Fixed-point Theorem for several reasons.

1. Fixed-point theorems are pretty rare and generally interesting.

2. This is the next case of a theorem first looked at in our study of connectivity, showing a bit more about the relationship of (path-) connectivity to simple-connectivity. Higher-dimensional analogs are easy to formulate and can be

[9] Okay, so this isn't a very precise proof that r is continuous. It's actually pretty easy to use the usual metric on $\mathbb{R}^2$ to write down an "epsilon–delta" proof that r is continuous, given what f is, and you're encouraged to do so if you're bothered by the quotation marks around the word "near."

proved in an almost identical manner, once one defines the higher homotopy groups $\pi_n(X)$.

3. The proof is absolutely beautiful! It shows quite elegantly how the algebraic ideas can be exploited to establish rather geometric facts.

Exercises

1. Consider the covering space $exp \times exp : \mathbb{R}^2 \to T^2$:

   (a) sketch the path $\alpha(s) = (\cos(2\pi s), \sin(2\pi s))$ in T^2, then sketch its lifting to the total space.

   (b) Do the same for the path $\beta(s) = (\cos(2\pi s), \sin(4\pi s))$.

2. Prove Theorem 11.6.4: If $r : X \to A$ is a strong deformation retract of X onto its subspace A, and if $a_0 \in A$, then both $i : A \hookrightarrow X$ and $r : X \to A$ induce isomorphisms:

$$i_* : \pi_1(A, a_0) \xrightarrow{\cong} \pi_1(X, a_0)$$

   and

$$r_* : \pi_1(X, a_0) \xrightarrow{\cong} \pi_1(A, a_0).$$

3. Prove Theorem 11.6.2: Let $p : E \to B$ be a covering space, and let $e_0 \in p^{-1}(b_0)$. Define $f : \pi_1(B, b_0) \to p^{-1}(b_0)$ by $f([\alpha]) = \tilde{\alpha}(1)$, where $\tilde{\alpha}$ is the lifting of α that starts at e_0. Then f is onto. Further, if E is simply-connected, then f is a bijection.

4. Note that Theorem 11.6.2 often gives you a bound on the size of the fundamental group of the base space of a classifying space. For example, say that $p : E \to B$ is a 3-sheeted cover, with B path-connected and E simply-connected. What can you say about $\pi_1(B)$? What if E is not simply-connected?

11.7 THE FUNDAMENTAL GROUPS OF SURFACES

In this section, we compute the fundamental groups of several of the surfaces we've come to know so well. Unfortunately, the relationship between the fundamental group and the connected sum operation on surfaces is not easily understood, so our work here will not yield the fundamental groups of all surfaces. We begin by looking at the fundamental groups of (finite) products of spaces.

Theorem 11.7.1. *If $x_0 \in X$ and $y_0 \in Y$, then the fundamental group of the product space $X \times Y$ is given by*

$$\pi_1(X \times Y, (x_0, y_0)) \cong \pi_1(X, x_0) \oplus \pi_1(Y, y_0).$$

In words, the fundamental group of the product of two spaces is the direct sum of the fundamental groups of the spaces. The notation here is the usual for the direct sum of two groups; if G and H are groups, then $G \oplus H$ is a group with the underlying set $G \times H$ and the operation given by

$$(g_1, h_1)(g_2, h_2) = (g_1 g_2, h_1 h_2),$$

with the operation in the last ordered pair occurring in G on the first coordinate and H on the second.

Proof: Let $x_0 \in X$ and $y_0 \in Y$ be any basepoints. We will build a homomorphism

$$f : \pi_1(X \times Y, (x_0, y_0)) \to \pi_1(X, x_0) \oplus \pi_1(Y, y_0)$$

using the projection maps, $p_X : X \times Y \to X$ and $p_y : X \times Y \to Y$. Given a path-homotopy class $[\alpha] \in \pi_1(X \times Y, (x_0, y_0))$, let

$$f([\alpha]) = (p_X)_*([\alpha]) \times (p_Y)_*([\alpha]).$$

In short, we let $f = (p_X)_* \times (p_Y)_*$. Then f is a group homomorphism by definition, since it's the direct sum of two homomorphisms. Now we can show that f is an isomorphism.

onto) Let $\beta : I \to X$ be any loop based at x_0 and $\gamma : I \to Y$ be any loop at y_0. We need to find a loop $\alpha : I \to X \times Y$ based at (x_0, y_0) whose path-homotopy class hits $[\beta] \times [\gamma] \in \pi_1(X, x_0) \oplus \pi_1(Y, y_0)$. Define α as follows: $\alpha(s) = (\beta(s), \gamma(s)) \in X \times Y$ for all $s \in I$. Note that α is continuous as a map into the product space $X \times Y$, since its compositions with the projection maps are given by $p_X \circ \alpha = \beta$ and $p_Y \circ \alpha = \gamma$, both of which are continuous. (Remember the "Big Theorem on Maps into Products"?) Then

$$f([\alpha]) = (p_X)_*([\alpha]) \times (p_Y)_*([\alpha]) = [\beta] \times [\gamma],$$

as we wished.

1-1) The easiest way to show that a group homomorphism is one-to-one is to show that its kernel is the trivial subgroup. Let $\delta : I \to X \times Y$ be a path at (x_0, y_0) such that

$$f([\delta]) = 0 = [c_{x_0}] \times [c_{y_0}].$$

Then we know that $p_X \circ \delta \simeq_p c_{x_0}$, via some path-homotopy F, say, and $p_Y \circ \delta \simeq_p c_{y_0}$, via some path-homotopy G. Then the map

$$H = F \times G : I \times I \to X \times Y$$

is a path-homotopy from

$$f([\delta]) = (p_X)_*([\delta]) \times (p_Y)_*([\delta])$$

to $c_{x_0} \times c_{y_0}$, as we wished. □

Corollary 11.7.2. *The fundamental group of the torus is the direct sum of $\mathbb{Z}$ with itself. That is, for any point $z \in T^2$, $\pi_1(T^2, z) \cong \mathbb{Z} \oplus \mathbb{Z}$.*

Proof: Recall that the exponential map $\exp : I \to S^1$ yields a homeomorphism $T^2 \cong S^1 \times S^1$ and apply the theorem. □

To continue with our computation of the fundamental groups of surfaces, we'll need some heavier artillery. The following theorem is a special case of a result due independently to Seifert and Van Kampen in 1932. Their full theorem is much more powerful, but it requires that the reader know about free products of groups and their quotients, which are best described in terms of "universal diagrams" of groups. (See Ref. [12] for full details.) We need only the special case given here for our purposes, so we'll skip this chance to learn about groups defined by universal properties. Maybe next time...

Theorem 11.7.3. *(Special Case of Seifert-Van Kampen Theorem)*

Let $X = U \cup V$, where U and V are both open in X and both simply-connected (and hence both path-connected), with $U \cap V$ also path-connected. Then the whole space X is also simply-connected.

Again, one can prove much more general results than this, but all of them require that the the component spaces U and V be path-connected and, at least, that they not be closed in X. Our proof is actually not too difficult because of our stronger hypotheses.

Proof: We start with our space X written as a union $X = U \cup V$, where U and V are both open in X and both simply-connected, with $U \cap V$ also path-connected. Let $x_0 \in U \cap V$. We will show that $\pi_1(X, x_0) = 0$, so we need to prove that any loop $\alpha : I \to X$ based at x_0 is path-homotopic to the constant loop. Since we're told that U and V are simply-connected, we will be finished if we can demonstrate that the loop α is path-homotopic to a loop that lies entirely in U or entirely in V. To make an arbitrary choice, we'll show how to deform α into a loop entirely in U.

First, since the U and V are open and since $\alpha : I \to X = U \cup V$ has a compact image, we know that there exists a partition

$$0 = s_0 < s_1 < s_2 < \cdots < s_i < \cdots < s_n = 1$$

so that for each i, the image $\alpha[s_{i-1}, s_i]$ is a subset that lies entirely in U or entirely in V. Further, we can choose such a partition to be minimal, in that the value of each partition point must lie in the intersection :

$$\alpha(s_j) \in U \cap V \text{ for all } j = 1, \ldots, n.$$

(Why? If α applied to any of the partition points were just in U but not V, say, then we could eliminate that particular partition point to get a strictly smaller partition satisfying the defining property that α applied to each subinterval lies entirely in exactly one of the two open sets.)

For each $i = 1, \ldots, n$, consider the restriction $\alpha|_{[s_{i-1}, s_i]}$, whose image lies entirely in U or entirely in V. After suitable reparametrization [10] the restriction $\alpha|_{[s_{i-1}, s_i]}$ is a path from $\alpha(s_{i-1})$ to $\alpha(s_i)$, with both endpoints lying in the intersection $U \cap V$. We'll now demonstrate that this restriction $\alpha|_{[s_{i-1}, s_i]}$ is path-homotopic to a path lying entirely in U.

Of course, the two choices are that the restriction $\alpha|_{[s_{i-1}, s_i]}$ lies entirely in U or entirely in V, so that half the time we won't have to do anything! If the restriction $\alpha|_{[s_{i-1}, s_i]}$ lies entirely in V, here's how we'll construct the desired path-homotopy. Since the endpoints $\alpha(s_{i-1})$ and $\alpha(s_i)$ lie in $U \cap V$, which is path-connected, there must exist paths β and γ in $U \cap V$, where β runs from x_0 to $\alpha(s_{i-1})$ and γ runs from x_0 to $\alpha(s_i)$. Then the product $\beta * \alpha * \tilde{\gamma}$ is a loop, based at x_0, lying entirely in V. Since V is simply-connected, we obtain

$$\beta * \alpha * \tilde{\gamma} \simeq_p c_{x_0}.$$

By simple "algebra" of paths, then, we have

$$\alpha \simeq_p \tilde{\beta} * c_{x_0} * \gamma \simeq_p \tilde{\beta} * \gamma.$$

Thus $\alpha|_{[s_{i-1}, s_i]}$ is path homotopic to a path from $\alpha(s_{i-1})$ to $\alpha(s_i)$ which *lies entirely in the intersection* $U \cap V$. The following diagram illustrates this idea:

[10]The reparametrization uses the homeomorphism $[s_{i-1}, s_i] \cong [0, 1]$, given by $g(t) = s_{i-1} + (s_i - s_{i-1})t$, so that we should replace the restriction $\alpha|_{[s_{i-1}, s_i]}$ with a new path δ defined by $\delta(t) = \alpha((1 - t)s_{i-1} + ts_i)$, if we want to be quite precise (meaning that our paths have domain [0,1] rather than an arbitrary closed interval like $[s_{i-1}, s_i]$.) We'll be somewhat more relaxed and refer to the restriction $\alpha|_{[s_{i-1}, s_i]}$ as a path.

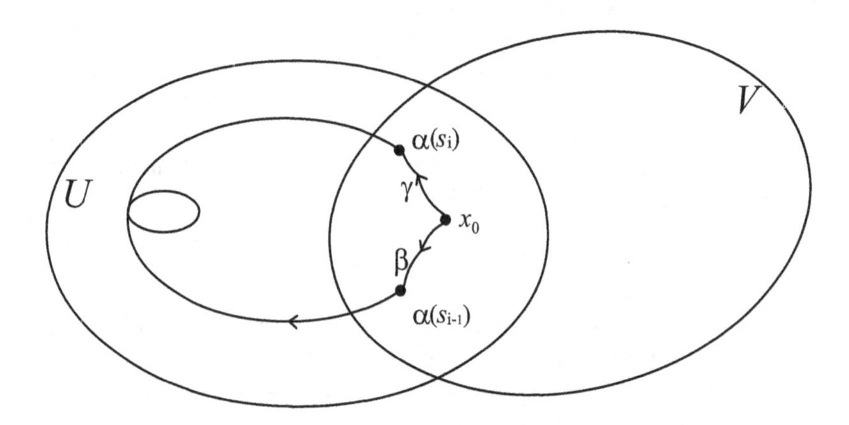

Doing this construction for (half of) the subintervals in the partition shows that the entire path α is path-homotopic to a path that lies entirely in U. Since U is simply-connected, this shows that $\alpha \simeq_p c_{x_0}$, as we wished. □

We now apply this special case of the Seifert–Van Kampen Theorem to compute the fundamental group of the 2-sphere.

Theorem 11.7.4. *The two-sphere S^2 is simply-connected; that is, $\pi_1(S^2, x) = 0$ for any $x \in S^2$.*

Proof: To apply the special case of the Seifert-Van Kampen Theorem requires that we write S^2 as a union of two open subsets, both simply-connected, such that the overlap of the two subsets is also path-connected. One might hope to use the upper and lower hemispheres, but these are not open is S^2. Instead, we "fatten" up these hemispheres, so that we get an open subset containing each. In fact, we previously defined an interesting function that allows us to fatten up the hemispheres as much as possible; let $U = S^2 \setminus \{(0,0,1)\}$ and $V = S^2 \setminus \{(0,0,-1)\}$, the sphere with the north and south poles, respectively, deleted. We recall that $S^2 \setminus \{(0,0,1)\}$ is homeomorphic to $\mathbb{R}^2$ by the homeomorphism known as *stereographic projection*

$$s : S^2 \setminus \{(0,0,1)\} \to \mathbb{R}^2$$

defined as in the following diagram:

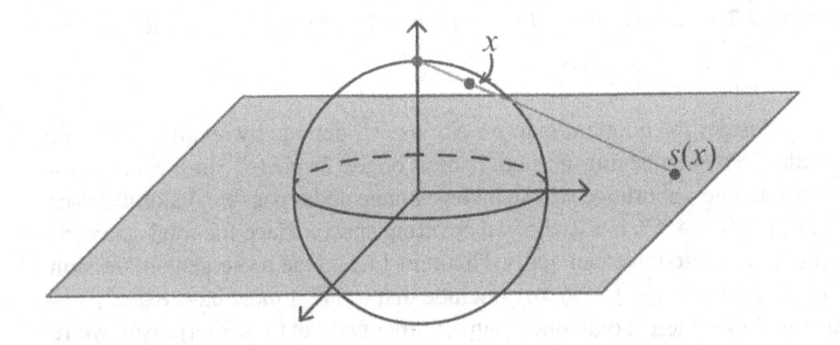

Note that the map is given by $s(x, y, z) = \frac{1}{1-z}(x, y, 0)$, which is, indeed, a continuous and open bijection. Since the plane $\mathbb{R}^2$ is simply-connected, and since the intersection $U \cap V = S^2 \setminus \{NP, SP\}$ is path connected, we've fulfilled the hypotheses of the theorem. The Seifert-Van Kampen Theorem thus shows that S^2 is simply-connected. □

In fact, we can define higher-dimensional analogs of this stereographic projection to prove the following generalization:

Theorem 11.7.5. *For $n > 1$, the n-sphere S^n is simply-connected; that is, $\pi_1(S^n, x) = 0$ for any $x \in S^n$.*

We exploit this to present a simple proof of the following fact:

Corollary 11.7.6. *For $n \geq 3$, the usual n-space $\mathbb{R}^n$ is not homeomorphic to the usual plane $\mathbb{R}^2$.*

Proof: First, note that for any $n \geq 1$, the $n-1$-sphere S^{n-1} is a strong deformation retract (see Definition 11.6.1) of the punctured n-space $\mathbb{R}^n \setminus \{\vec{0}\}$, where the retraction is $r(\vec{x}) = \frac{1}{||\vec{x}||}\vec{x}$. Thus $\pi_1(\mathbb{R}^n \setminus \{\vec{0}\}, x) = 0$ for all xs, if $n \geq 3$,
but $\pi_1(\mathbb{R}^2 \setminus \{\vec{0}\}, x) = \mathbb{Z}$, since the circle is an SDR of the punctured plane. Hence, deleting a point from $\mathbb{R}^n$ leaves a simply-connected space, if $n \geq 3$, but deleting a point from $\mathbb{R}^2$ doesn't, so that $\mathbb{R}^n$ and $\mathbb{R}^2$ can't be homeomorphic if $n \geq 3$. □

Note that this is similar in nature to our proof that that the line $\mathbb{R}$ is not homeomorphic to the plane, Proposition 6.4.3, which used path-connectivity rather than simple-connectivity.

We can also use our calculation of $\pi_1(S^2)$ to compute the fundamental group of the projective plane.

Theorem 11.7.7. *The fundamental group of the projective plane has order 2. Hence, $\pi_1(P^2, x) = \mathbb{Z}_{/2}$ for all $x \in P^2$.*

Proof: Consider the quotient map $p : S^2 \to P^2$, defined by recalling that one can build P^2 by identifying antipodal points on the 2-sphere. Since every point $x \in P^2$ has a neighborhood whose inverse image under p is two disjoint copies of itself, $p : S^2 \to P^2$ is a 2-sheeted covering space. Since the total space S^2 is simply-connected, we can apply Theorem 11.6.2, the more general version of the proof that $\pi_1(S^1) \cong \mathbb{Z}$, to conclude that $\pi_1(P^2)$ must have exactly two elements. Since there's only one group of order two, up to isomorphism, we're done. □

We now know that all three of the "basic" surfaces must be distinct:

Corollary 11.7.8. *The surfaces S^2, T^2 and P^2 are all topologically distinct.*

Proof: Look at their fundamental groups. Spaces with different fundamental groups can't be homeomorphic – they can't even be homotopy equivalent! □

Finally, we'll attempt to get some information about the fundamental group of the connected sum of two surfaces. Since any surface X connected sum with a sphere is homeomorphic to X back again, the first nontrivial case is $2T^2 = T^2 \# T^2$. We can't get complete information on its fundamental group with the technology that we have, but we can make some progress. First, we consider a simpler space.

Theorem 11.7.9. *The fundamental group of the figure-8 $S^1 \vee S^1$ is nonabelian.*

Proof: Recall that the exponential map $exp : \mathbb{R} \to S^2$ is an infinite-sheeted cover. We'll use this to build a covering space for the figure-8. Let C be the subspace of $\mathbb{R}^2$ consisting of the x-axis, the y-axis, and small circles tangent to each axis at each nonzero integer point:

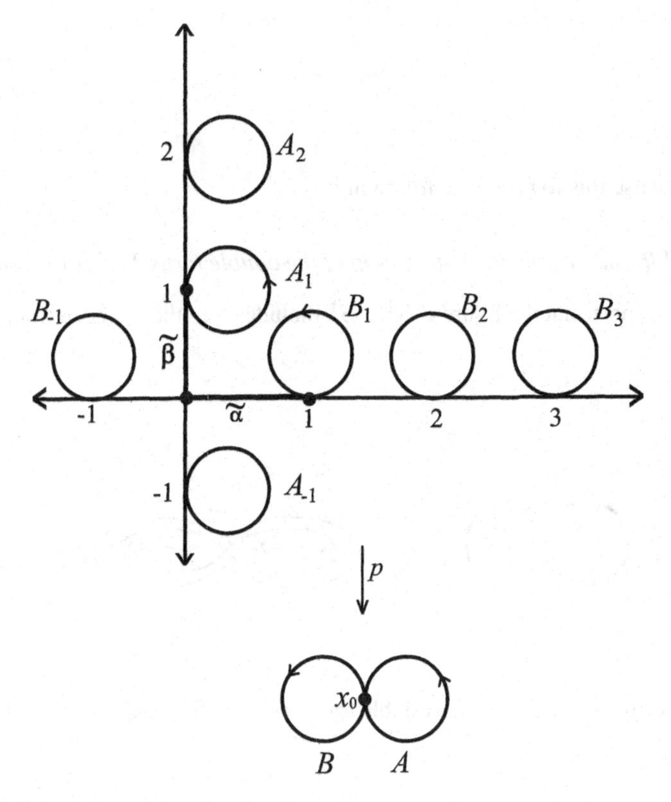

The map $p : C \to S^1 \vee S^1$ maps the x-axis around circle A and the y-axis around circle B by the exponential maps. Further, it takes the tangent circles "along for the ride": p maps each circle tangent to the x-axis homeomorphically to circle B and each circle tangent to the y-axis homeomorphically to circle A, as labeled in the diagram above. This is an infinite-sheeted covering space of $S^1 \vee S^1$.

Now we'll look at some paths in the total space and see what loop they map to in the base space. Let $\tilde{\alpha}$ be the path from $(0,0)$ to $(1,0)$ in C: $\tilde{\alpha}(s) = (s, 0)$. Let $\tilde{\beta}(s) = (0, s)$, so that $\tilde{\beta}$ goes from $(0,0)$ to $(0,1)$. Then composing these with the covering map $p : C \to S^1 \vee S^1$ yields α, which wraps around circle A once, counterclockwise, and β, which wraps around circle B once, counterclockwise. Here's the punchline: We claim that $\alpha * \beta$ is *not* path-homotopic to $\beta * \alpha$. Why? Consider the liftings to C: $\alpha * \beta$ lifts to a path in C that starts at $(0,0)$, goes along the x-axis to $(1,0)$, then wraps around the circle B_1 once, counterclockwise. One the other hand, $\beta * \alpha$ lifts to the path in C that starts at $(0,0)$, goes up the y-axis to $(0,1)$, then wraps around the circle A_1 exactly once. The two liftings cannot be path-homotopic, since they end at different terminal points. Hence the paths in the base space cannot be path-homotopic:

$$\alpha * \beta \not\simeq_p \beta * \alpha.$$

□

We now use this to prove the following:

Corollary 11.7.10. *The fundamental group of the double torus $2T^2$ is nonabelian.*

Proof: First, note that the figure-8 $S^1 \vee S^1$ includes into the connected sum of two tori:

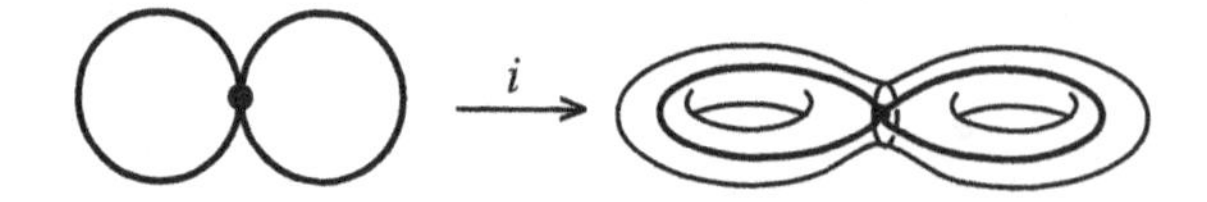

Now we form a retraction of the double torus onto the figure-8:

Now we use the usual algebraic property of retracts; the maps

$$8 \xrightarrow{i} 2T^2 \xrightarrow{r} 8$$

compose to the identity on the figure 8, and they induce

$$\pi_1(8) \xrightarrow{i_*} \pi_1(2T^2) \xrightarrow{r_*} \pi_1(8),$$

which is the identity homomorphism on $\pi_1(8)$. Thus $\pi_1(T^2)$ must contain a subgroup isomorphic to $\pi_1(8)$. □

So our techniques suffice to show that the double torus $2T^2$ is not homeomorphic to the basic surfaces S^2, P^2 or T^2. However, we need better machinery if we want to compute the fundamental groups of connected sums. See Ref. [12] for details, but be sure to brush up on understanding groups in terms of generators and relations before you do.

Exercises

1. Precisely define the higher-dimensional version of stereographic projection: $s : S^n \setminus \{(0,0,\ldots,1)\} \to \mathbb{R}^n$.

2. Find the fundamental group of $S^2 \times S^1$.

3. Find the fundamental group of $S^2 \vee S^1$. (*Hint*: you can construct a covering space along the lines of Theorem 11.7.9.)

4. Find the fundamental group of the "solid torus" $S^1 \times D^2$. Give an intuitive explanation to go along with your precise proof.

REFERENCES

1. Edwin Abbott, *Flatland, a Romance in Many Dimensions*, 1880; reprinted by Dover, 1992.

2. Crump W. Baker, *Introduction to Topology*, Krieger Publishing, 1991.

3. Ralph Boas, Bourbaki and Me, *Mathematical Intelligencer* **8** (4) 644–5 (1986).

4. Carl B Boyer, *A History of Mathematics*, Wiley Publishing, 1968 (more recent editions are jointly published with Merzbach, but the older editions have more of what you want).

5. Neil Cornish and Jeff Weeks, "Measuring the Shape of the Universe," *Notices of the American Mathematical Society* **45** 1463–1471 (Dec. 1998).

6. Jean Dieudonne, *A History of Algebraic and Differential Topology, 1900–1960*, Birkhauser, Boston, 1989.

7. Daniel Gorenstein, *Finite Groups*, American Mathematical Society, 1968.

8. Daniel Gorenstein, Richard Lyons and Ronald Solomon, *The Classification of Finite Simple Groups* Vols 1-5, American Mathematical Society, 1994-2002. (Volumes 1 and 2 are available free of charge online at `www.ams.org`.)

9. Paul Halmos, *Naive Set Theory*, Springer-Verlag, 1974

10. I. N. Herstein, *Abstract Algebra*, 3rd edition, Prentice-Hall, 1996

11. Janna Levin, *How the Universe Got Its Spots*, First Anchor Books, 2002.

12. William Massey, *Algebraic Topology, an Introduction*, Springer-Verlag, 1967.

13. James Munkres, *Topology, a First Course*, 2nd edition, Prentice-Hall, 2001.

14. *New York Times* article on Weeks, et al., Nov., 2003.

15. Heinrich Reitberger, "Obituary for Vietoris," *Notices of the American Mathematical Society* 1232, Nov., 2002.

16. Arthur Seebach and Lynn Steen, *Counterexamples in Topology*, Dover Publications, 1995.

17. H. Seifert and W. Threlfall, *Lehrbuch der Topologie*, Academic Press, 1980.

18. The MacTutor History of Mathematics archive, website hosted by the University of St. Andrews, Scotland.

19. Jeff Weeks, *The Shape of Space*, 2nd edition, Marcel Dekker, 2002.

20. Hans Wussing, *The Genesis of the Abstract Group Concept* (translated by A. Shenitzer), M.I.T. Press, 1984.

INDEX

PURE AND APPLIED MATHEMATICS

A Wiley-Interscience Series of Texts, Monographs, and Tracts

ADÁMEK, HERRLICH, and STRECKER—Abstract and Concrete Catetories
ADAMOWICZ and ZBIERSKI—Logic of Mathematics
AINSWORTH and ODEN—A Posteriori Error Estimation in Finite Element Analysis
AKIVIS and GOLDBERG—Conformal Differential Geometry and Its Generalizations
ALLEN and ISAACSON—Numerical Analysis for Applied Science
*ARTIN—Geometric Algebra
AUBIN—Applied Functional Analysis, Second Edition
AZIZOV and IOKHVIDOV—Linear Operators in Spaces with an Indefinite Metric
BASENER—Topology and Its Applications
BERG—The Fourier-Analytic Proof of Quadratic Reciprocity
BERMAN, NEUMANN, and STERN—Nonnegative Matrices in Dynamic Systems
BERKOVITZ—Convexity and Optimization in $\mathbb{R}^n$
BOYARINTSEV—Methods of Solving Singular Systems of Ordinary Differential Equations
BRIDGER—Real Analysis: A Constructive Approach
BURK—Lebesgue Measure and Integration: An Introduction
*CARTER—Finite Groups of Lie Type
CASTILLO, COBO, JUBETE, and PRUNEDA—Orthogonal Sets and Polar Methods in Linear Algebra: Applications to Matrix Calculations, Systems of Equations, Inequalities, and Linear Programming
CASTILLO, CONEJO, PEDREGAL, GARCIÁ, and ALGUACIL—Building and Solving Mathematical Programming Models in Engineering and Science
CHATELIN—Eigenvalues of Matrices
CLARK—Mathematical Bioeconomics: The Optimal Management of Renewable Resources, Second Edition
COX—Galois Theory
†COX—Primes of the Form $x^2 + ny^2$: Fermat, Class Field Theory, and Complex Multiplication
*CURTIS and REINER—Representation Theory of Finite Groups and Associative Algebras
*CURTIS and REINER—Methods of Representation Theory: With Applications to Finite Groups and Orders, Volume I
CURTIS and REINER—Methods of Representation Theory: With Applications to Finite Groups and Orders, Volume II
DINCULEANU—Vector Integration and Stochastic Integration in Banach Spaces
*DUNFORD and SCHWARTZ—Linear Operators
Part 1—General Theory
Part 2—Spectral Theory, Self Adjoint Operators in Hilbert Space
Part 3—Spectral Operators

*Now available in a lower priced paperback edition in the Wiley Classics Library.
†Now available in paperback.

FARINA and RINALDI—Positive Linear Systems: Theory and Applications
FATICONI—The Mathematics of Infinity: A Guide to Great Ideas
FOLLAND—Real Analysis: Modern Techniques and Their Applications
FRÖLICHER and KRIEGL—Linear Spaces and Differentiation Theory
GARDINER—Teichmüller Theory and Quadratic Differentials
GILBERT and NICHOLSON—Modern Algebra with Applications, Second Edition
*GRIFFITHS and HARRIS—Principles of Algebraic Geometry
GRILLET—Algebra
GROVE—Groups and Characters
GUSTAFSSON, KREISS and OLIGER—Time Dependent Problems and Difference Methods
HANNA and ROWLAND—Fourier Series, Transforms, and Boundary Value Problems, Second Edition
*HENRICI—Applied and Computational Complex Analysis
Volume 1, Power Series—Integration—Conformal Mapping—Location of Zeros
Volume 2, Special Functions—Integral Transforms—Asymptotics—Continued Fractions
Volume 3, Discrete Fourier Analysis, Cauchy Integrals, Construction of Conformal Maps, Univalent Functions
*HILTON and WU—A Course in Modern Algebra
*HOCHSTADT—Integral Equations
JOST—Two-Dimensional Geometric Variational Procedures
KHAMSI and KIRK—An Introduction to Metric Spaces and Fixed Point Theory
*KOBAYASHI and NOMIZU—Foundations of Differential Geometry, Volume I
*KOBAYASHI and NOMIZU—Foundations of Differential Geometry, Volume II
KOSHY—Fibonacci and Lucas Numbers with Applications
LAX—Functional Analysis
LAX—Linear Algebra
LOGAN—An Introduction to Nonlinear Partial Differential Equations
MARKLEY—Principles of Differential Equations
MORRISON—Functional Analysis: An Introduction to Banach Space Theory
NAYFEH—Perturbation Methods
NAYFEH and MOOK—Nonlinear Oscillations
PANDEY—The Hilbert Transform of Schwartz Distributions and Applications
PETKOV—Geometry of Reflecting Rays and Inverse Spectral Problems
*PRENTER—Splines and Variational Methods
RAO—Measure Theory and Integration
RASSIAS and SIMSA—Finite Sums Decompositions in Mathematical Analysis
RENELT—Elliptic Systems and Quasiconformal Mappings
RIVLIN—Chebyshev Polynomials: From Approximation Theory to Algebra and Number Theory, Second Edition
ROCKAFELLAR—Network Flows and Monotropic Optimization
ROITMAN—Introduction to Modern Set Theory
ROSSI—Theorems, Corollaries, Lemmas, and Methods of Proof
*RUDIN—Fourier Analysis on Groups
SENDOV—The Averaged Moduli of Smoothness: Applications in Numerical Methods and Approximations
SENDOV and POPOV—The Averaged Moduli of Smoothness
SEWELL—The Numerical Solution of Ordinary and Partial Differential Equations, Second Edition
SEWELL—Computational Methods of Linear Algebra, Second Edition

*Now available in a lower priced paperback edition in the Wiley Classics Library.
†Now available in paperback.

SHICK—Topology: Point-Set and Geometric
*SIEGEL—Topics in Complex Function Theory
Volume 1—Elliptic Functions and Uniformization Theory
Volume 2—Automorphic Functions and Abelian Integrals
Volume 3—Abelian Functions and Modular Functions of Several Variables
SMITH and ROMANOWSKA—Post-Modern Algebra
ŠOLÍN–Partial Differential Equations and the Finite Element Method
STADE—Fourier Analysis
STAKGOLD—Green's Functions and Boundary Value Problems, Second Editon
STAHL—Introduction to Topology and Geometry
STANOYEVITCH—Introduction to Numerical Ordinary and Partial Differential Equations Using MATLAB®
*STOKER—Differential Geometry
*STOKER—Nonlinear Vibrations in Mechanical and Electrical Systems
*STOKER—Water Waves: The Mathematical Theory with Applications
WATKINS—Fundamentals of Matrix Computations, Second Edition
WESSELING—An Introduction to Multigrid Methods
†WHITHAM—Linear and Nonlinear Waves
ZAUDERER—Partial Differential Equations of Applied Mathematics, Third Edition

*Now available in a lower priced paperback edition in the Wiley Classics Library.
†Now available in paperback.